Precalculus

A Functional Approach to Algebra
and Trigonometry

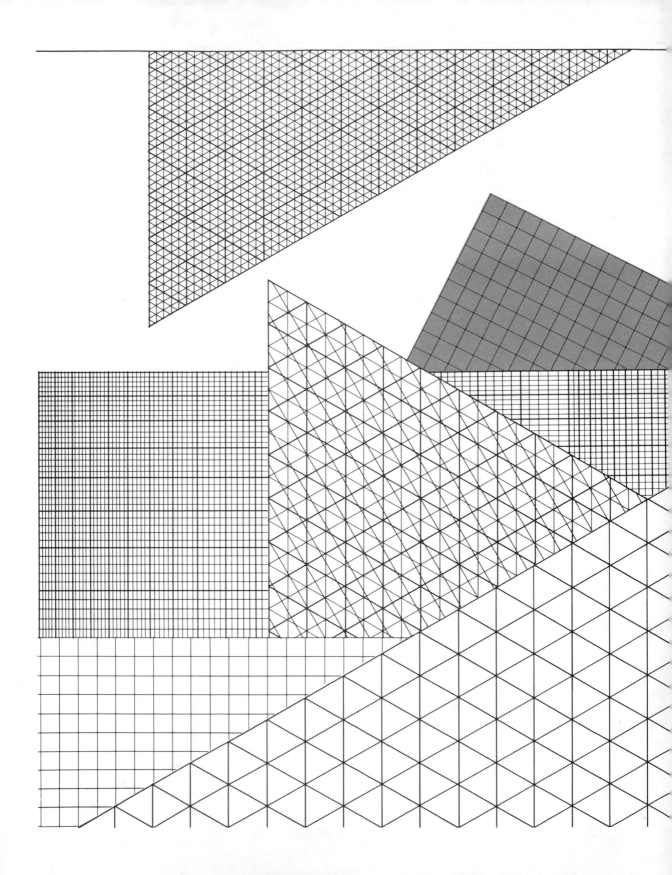

Precalculus

A Functional Approach to Algebra
and Trigonometry

Peter Evanovich   Lafayette College

Martin Kerner   Lafayette College

 Holden-Day, Inc.
San Francisco

This book was designed by Victoria Ann Philp. The cover design is by
Leslie Berman. Illustrations were rendered by J & R Services. The
composition is by Allservice Phototypesetting Company of Arizona.
Printing and binding were done by The Book Press.

Precalculus

A Functional Approach to Algebra and Trigonometry

Copyright © 1982 by Holden-Day, Inc.
500 Sansome Street, San Francisco, CA 94111

Library of Congress Catalog Card Number: 79-91354

ISBN: 0-8162-2715-2

Printed in the United States of America

9 8 7 6 5 4 3 2 1

To Janet, Alexandra, and Peter
P.E.

For Susan and Andrew
M.K.

Contents

Preface

This book is designed for a one-semester course in precalculus, and assumes that students have taken elementary algebra and plane geometry. It contains enough material to be used in a variety of ways. At Lafayette College we have used it in a one-semester course meeting three times each week for fourteen weeks, as well as in a six-week summer session course. In these courses we covered the material in Chapters 1–4, Sections 5.1–5.3, 6.1–6.8, 7.1–7.6, and Chapter 8. By including other sections and the appendices, the text is suitable for one-semester courses meeting four or five times each week.

Planning a text for a one-semester course involved difficult choices. We tried to limit topics to those that would be most helpful for the study of calculus, and within those topics we emphasized the skills and concepts that the student will use most frequently in a calculus course. The text has a number of features that we feel make it particularly suitable for a one-semester course:

We have kept preliminaries to a minimum. Rather than cover elementary algebra and geometry in a separate chapter, we review them as needed to understand new concepts.

We have used minimal notation in the text.

In almost every section we have included exercises that prepare students for work in future sections.

Our treatment of functions is more detailed than that usually found in precalculus texts. We have addressed the ambiguities of functional notation and have consistently applied the theory developed in the chapter on functions to specific functions introduced later in the text.

Rather than using the wrapping function, our treatment of trigonometry uses reference angles, an approach that students with no background in trigonometry find easier to understand and one that parallels the right-triangle trigonometry many students in this course have already seen.

We make extensive use of point plotting to draw graphs, taking advantage of the hand-held calculators readily available to students.

Our concluding chapter introduces limits as an extension of the work we have done with functions and graphs—not as a formal study. Applications in this chapter reinforce the properties of trigonometric, logarithmic, and exponential functions presented earlier.

Aware that students in a one-semester precalculus course have a great amount of material to learn in a short time, we tried to be sensitive to this when deciding among alternative ways to present a topic. Our approach to trigonometry, in particular, is more concrete and less abstract than that taken by many other texts, and is the approach with which our students have had the most success. We have made a particular effort to introduce each chapter with a discussion or example that motivates students. New concepts are always presented along with interesting examples illustrating how the concepts are applied in fields like engineering, physics, and business. We have tried to achieve a balance of motivation and rigor while keeping the book as readable as possible.

Answers to odd-numbered exercises are included to give students a chance to check their work, but we have omitted answers to those odd-numbered exercises that ask for proofs (hints for these are usually provided in the exercise) and/or those asking for graphs that may appear in other parts of the text. In part, we omitted these graphs from the answer section thinking it especially important that the student go through the process of point plotting graphs.

Because many students have hand-held calculators, we have indicated with the symbol ▦ the exercises for which a calculator might be helpful. Consider the use of a calculator a suggestion, not a requirement. A calculator will save considerable time in those exercises that ask students to plot graphs, and in other exercises it will save students the trouble of repeatedly looking up entries in the tables at the back of the book. However, the calculator symbol means that a calculator is helpful, not that it is absolutely necessary. Students without calculators should not hesitate to try these problems.

For students who want additional review, the *Study Guide* outlines the text section by section and presents additional solved examples and exercises. The *Instructor's Manual* includes notes on what we tried to accomplish in each section of the text, as well as worked-out solutions to all text exercises.

ACKNOWLEDGMENTS

Many people have helped us with this text. We are grateful to the following, who read and commented on various drafts of the manuscript:

Vincent J. Bruno, *San Francisco State University*

Norbert L. Ellman, *Long Beach Community College*

Adelaide T. Harmon-Elliott, *California Polytechnic State University*

James F. Hurley, *University of Connecticut*

Thomas A. McCready, *California State University, Chico*

Marcus A. McWaters, *University of South Florida*

To all the above, we'd like to extend our thanks, adding a special note of appreciation for the suggestions and encouragement of Professor James F. Hurley.

We would also like to thank those of our students who struggled through preliminary versions. Many of their suggestions found their way into this edition. Our thanks also go to Barbara Walen and Randy Stonesifer for their help, and to the staff at Holden-Day—particularly to Carol Pritchard-Martinez. Finally, our thanks to Claudette Dahlinger for typing the final manuscript.

October, 1981   P.E.
M.K.

Precalculus

A Functional Approach to Algebra
and Trigonometry

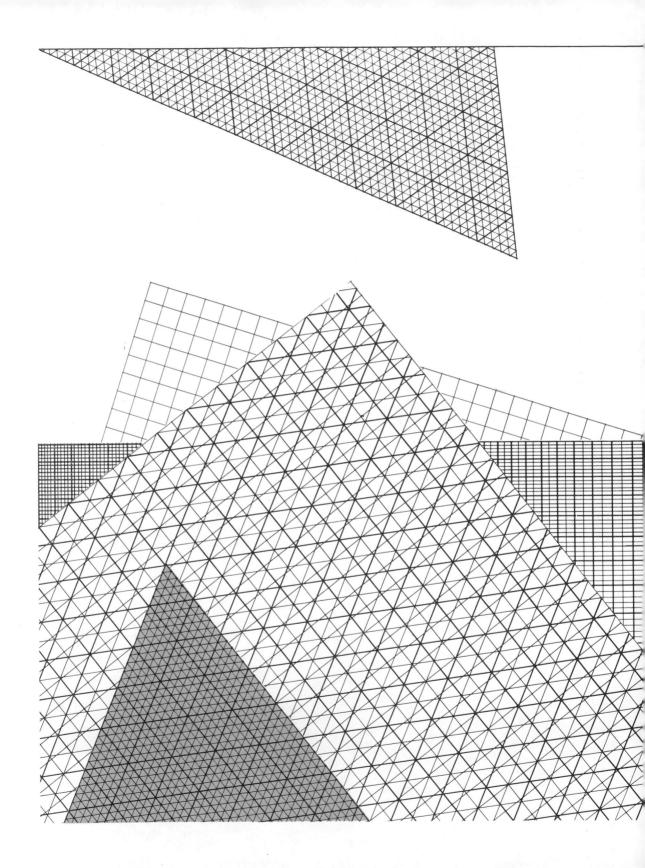

1

Preliminaries

INTRODUCTION

In this short chapter we will review topics in elementary algebra and introduce notation that we will use throughout this text.

INEQUALITIES

The real numbers can be thought of as all the points on a scaled, directed line as pictured in Figure 1.2.1.

Figure 1.2.1

The points to the right of the 0 point in Figure 1.2.1 are **positive real numbers;** the points to the left of 0 are **negative numbers;** 0 itself is neither positive nor negative.

The symbols $<, \leq, >, \geq$, called **inequality signs,** are used to describe the relative positions of any two real numbers. To be precise, if x and y are real numbers,

$$x < y$$

means x lies to the left of y on the number line and we say x **is less than** y. For example, $2 < 3$, $-2 < 0$, and $-3 < 2$.

When we write

$$x \leq y$$

(read: x **is less than or equal to** y) we mean $x < y$ or $x = y$. For example, $2 \leq 2$, $3 \leq 4$, and $-3 \leq 1$.

The meanings of $x > y$ and $x \geq y$ are analogous to those of $x < y$ and $x \leq y$. In particular, $x > y$ (read: x **is greater than** y) means x lies to the right of y on the number line and $x \geq y$ (read: x **is greater than or equal to** y) means $x > y$ or $x = y$. For example, $3 > 2$, $3 \geq -2$, $-2 \geq -2$, and $-2 > -3$.

Since "x is to the right of y" is equivalent to "y is to the left of x,"

$$x > y \quad \text{is equivalent to} \quad y < x$$

and

$$x \geq y \quad \text{is equivalent to} \quad y \leq x$$

Inequalities will be frequently used to replace statements such as "x is a positive number." In particular,

$x > 0$ means x is positive.

$x \geq 0$ means x is non-negative (i.e., either positive or zero).

$x < 0$ means x is negative.

$x \leq 0$ means x is non-positive.

Example 1

The collection of all real numbers, x, which satisfies $x < 3$, can be graphically described as the set of points to the left of 3 on the number line. Note in Figure 1.2.2 that the open circle at 3 indicates 3 is not a member of this collection.

Figure 1.2.2   **$x < 3$**

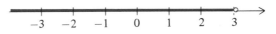

Example 2

The collection of real numbers x for which $x \geq -2$ is depicted in Figure 1.2.3. The solid circle at -2 indicates -2 is in this collection.

Figure 1.2.3   **$x \geq -2$**

Frequently, a number satisfies two inequalities simultaneously—for example, $-1 < 2$ and $2 \leq 4$. In this case, we condense the phrase $-1 < 2$ and $2 \leq 4$ to $-1 < 2 \leq 4$. In general, when we write $x * y \ \square \ z$, where $*$ and $\square$ are $<$, $\leq$, $>$, or $\geq$, we mean $x * y$ **and** $y \ \square \ z$.

Example 3

The inequality $-2 \leq x < 3$ means $-2 \leq x$ and $x < 3$. The collection of real numbers x satisfying these inequalities is shown in Figure 1.2.4.

Figure 1.2.4   **$-2 \leq x < 3$**

Example 4

> The inequality $-2 > x > 3$ means $-2 > x$ and $x > 3$. No real numbers will satisfy both of these inequalities simultaneously since no number can lie to the right of 3 and to the left of -2.

Example 5

> The inequality $2 < x < 4$ means $2 < x$ and $x < 4$ as shown in Figure 1.2.5.
>
> **Figure 1.2.5** $2 < x < 4$
>
>

Example 6

> The inequality $2 > x > 4$ means $2 > x$ and $x > 4$. No real numbers satisfy this pair of inequalities since no number is larger than 4 and less than 2.

Exercises 1.2

In Exercises 1–12, indicate whether the inequality is true or false.

1. $2 \leq 5$	**2.** $2 < 5$	**3.** $-2 < 5$
4. $-2 \leq -5$	**5.** $\pi > 3.14$	**6.** $-3 \leq -2$
7. $-\pi \leq 0$	**8.** $-\frac{1}{2} < -1$	**9.** $2 < -1$
10. $1.01 \leq 1.001$	**11.** $-1.01 \leq 1.001$	**12.** $\sqrt{2} \leq \sqrt{3}$

In Exercises 13–22, place one of the symbols $<$ or $>$ between the numbers to make a true statement.

13. $2 \underline{\hspace{1cm}} 5$	**14.** $-2 \underline{\hspace{1cm}} -5$
15. $-2 \underline{\hspace{1cm}} 5$	**16.** $2 \underline{\hspace{1cm}} -5$
17. $0 \underline{\hspace{1cm}} -1$	**18.** $100 \underline{\hspace{1cm}} -1000$
19. $\pi^2 \underline{\hspace{1cm}} \pi$	**20.** $(\frac{1}{2})^2 \underline{\hspace{1cm}} (\frac{1}{4})^2$
21. $(-\frac{1}{2})^2 \underline{\hspace{1cm}} (-\frac{1}{4})^2$	**22.** $-\frac{1}{2} \underline{\hspace{1cm}} -\frac{1}{4}$

In Exercises 23–34, sketch the graph of the real numbers x satisfying the given inequality. (See Examples 1–4.)

23. $x \leq 2$	**24.** $x > -1$	**25.** $x \geq \sqrt{2}$
26. $-3 < x \leq 2.5$	**27.** $5 < x < 4$	**28.** $4 < x < 5$
29. $-1 \leq x < 5$	**30.** $2 > x < 4$	**31.** $-3 \leq x \leq -1$
32. $-2 \leq x \leq 6$	**33.** $x \geq \sqrt{3}$	**34.** $x < -\sqrt{5}$

35. The inequalities $<, \le, >, \ge$ can also be defined as follows:

$x < y$ if and only if $y - x$ is positive.

$x \le y$ if and only if $y - x$ is non-negative.

$x > y$ if and only if $x - y$ is positive.

$x \ge y$ if and only if $x - y$ is non-negative.

For example, $2 < 5$ since $5 - 2 = 3$ is positive, and $-2 > -3$ since $-2 - (-3) = -2 + 3 = 1$ is positive. Use the above definitions to verify the following inequalities:

a. $\frac{1}{2} < 1$ b. $-\frac{1}{2} > -\frac{3}{2}$ c. $5 \le 7$
d. $-3 \le 1$ e. $-1 \ge -4$ f. $1 \ge 0$
g. $2 > -3$ h. $4 \le 4$

36. If $x < y$, what inequality describes the relative positions of $x + 2$ and $y + 2$ on the number line?

37. If $x \ge y$, what inequality describes the relative positions of $x - 2$ and $y - 2$ on the number line?

38. Place $<$ or $>$ between the following pairs of numbers to make a true statement.

a. $2 \underline{\hspace{1cm}} 3$ b. $2 \cdot 2 \underline{\hspace{1cm}} 3 \cdot 2$
c. $2(-2) \underline{\hspace{1cm}} 3(-2)$ d. $3 \underline{\hspace{1cm}} -5$
e. $3 \cdot 4 \underline{\hspace{1cm}} (-5)(4)$ f. $3(-4) \underline{\hspace{1cm}} (-5)(-5)$
g. $3(-4) \underline{\hspace{1cm}} (-5)(-4)$ h. $-3 \underline{\hspace{1cm}} -2$
i. $(-3)(-5) \underline{\hspace{1cm}} (-2)(-5)$

39. Suppose $x < y$. What inequality describes the relative position of $2x$ and $2y$ on the number line? What inequality describes the relative positions of $-2x$ and $-2y$ on the number line? (*Hint:* Examine your answers to questions in Exercise 38.)

SECTION 1.3   **INTERVALS**

The real numbers 2 and 4 satisfy $2 < 4$. The set of real numbers x with $2 < x < 4$ is called an **interval** and is denoted by $(2, 4)$. The graph of $(2, 4)$ is shown in Figure 1.3.1.

Figure 1.3.1   **The Open Interval (2, 4)**

Note that if u and v are points in (2, 4) and $u < v$, then any number x satisfying $u < x < v$ must also be in (2, 4). This means that there are no gaps or holes in the interval (2, 4), an idea that characterizes intervals.

There are other types of intervals used frequently in mathematics. Here is a complete list of intervals together with corresponding notation, names, and sketches of their graphs.

Let a and b be real numbers with $a < b$.

A.   The **open interval** (a, b) (shown in Figure 1.3.2) is the set of real numbers x such that $a < x < b$.

Figure 1.3.2   The Open Interval (a, b)

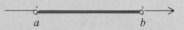

B.   The **closed interval** $[a, b]$ (shown in Figure 1.3.3) is the set of real numbers x such that $a \leq x \leq b$.

Figure 1.3.3   The Closed Interval $[a, b]$

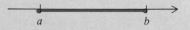

C.   The interval $[a, b)$ is the set of real numbers x for which $a \leq x < b$. Similarly, $(a, b]$ is the set of real numbers x for which $a < x \leq b$. These intervals are called **half open** (or **half closed**), and are shown in Figures 1.3.4 (a) and (b), respectively.

Figure 1.3.4
(a) The Half-Open Interval $[a, b)$

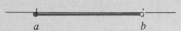

(b) The Half-Open Interval $(a, b]$

D.   The symbols $+\infty$, $-\infty$ are used to denote open-ended intervals. In particular, the open interval $(-\infty, a)$, shown in Figure 1.3.5 (a), is the set of real numbers x for which $x < a$ and the open interval $(a, +\infty)$ (Figure 1.3.5 (b)) is the set of real numbers x for which $x > a$.

Figure 1.3.5
(a) The Open Interval $(-\infty, a)$

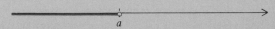

(b) The Open Interval $(a, +\infty)$

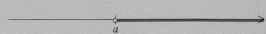

E.   The closed interval $(-\infty, a]$ is the set of real numbers x for which $x \le a$ and the closed interval $[a, -\infty)$ is the set of real numbers x for which $x \ge a$. (See Figure 1.3.6.)

Figure 1.3.6
(a) The Closed Interval $(-\infty, a]$

(b) The Closed Interval $[a, +\infty)$

F.   The symbol $(-\infty, +\infty)$ is used to denote the set of all real numbers, which is illustrated in Figure 1.3.7.

Figure 1.3.7   The Interval $(-\infty, +\infty)$

Example 1

Figure 1.3.8
(a) The Open Interval $(-1, 2)$

(b) The Closed Interval $[1, 3]$

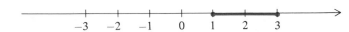

(c) The Half-Open Interval $[-2, 2)$

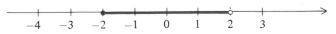

(d) The Closed Interval [2, +∞]

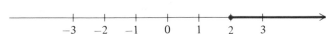

(e) The Open Interval (−∞, 1)

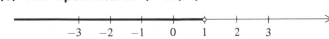

Note that neither symbol $-\infty$ nor $+\infty$ represents a real number. Many students would like to regard $+\infty$ as a very large number. But $+\infty$ is *not* a number, large or otherwise. In particular, we have no way of adding, subtracting, multiplying, or dividing with $+\infty$ or $-\infty$. We have introduced these symbols to indicate that the numbers in a particular interval have either no lower boundary point or no upper boundary point.

Exercises 1.3

In Exercises 1–7, sketch the graph of the given interval.

  1. $[-2, 1)$ **2.** $(-2, 1]$

  3. $[-\sqrt{2}, \frac{1}{2}]$ **4.** $(\pi, +\infty)$

  5. $(-\infty, -\pi]$ **6.** $(3, 5)$

  7. $[-3, 2)$

In Exercises 8–12 describe each set of points using interval notation.

8.

9.

10.

11.

12.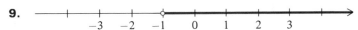

Each of the sets of numbers described in Exercises 13–22 is an interval. Describe each set using interval notation.

13. The set of real numbers x for which $2 < x \le 3$

14. The set of real numbers x for which $x \ge -2$

15. The set of non-negative real numbers less than 5

16. The set of real numbers that are less than 4 units from the origin (i.e., the point marked 0)
17. The set of real numbers whose squares are less than 4 units from the origin
18. The set of non-negative real numbers whose square roots are less than 4 units from the origin
19. The set of real numbers x satisfying $1 \leq x < 5$ and $2 \leq x \leq 7$
20. The set of real numbers x satisfying $x \geq 2$ and $x < 7$
21. The set of real numbers x satisfying $x \leq 7$, $x < -2$, and $x < 0$
22. The set of real numbers x satisfying $x \leq 7$, $x < 3$, $x > -1$, and $x \geq 0$

In Exercises 23–32 a set of numbers is described. In each case determine whether the set is an interval. If a set is an interval, describe it using interval notation.

23. The set of numbers x such that $x > 2$ or $x < -2$
24. The set of numbers x such that $x < 2$ and $x \geq -2$
25. The set of numbers consisting of 1 and 2
26. The set of numbers x such that $x^2 > 4$
27. The set of numbers x such that $x^3 > 27$
28. The set of numbers x such that $x^2 \leq 4$
29. The set of numbers x such that $x^3 \leq 27$
30. The set of numbers x such that $x^2 \geq -1$
31. The set of numbers x such that $x^2 \geq 0$
32. The set of numbers x such that $x^2 > 0$

SECTION 1.4   **SETS OF REAL NUMBERS**

Some sets of real numbers are referred to and used so often that it has been convenient to develop special names and symbols for these sets. Here are three of these sets together with their generally accepted names and notational representations.

> The set of **natural (or counting) numbers** consists of the numbers 1, 2, 3, 4, ... and is denoted by **N**.

> The set of **integers** consists of the numbers ..., $-4, -3, -2, -1, 0, 1, 2, 3, 4, \ldots$ and is denoted by **Z** (from the German word for numbers, Zahlen). Frequently, the set of natural numbers is called the set of **positive integers**.

> The set of **rational numbers** consists of the set of numbers that can be expressed in the form m/n where m and n are integers and $n \neq 0$. Since each rational number is a quotient, the set of rational numbers will be denoted by **Q**.

Each **real number** can be represented in decimal form. For example, $\frac{3}{2} = 1.5000 \ldots$ and $\frac{1}{3} = 0.3333 \ldots$. (The converse is also true: every infinite decimal represents a real number.) The rational numbers may be characterized as those numbers whose decimal representations terminate (have all zeros after a certain point) or are repeating. For example, $0.3575757575 \ldots$ is a repeating decimal, a repeating factor being 75. The number $r = 0.10100100010000 \ldots$ is not a repeating decimal since the same block of digits does not repeat itself. Thus r is not a rational number. The set of real numbers that are not rational is called the set of **irrational numbers.** Many well-known mathematical constants—π and $\sqrt{2}$, for example—are irrational.

> Together, the rational and irrational numbers make up the **real numbers**—the points on the number line. The set of real numbers is denoted by **R**.

Exercises 1.4

In Exercises 1–8, list the integers that satisfy the given inequality.

1. $-\sqrt{2} < x < 5.9$ 2. $0 \leq x < 5.5$
3. $-3 < x \leq 2$ 4. $-3 < x < -2$
5. $-3 \leq x < -1$ 6. $3 \leq x \leq 4$
7. $3 > x \geq 7$ 8. $-3 < x$

In Exercises 9–14, list the integers in the given interval.

9. $[-3, 2)$ 10. $(2, 3)$ 11. $(-\infty, 2]$
12. $(0, +\infty)$ 13. $[1, 2)$ 14. $[1, 2]$

In Exercises 15–22, list the natural numbers that satisfy the given inequality.

15. $-\sqrt{2} < x < 5.9$ 16. $0 \leq x < 5.5$ 17. $-3 < x \leq 2$
18. $-3 < x < -2$ 19. $-3 \leq x \leq 1$ 20. $3 \leq x \leq 4$
21. $x < 2$ 22. $x > 2$

In Exercises 23–29, list the natural numbers in the given interval.

23. $[-\pi, \pi)$ 24. $[-3, 2)$ 25. $(2, 3)$ 26. $[2, 3)$
27. $(-\infty, 4]$ 28. $(4, +\infty)$ 29. $[1, 5]$

30. If $x = 0.3121212\ldots$ then x is a rational number since it is represented by a repeating decimal. The following technique can be used to express x in the form m/n where m, n are integers and $n \neq 0$.

$$1000x = 312.121212\ldots$$
$$10x = 3.121212\ldots$$

Subtracting,

$$990x = 309$$
$$x = \frac{309}{990}$$

Use this technique to find fractional representations of

a. $x = 3.217217217\ldots$ **b.** $x = 0.11111\ldots$
c. $x = 0.999999\ldots$

31. a. Find a decimal representation of $4/7$ and identify the repeating factor. (*Hint:* Use the long division algorithm to divide 4 by 7.)
    b. Find a decimal representation of $22/7$. (This number is frequently used as a rational approximation of π. Problem 32 gives a much better rational approximation of π.)

32. Find the first 6 digits in the decimal representation of $355/113$.

33. A property of the rational numbers is that, given any two real numbers a and b with $a < b$, there is a rational number r such that $a < r < b$. In each exercise below find a rational number r such that $a < r < b$.

    a. $a = 1.5$, $b = 1.51$ **b.** $a = 1.5$, $b = 1.501$
    c. $a = 1.5$, $b = 1.5001$ **d.** $a = \sqrt{2}$, $b = \sqrt{3}$
    e. $a = \pi$, $b = (\pi + 4)/2$

34. Is every integer a rational number?
35. Is every rational number an integer?
36. Give an example of two irrational numbers whose sum is rational.
37. Give an example of two irrational numbers whose product is rational.
38. Is there a smallest real number x such that $x > 0$? Support your answer.

Review Exercises
Chapter 1

In Exercises 1–10, place either $<$ or $>$ between the two numbers to make a true statement.

1. -2 _____ -1 **2.** 2 _____ -1

3. 1 _____ -2 **4.** 2 _____ 1

5. $\sqrt{2}$ _____ 1 **6.** $\sqrt{2}$ _____ -2

7. π _____ 3.14 **8.** 2^2 _____ 3^2

9. $\dfrac{1}{2^2}$ _____ $\dfrac{1}{3^2}$ **10.** $-\left(\dfrac{1}{2}\right)^2$ _____ $-\left(\dfrac{1}{3}\right)^2$

In Exercises 11–20, sketch a graph of the real numbers x satisfying the given inequality (or inequalities). If your graph is an interval, use interval notation to describe it.

11. $x \geq 5$ **12.** $x < -3$

13. $2 \leq x < 5$ **14.** $-3 < x \leq 4$

15. $-2 \leq x \leq 0$ **16.** $x^2 > 4$

17. $x^2 \leq 9$ **18.** $4 \geq x \leq 3$

19. $-3 \leq x \geq 3$ **20.** $x - 2 \leq 3$

Sketch the graphs of the sets of real numbers described in Exercises 21–30. If a set is an interval, describe it using interval notation.

21. The set of real numbers x such that $-1 \leq x < 3$ and $0 < x \leq 4$
22. The set of real numbers x such that $x^2 > 9$
23. The set of real numbers x such that $x^3 > 64$
24. The set of integers x such that $-3 \leq x < 3$
25. The set of integers x such that $-3 < x < 3$
26. The set of natural numbers x such that $-5 \leq x \leq 1$
27. The set of integers whose distances from the point 1 on the number line are not more than 3 units
28. The set of natural numbers whose distances from the point 1 on the number line are not more than 3 units
29. The set of real numbers x whose distances from the point 1 on the number line are not more than 3 units
30. The set of real numbers x that are less than 3 units from the point 2 and more than 4 units from the point -2

In Exercises 31–33, represent the given rational number in decimal form. Identify the repeating sequence of digits in your representation.

31. $\dfrac{2}{7}$ **32.** $\dfrac{3}{11}$ **33.** $\dfrac{13}{9}$

In Exercises 34–36, a rational number is given in decimal form. Express each rational number as the ratio of two integers. (See Exercise 30, Section 1.4.)

34. $3.21767676\ldots$
35. $2.517000\ldots$
36. $-1.7152152152152\ldots$

37. Is every non-negative integer a natural number?
38. Is every non-negative integer a rational number?
39. Is the sum of two integers always an integer?
40. Is the sum of two natural numbers always a natural number?
41. Is the sum of two rational numbers always a rational number?
42. Is the sum of two irrational numbers always an irrational number?
43. Is the reciprocal (the reciprocal of a number x is $1/x$) of a natural number always a natural number?
44. Is the reciprocal of a non-zero rational number always a rational number?
45. Is there a largest integer less than 0? If so, what is it?
46. Is there a largest rational number less than 0? If so, what is it?

In Exercises 47–50, enter one of the symbols $<$ or $>$ at the indicated point to make a true sentence. (See Exercises 36, 37, and 39, Section 1.2.)

47. If $c > 0$ then $2 + c$ _____ $3 + c$
48. If $c < 0$ then $2 + c$ _____ $3 + c$
49. If $c > 0$ then $2 \cdot c$ _____ $3 \cdot c$
50. If $c < 0$ then $2 \cdot c$ _____ $3 \cdot c$

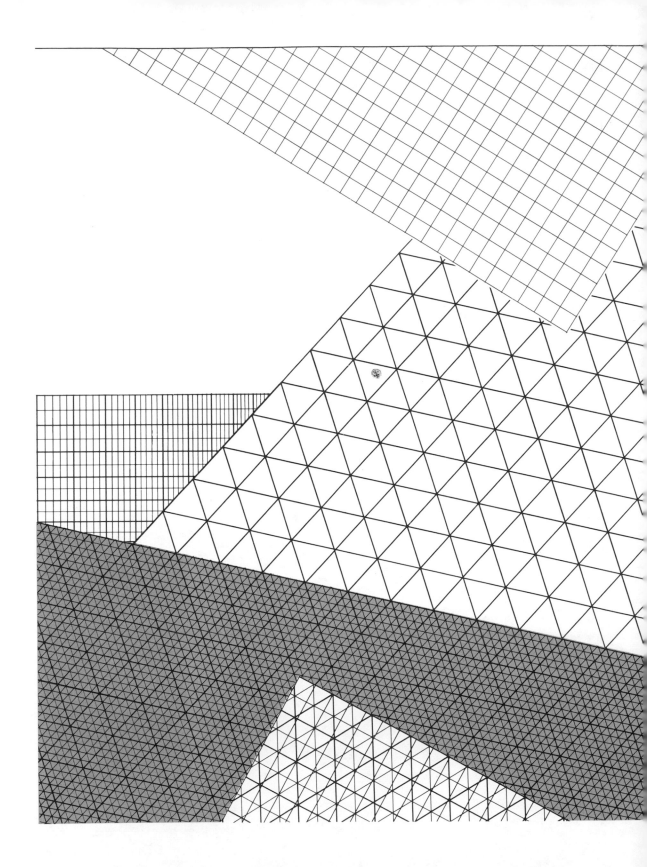

2 Polynomials

SECTION 2.1

INTRODUCTION

This chapter is devoted to a study of polynomials and their arithmetic properties—sums, differences, and products. Many of the arithmetic properties of polynomials are identical to the arithmetic properties of real numbers. For example, we know that if a and b are two real numbers then $a + b = b + a$. This property will also hold if a and b are polynomials. As we will see, there are many other similarities.

This chapter is important for two reasons. The first is that polynomials are an important tool for studying functions, a topic we will spend much time on later in this text. The second is that while we are studying polynomials, we will be reviewing important topics from elementary algebra.

SECTION 2.2

DEFINITIONS AND PROPERTIES OF POLYNOMIALS

A **polynomial in x (the variable or unknown)** is an expression of the form

$$a_0 + a_1 x + a_2 x^2 + \cdots + a_n x^n$$

where $a_0, a_1, \ldots, a_n$ are real numbers and n is a non-negative integer. The a_i's are called **coefficients** and, in particular, a_i is called the coefficient of x^i. The coefficient a_0 is called the **constant coefficient** and since $x^0 = 1$, it is sometimes viewed as the coefficient of x^0. When the coefficient of x^i is 1 we usually write x^i rather than $1x^i$. For example, the coefficient of x^2 in $3x^3 + x^2 + 5$ is 1. A common error is to think x^i has no coefficient or a coefficient of 0 when a coefficient isn't written explicitly.

Some examples of polynomials are

A. $x^5 + 3x^4 - x^3 + x + \pi$
B. $3x - 2$
C. 5
D. $x^{1000} - 1$
E. 0

The polynomial given in (E) is called the **zero polynomial**.

In Example (A), the coefficient of x^5 is 1, the coefficient of x^4 is 3, the coefficient of x^3 is -1, the coefficient of x is 1, and the constant coefficient is π. Since the coefficient of x^3 is -1, (A) could be (more laboriously) written $1x^5 + 3x^4 + (-1)x^3 + 1x + \pi$.

Notice that an expression such as $3x^{-2} + 5x + 6$ is not a polynomial since one of the exponents is a negative integer.

Usually if a polynomial contains a term $a_k x^k$ and $a_k = 0$, the term $a_k x^k$ is not written. For example, $x^2 + 0x + 1$ is usually written $x^2 + 1$, and $0x^2 + 3x - 2$ is usually written $3x - 2$. Furthermore, the order in which the terms appear is immaterial. So, for example, $x^5 + 7x^2 + x + 3$ is the same polynomial as $7x^2 + x^5 + 3 + x$. However, we will usually write polynomials with exponents appearing in either ascending or descending order.

Now suppose

$$a_0 + a_1 + a_2 x^2 + \cdots + a_n x^n$$

and

$$b_0 + b_1 x + b_2 x^2 + \cdots + b_n x^n$$

are two polynomials.

We say that these polynomials are **equal** if and only if

$$a_0 = b_0, a_1 = b_1, a_2 = b_2, \ldots, a_n = b_n$$

Example 1

$$a_0 + a_1 x + x^2 = 2 + x^2$$

if and only if

$$a_0 = 2 \quad \text{and} \quad a_1 = 0$$

since

$$2 + x^2 = 2 + 0x + x^2$$

Example 2

It is impossible to choose a_0 and a_1 so that

$$a_0 + a_1 x + x^2 = 2 - 3x + 2x^2$$

since the coefficient of x^2 is 1 in one of the polynomials and 2 in the other.

Example 3

$$1 - 2x + x^2 \neq 1 - 2x$$

since

$$1 - 2x = 1 - 2x + 0x^2$$

and the coefficients of x^2 differ.

Given a nonzero polynomial, a polynomial with at least one nonzero coefficient, the **degree** of this polynomial is the largest exponent of x appearing in a term with nonzero coefficient. The degree of the polynomials given in A–D above are 5, 1, 0, and 1000, respectively. Note that $0x^3 + 2x^2 + x + 1 = 2x^2 + x + 1$. The degree of this polynomial is 2 even though x^3 appears in one version of the polynomial.

We can now restate the definition of equal polynomials: **Two polynomials are equal if and only if they have the same degree and if coefficients of like powers of x are equal.**

Suppose $P(x) = a_0 + a_1 x + a_2 x^2 + \cdots + a_n x^n$ is a polynomial. The **value of $P(x)$ at a real number r** is the number obtained when the variable x is replaced by the real number r. The value of $P(x)$ at r is denoted by $P(r)$. For example if $P(x) = x^2 + 3x - 3$ the value of $P(x)$ at 2 is $(2)^2 + 3(2) - 3 = 7$. Equivalently, $P(2) = 7$. The value of $P(x)$ at -3 is $(-3)^2 + 3(-3) - 3 = -3$. Equivalently, $P(-3) = -3$.

Example 4

> If
> $$P(x) = x^3 - 3x^2 - x + 3$$
> then
> $$P(-2) = (-2)^3 - 3(-2)^2 - (-2) + 3$$
> $$= -8 - 12 + 2 + 3 = -15$$
> $$P(-1) = (-1)^3 - 3(-1)^2 - (-1) + 3$$
> $$= -1 - 3 + 1 + 3 = 0$$
> $$P(0) = 0^3 - 3(0)^2 - 0 + 3 = 3$$
> $$P(1) = 1^3 - 3(1)^2 - 1 + 3 = 0$$

Polynomials are often used to describe relationships between two measurable quantities. For example, the polynomial $P(x) = x^2$ can be thought of as describing the relationship between the area of a square and the length of a side of that square. To be precise, we mean that if r is a positive real number then $P(r)$, the value of $P(x)$ at r, is the area of a square of side r.

The following examples also illustrate this use of polynomials.

Example 5

> The polynomial $P(x) = 4x$ describes the relationship between the side of a square and the perimeter of that square. For example, if a square's side measures 10 units, its perimeter is $P(10) = 4 \times 10 = 40$. In general, if r is the measure of the side of a square, then $P(r)$ is the perimeter of the square.

Example 6

> The volume of a right circular cone (shown in Figure 2.2.1) with height h and base radius r is given by
> $$V = \frac{1}{3} \pi r^2 h$$

Figure 2.2.1

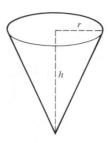

Suppose we have a water tank in the shape of a right circular cone with height 25 feet, and it has the property that, when the water in the tank is h feet high, the base radius of the right circular cone formed by the water is $\frac{1}{2}h$ feet. (See Figure 2.2.2.) Then, the volume of the water in the tank when the height of the water is h feet is

$$\frac{1}{3}\pi\left(\frac{1}{2}h\right)^2 h = \frac{\pi}{12}h^3$$

Figure 2.2.2

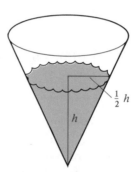

The polynomial

$$P(x) = \frac{\pi}{12}x^3$$

describes the relationship between the height and volume of water in the tank. Of course, physical considerations tell us that, when using $P(x)$ to describe this relationship, we must use the values $P(h)$ where $0 \leq h \leq 25$ since a negative height is meaningless and the maximum height of the water is 25 feet.

Exercises 2.2

In Exercises 1 and 2, determine which of the expressions in x are polynomials and give the degrees of each of the polynomials. If an expression is a polynomial of degree n, give the coefficients of x^n, $x^{n-1}, \ldots, x^2$, x, and the constant coefficient.

1. **a.** $x^5 + x^4 + 2x - 3$
   b. $x^{1/2} + 1$
   c. $x^6 + x^{100} + x^{599} - 7000$
   d. $2x^3 + (1/x) - 3$

2. **a.** $\sqrt{x^2} + x + 1$
   b. $\sqrt{2x^3} + x^2 + x - 3$
   c. $-3x^4 + 2x^5 + 3x^3 - 2x^2 + x - 5$
   d. $\sqrt{2}x^{10} + 7x^4 - 3x + \sqrt{3}/2$

In Exercises 3–8, evaluate the given polynomial at the indicated points.

3. $x^2 - 3x + 2$ at $-2, -1, 0, 1, 2$, and 3
4. $x^3 - 1$ at $-1, 0, 1$, and $\sqrt[3]{3}$
5. $x^4 - 1$ at $-1, 0, 1$, and $\sqrt{2}$
6. $2x^3 + x - 3$ at $-2, 0, 2$, and 3
7. $3x^2 - 10x + 10$ at $1, 3$, and 5
8. $-x^4 + 3$ at $-2, 0, 2, -\sqrt[4]{3}$, and $\sqrt[4]{3}$

9. Find constants a, b, and c such that

$$-ax^3 + x^2 + bx - c = 3x^3 + x^2 - 4$$

10. Find constants a, b, and c such that

$$ax^2 - bx + c = 2x^2 + 3x - 7$$

11. Find constants a, b, c, d, and e such that

$$ax^4 + bx^3 + cx^2 + dx + e = x^4 - 1$$

12. Find constants a, b, c, d, and e such that

$$ax^4 + bx^3 + cx^2 + dx + e = -x^3 + x^2 - x + 1$$

13. Can two polynomials of different degrees be equal?

14. A fact about polynomials is that two polynomials are equal if and only if they have the same values at each real number. Show that the following polynomials are not equal by finding a real number r where the polynomials have different values.

   a. $x^3 - x$ $\quad 3x^3 - 3x$
   b. $x^5 - 5x^3 + 4x$ $\quad x^6 - 3x^5 - 5x^4 + 15x^3 + 4x^2 - 12x$

15. Give a polynomial whose value at a positive number r is the volume of a cube of side r. Evaluate your polynomial at 2 and 4 to find the volumes of cubes of sides 2 and 4, respectively.

16. An open tin box is formed by cutting congruent squares from the corners of a rectangular sheet and folding up the sides. If the sheet of tin measures 10×20 inches, give a polynomial whose value at r, $0 \le r \le 5$, is the volume of the box obtained by cutting squares of side r in from the corners. Find the volume of the box where the squares cut out have sides 2 inches long. Find the volume when the squares cut out have sides 3 inches long.

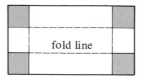

fold line

SECTION 2.3   THE ARITHMETIC OF POLYNOMIALS

Two polynomials can be either added or multiplied together. We suggested in the introduction to this chapter that the arithmetic properties of polynomials are very often identical to the arithmetic properties of the real numbers. In fact, addition and multiplication of polynomials are carried out using the same arithmetic rules that we have been using since we were introduced to numbers in grammar school. It will be useful to keep this in mind even though we will introduce some mathematical formalism when we give definitions for the sum and product of polynomials.

Suppose

$$P(x) = a_0 + a_1 x + a_2 x^2 + \cdots + a_n x^n$$
$$Q(x) = b_0 + b_1 x + b_2 x^2 + \cdots + b_n x^n$$

The **sum** of the polynomials $P(x)$ and $Q(x)$ is given by

$$P(x) + Q(x) = (a_0 + b_0) + (a_1 + b_1)x$$
$$+ (a_2 + b_2)x^2 + \cdots + (a_n + b_n)x^n$$

Example 1

A. $(x^2 + 2x + 3) + (x^3 + 5x + 1)$
   $= (0x^3 + 1x^2 + 2x + 3) + (1x^3 + 0x^2 + 5x + 1)$
   $= (0 + 1)x^3 + (1 + 0)x^2 + (2 + 5)x + (3 + 1)$
   $= x^3 + x^2 + 7x + 4$

B. $(x^3 + 2x^2 - 3x + 7) + (2x^3 + x^2 - 3x - 7)$
   $= (1 + 2)x^3 + (2 + 1)x^2 + [-3 + (-3)]x + [7 + (-7)]$
   $= 3x^3 + 3x^2 - 6x$

An important point to remember is that when adding two polynomials, terms with like powers of x, such as $a_k x^k$ and $b_k x^k$, are condensed into a single term,

$(a_k + b_k)x^k$. A common error in adding polynomials is to condense terms with unlike powers of x into a single term. For example, if $P(x) = 2x^2$ and $Q(x) = 5x^3$, many students write $P(x) + Q(x) = 2x^2 + 5x^3$ as $7x^5$ or $7x^6$, neither of which is the sum of $P(x)$ and $Q(x)$. If you have difficulty, write $P(x)$ and $Q(x)$ with the same number of terms and follow the rule outlined in our definition of addition. For example,

$$P(x) + Q(x) = (0x^3 + 2x^2 + 0x + 0) + (5x^3 + 0x^2 + 0x + 0)$$
$$= 5x^3 + 2x^2$$

Although we have not defined how to take the product of two polynomials, we will illustrate how to take the product of $P(x) = 2x + 3$ and $Q(x) = x^3 + 4x^2 + x - 5$ assuming we can use the usual rules of arithmetic.

In particular, we will use the distributive law, which says: If a, b, c are real numbers, then $a(b + c) = ab + ac$. For example, $5(3 + 2 + 7) = 5(3) + 5(2) + 5(7)$. We will also use the following fact from elementary algebra:

$$a_j x^j \cdot b_k x^k = a_j b_k x^{j+k} \tag{1}$$

Now,

$$P(x)Q(x) = (2x + 3)(x^3 + 4x^2 + x - 5)$$
$$= (2x + 3)x^3 + (2x + 3)4x^2$$
$$\qquad + (2x + 3)x + (2x + 3)(-5)$$
$$= (2x)x^3 + 3x^3 + (2x)4x^2 + 3(4x^2)$$
$$\qquad + (2x)x + 3x + (2x)(-5) + 3(-5)$$
$$= 2x^4 + (3x^3 + 8x^3) + (12x^2 + 2x^2)$$
$$\qquad + [3x + (-10)x] + (-15)$$
$$= 2x^4 + 11x^3 + 14x^2 - 7x - 15$$

The above example illustrates that $P(x)Q(x)$ is the sum of all products of terms in $P(x)$ with terms in $Q(x)$. Since we know how to add polynomials, understanding multiplication only requires that we know how to take the product of arbitrary terms. Of course, (1) above tells us how to take these products.

Example 2

$$5(x^2 + 3x - 3) = 5x^2 + 15x - 15$$

Example 3

$$(x + 5)(x^2 + 3x - 2) = (x + 5)x^2 + (x + 5)(3x)$$
$$\qquad + (x + 5)(-2)$$
$$= x^3 + 5x^2 + 3x^2 + 15x - 2x - 10$$
$$= x^3 + 8x^2 + 13x - 10$$

The following example illustrates a multiplication technique that is frequently useful when taking the product of polynomials with many terms. This technique is based on the fact that if

$$P(x) = a_0 + a_1 x + \cdots + a_n x^n$$

and

$$Q(x) = b_0 + b_1 x + \cdots + b_m x^m$$

then

$$P(x)Q(x) = P(x)(b_0 + b_1 x + \cdots + b_m x^m)$$
$$= P(x)b_0 + P(x)b_1 x + \cdots + P(x)b_m x^m$$

That is, $P(x)Q(x)$ is the sum of the products of $P(x)$ and the terms of $Q(x)$.

Example 4

The product $(x^4 + 3x^2 - 2x - 3)(-x^3 + 2x^2 + x - 5)$ can be found as follows:

$$x^4 + 3x^2 - 2x - 3$$
$$\underline{-x^3 + 2x^2 + x - 5}$$

$-x^7 \qquad\quad - 3x^5 + 2x^4 + 3x^3$	(2)
$2x^6 \qquad\quad + 6x^4 - 4x^3 - 6x^2$	(3)
$x^5 \qquad\quad + 3x^3 - 2x^2 - 3x$	(4)
$- 5x^4 \qquad\quad - 15x^2 + 10x + 15$	(5)
$\overline{-x^7 + 2x^6 - 2x^5 + 3x^4 + 2x^3 - 23x^2 + 7x + 15}$	(6)

If $P(x) = x^4 + 3x^2 - 2x - 3$ and $Q(x) = -x^3 + 2x^2 + x - 5$, then the polynomials (2)–(5) are the products of $P(x)$ with the terms of $Q(x)$. In particular, (2), (3), (4), and (5) are the products $P(x) \cdot (-x^3)$, $P(x) \cdot (2x^2)$, $P(x) \cdot (x)$, and $P(x) \cdot (-5)$, respectively. The polynomial (6) is the sum of (2)–(5) and, of course, is our product.

If

$$P(x) = a_0 + a_1 x + a_2 x^2 + \cdots + a_n x^n$$

and

$$Q(x) = b_0 + b_1 x + b_2 x^2 + \cdots + b_n x^n$$

then we define the **difference** $P(x) - Q(x)$ as follows:

$$P(x) - Q(x) = P(x) + (-1)Q(x)$$

For example,

$$(2x^2 + 3x - 7) - (x^2 - 4x + 2)$$
$$= (2x^2 + 3x - 7) + (-1)(x^2 - 4x + 2)$$
$$= (2x^2 + 3x - 7) + (-x^2 + 4x - 2)$$
$$= x^2 + 7x - 9$$

Now,

$$(-1)(b_0 + b_1 x + b_2 x^2 + \cdots + b_n x^n)$$
$$= -b_0 + (-b_1)x + \cdots + (-b_n)x^n$$

and $(-1)P(x)$ is most often written as $-P(x)$. For example,

$$-(x^5 + 7x^2 - 41) = -x^5 - 7x^2 + 41$$

In general,

$$(a_0 + a_1 x + a_2 x^2 + \cdots + a_n x^n) - (b_0 + b_1 x + b_2 x^2 + \cdots + b_n x^n)$$
$$= (a_0 - b_0) + (a_1 - b_1)x + (a_2 - b_2)x^2 + \cdots + (a_n - b_n)x^n$$

Example 5

> A. $(x^3 + 3x^2 - 6x + 2) - (2x^2 + 5x - 7)$
> $$= x^3 + (3 - 2)x^2 + (-6 - 5)x + (2 + 7)$$
> $$= x^3 + x^2 - 11x + 9$$
>
> B. $(x^2 + 5x - 3) - (2x^3 + x^2 + 7x - 5)$
> $$= -2x^3 - 2x + 2$$

We have assumed throughout this section that polynomial addition and multiplication obey many of the rules that real numbers obey. We have, in fact, used these rules in developing and motivating our definitions. To be complete, these properties are shown below.

If $P(x)$, $Q(x)$, and $R(x)$ are any three polynomials in x, then

$$P(x) + Q(x) = Q(x) + P(x) \tag{7}$$

$$[P(x) + Q(x)] + R(x) = P(x) + [Q(x) + R(x)] \tag{8}$$

$$P(x) + 0 = P(x) \tag{9}$$

There is a polynomial called $-P(x)$ such that

$$P(x) + [-P(x)] = 0 \tag{10}$$

$$P(x)Q(x) = Q(x)P(x) \tag{11}$$

$$[P(x)Q(x)]R(x) = P(x)[Q(x)R(x)] \tag{12}$$

$$P(x) \cdot 1 = P(x) \tag{13}$$

$$P(x)[Q(x) + R(x)] = P(x)Q(x) + P(x)R(x) \tag{14}$$

Exercises 2.3   In Problems 1–10, find $P(x) + Q(x)$, $P(x) - Q(x)$, and $P(x) \cdot Q(x)$.

1. $P(x) = x + 3$ $Q(x) = 3x - 4$
2. $P(x) = 5$ $Q(x) = x^4 + 7x^3 + 6x^2 + 2x - 3$
3. $P(x) = 0$ $Q(x) = x^5 + x^3 + 1$
4. $P(x) = x^5 + x^3 + 1$ $Q(x) = 1$
5. $P(x) = x^4 + x^3 + x^2 + x + 1$ $Q(x) = x - 1$
6. $P(x) = x^3 + x^2 + x + 1$ $Q(x) = x^3 + 2x^2 + 3x - 5$
7. $P(x) = x^2 + 3x - 2$ $Q(x) = 2x^2 - x + 4$
8. $P(x) = 2x^3 + x^2 - 3x - 6$ $Q(x) = -x^2 + 3x + 2$
9. $P(x) = x^{40} + x + 1$ $Q(x) = 3x^2 - 2x - 2$
10. $P(x) = x - 1$ $Q(x) = x^4 + x^3 + x^2 + x + 1$

11. If $P(x)$ and $Q(x)$ are polynomials of degree 1, $P(x) = ax + b$ and $Q(x) = cx + d$ ($a \neq 0$ and $c \neq 0$), then $P(x)Q(x) = (ax + b)(cx + d) = acx^2 + (ad + bc)x + bd$. Four multiplications are involved in computing this product.

First Terms $(ax \cdot cx)$

Outer Terms $(ax \cdot d) = (ad)x$

Inner Terms $(b \cdot cx) = (bc)x$

Last Terms $(b \cdot d)$

As an example:

$$(2x + 4)(x + 3) = 2x^2 + 6x + 4x + 12 = 2x^2 + 10x + 12$$

$$\underset{\text{First}}{\uparrow} \quad \underset{\text{Outer}}{\uparrow} \quad \underset{\text{Inner}}{\uparrow} \quad \underset{\text{Last}}{\uparrow}$$

Use this method to find the following products:

a. $(3x + 1)(2x - 3)$ b. $(x - 4)(x + 5)$
c. $(2x - 2)(-3x + 1)$ d. $(-2x + 2)(7x + 4)$
e. $(3x - 10)(x - 8)$ f. $(\sqrt{2}x - 1)(x + \pi)$

12. Let $P(x)$ and $Q(x)$ be polynomials of degrees m and n, respectively. What can be said about the degrees of

a. $P(x) + Q(x)$?   b. $P(x)Q(x)$?

13. Show that properties (7)–(14) hold if $P(x)$, $Q(x)$, and $R(x)$ are the polynomials defined by

$$P(x) = x^2 + 1 \qquad Q(x) = x^2 - 1 \qquad R(x) = 3x + 2$$

If $R(x)$ is the product of the polynomials $P(x)$ and $Q(x)$, the value of $R(x)$ at a real number r can be found by replacing x by r in the polynomial representing the product or by taking the product of the values of $P(x)$ and $Q(x)$ at r. For example, if $P(x) = x + 5$ and $Q(x) = 2x - 1$, then $R(x) = P(x)Q(x) = 2x^2 + 9x - 5$. The value of $R(x)$ at 2 is $2(2)^2 + 9(2) - 5 = 21$. The values of $P(x)$ and $Q(x)$ at 2 are $2 + 5 = 7$ and $2(2) - 1 = 3$, respectively, and $P(2) \cdot Q(2) = 7 \cdot 3 = 21$, the value of $R(x)$ at 2. In problems 14–18, find the value of $P(x)Q(x)$ at the indicated value of r.

14. $P(x) = x - 1$ $Q(x) = x^2 + 5$ $r = -2$

15. $P(x) = x - 1$ $Q(x) = x^2 + 5$ $r = 1$

16. $P(x) = (x - 1)(x - 3)$
      $Q(x) = x^{40} + x^{39} + x^{38} + x^{15} + x^7 + x^6 + x + 1$ $r = 3$

17. $P(x) = (x + 3)(x - 2)$ $Q(x) = x^{1000} - 1$ $r = -3$

18. $P(x) = x^{500} - 1$ $Q(x) = x^{45} + 1$ $r = 1$

19.   **a.**   Prove that each of the following equalities is valid by taking the product of the polynomials on the left-hand side of the equality sign.

$$(x - a)(x + a) = x^2 - a^2$$
$$(x - a)(x^2 + ax + a^2) = x^3 - a^3$$
$$(x - a)(x^3 + ax^2 + a^2x + a^3) = x^4 - a^4$$
$$(x - a)(x^4 + ax^3 + a^2x^2 + a^3x + a^4) = x^5 - a^5$$

   b.   Find a polynomial $P(x)$ such that
$$(x - a)P(x) = x^6 - a^6$$

   c.   Find a polynomial $P(x)$ such that
$$(x - a)P(x) = x^n - a^n$$
where n is an integer larger than 1.

20.   **a.**   Prove that each of the following equalities is valid by taking the product of the polynomials on the left-hand side of the equality sign.

$$(x + a)(x^2 - ax + a^2) = x^3 + a^3$$
$$(x + a)(x^4 - ax^3 + a^2x^2 - a^3x + a^4) = x^5 + a^5$$

   b.   Find a polynomial $P(x)$ such that
$$(x + a)P(x) = x^7 + a^7$$

21. Show that there is *no* polynomial $P(x)$ such that

$$(x + 2)P(x) = x^2 + 2^2$$

(*Hint:* If $P(x)$ exists, it is of the form $cx + d$. Why?)

22. Show that there is no polynomial $P(x)$ such that

$$(x + 2)P(x) = x^4 + 2^4$$

(*Hint:* If $P(x)$ exists, it is of the form $cx^3 + dx^2 + ex + f$.)

23. Give a polynomial $P(x)$ whose value at a positive number r is the perimeter of a square of side r.

24. Give a polynomial $P(x)$ whose value at a positive number r is the area of a circle of radius r.

25. To purchase posters from a certain printer, you must pay a fixed charge of $10, which covers the cost of setting up the equipment to print the posters, and a charge of $.10 for each poster printed. Give a polynomial whose value at each positive integral value r gives the cost of purchasing r posters.

26. A Norman window is a window whose shape is a rectangle surmounted by a semicircle.

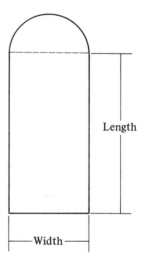

a. Assuming that the rectangular portion has a length that is twice its width, give a polynomial whose value at a positive number r gives the area of the window if the width of the rectangle is r units. Find the area of the window if the radius of the circle is 5 inches.

b. Assuming that the perimeter of the entire window is 100 inches, give a polynomial whose value at a positive number r is the area of the window if the width of the rectangle is r inches. Find the area of the window if the length is 2 inches.

SECTION 2.4   **ROOTS OF POLYNOMIALS**

In Section 2.2, we defined the **value** of a polynomial $P(x)$ at a real number r to be the numerical value obtained by replacing the variable x by the number r. We denote the value of P at r by $P(r)$. It is frequently important to know the real numbers at which a polynomial has value zero. We refer to numbers with this property so often that we have the following definition:

Definition

> A root (or zero) of a polynomial $P(x)$ is a real number r at which the value of the polynomial, $P(r)$, is 0.*

Example 1

> -1 and 2 are roots of $x^2 - x - 2$ since $(-1)^2 - (-1) - 2 = 0$ and $(2)^2 - 2 - 2 = 0$.
>
> The polynomial $x^4 - 3x^3 + x^2 - 3x$ has 0 and 3 as roots, since $0^4 - 3(0)^3 + (0)^2 - 3(0) = 0$ and $3^4 - 3(3)^3 + 3^2 - 3(3) = 0$.
>
> The polynomial $x^2 + 1$ has no roots, since if r is a real number, $r^2 \geq 0$ and hence $r^2 + 1 > 0$.

These examples illustrate that roots of polynomials are solutions of equations. For example, the roots of the polynomial

$$x^2 + x - 5 \tag{1}$$

are the solutions of the equation

$$x^2 + x - 5 = 0 \tag{2}$$

That is, the roots of (1) are those real numbers that, when substituted for x in (2), yield a true equality of numbers.

In general, finding the roots of

$$P(x) = a_0 + a_1 x + \cdots + a_n x^n$$

is equivalent to solving the equation

$$a_0 + a_1 x + \cdots + a_n x^n = 0$$

Polynomials appear frequently in scientific work. They have been used to describe phenomena studied in many disciplines including physics, chemistry, economics, and biology; and roots often give important information about the phenomena being studied.

For example, if an object is dropped from an initial height of 1000 feet and the effects of air resistance are ignored then the height of the object x seconds into its fall is given by the polynomial

$$P(x) = -16x^2 + 1000$$

To be specific, we mean that if t is a real number representing the number of

---

* Some students may have worked with complex numbers and know that they can be viewed as roots of polynomials. Except for the appendix to this chapter (Complex Numbers), we will study only the roots of polynomials that are real numbers.

seconds the object has been in free fall, then $P(t)$, the value of $P(x)$ at time t, is the height of the object.

To answer a question such as, "At what time is the object 500 feet high?" we must find a time t such that

$$500 = P(t) = -16t^2 + 1000$$

Equivalently,

$$0 = -16t^2 + 500$$

Solving this equation is equivalent to finding the roots of the polynomial $-16x^2 + 500$.

The remainder of this chapter is devoted to learning how to find roots of polynomials. We will begin by studying roots of **linear polynomials** (polynomials of degree 1) for two reasons. First, linear polynomials are quite useful, appearing frequently in scientific work. Second, roots of arbitrary polynomials can be found by finding roots of certain related linear polynomials.

2.4.1

Roots of Linear Polynomials

Let $P(x) = ax + b$ where $a \neq 0$. Then r is a root of $ax + b$ if and only if r satisfies the **linear equation** $ax + b = 0$. Now,

$$ar + b = 0$$

is equivalent to

$$ar = -b$$

which, in turn, is equivalent to

$$r = -\frac{b}{a}$$

The argument given above shows that each linear polynomial has exactly one root and tells us how to find the root.

In mathematics, the phrases "is equivalent to" and "if and only if" are used synonymously. In the next example and in subsequent work, we will use the symbol "$\Leftrightarrow$" instead of these phrases.

Example 2

To find the root of $2x + 3$ we proceed as follows:

$$2x + 3 = 0$$
$$\Leftrightarrow 2x = -3$$
$$\Leftrightarrow x = \frac{-3}{2}$$

Note that we did not replace x with r as in the discussion and calculations that preceded Example 2. It is customary in this context (and many others we will see throughout the text) to use a symbol like x to represent either a variable (or unknown) or a real number if no confusion arises from this practice.

Example 3

> The root of $-\pi x + \sqrt{2}$ is $\dfrac{\sqrt{2}}{\pi}$ since
>
> $$-\pi x + \sqrt{2} = 0$$
> $$\Leftrightarrow -\pi x = -\sqrt{2}$$
> $$\Leftrightarrow x = \frac{-\sqrt{2}}{-\pi} = \frac{\sqrt{2}}{\pi}$$

2.4.2

Roots of Quadratic Polynomials

A **quadratic polynomial** is a polynomial of degree 2. It has the form

$$ax^2 + bx + c \qquad \text{where} \quad a \neq 0$$

As we shall see, a second-degree polynomial has either 0, 1, or 2 roots, and there are several techniques available that can be used to either calculate the roots or conclude there are none. As in the general case, finding the roots of $ax^2 + bx + c$ is equivalent to solving the **quadratic equation**

$$ax^2 + bx + c = 0$$

Example 4 illustrates that it is possible for a quadratic polynomial to have either 0, 1, or 2 roots. Our discussion in Sections 2.4.2 and 2.4.3, showing how to calculate roots of quadratic polynomials, will prove that it is impossible for a quadratic polynomial to have more than two roots.

Example 4

> The polynomial $x^2 - 1$ has two roots. To find these roots, we must solve $x^2 - 1 = 0$. Since $x^2 - 1 = (x - 1)(x + 1)$, solving $x^2 - 1 = 0$ is equivalent to solving
>
> $$(x - 1)(x + 1) = 0$$
>
> A product of numbers $ab = 0$ if and only if $a = 0$ or $b = 0$, so
>
> $$x^2 - 1 = 0$$
> $$\Leftrightarrow x - 1 = 0 \quad \text{or} \quad x + 1 = 0$$
> $$\Leftrightarrow x = 1 \quad \text{or} \quad x = -1$$
>
> Hence $x^2 - 1$ has two roots, 1 and -1.

The polynomial $x^2 - 2x + 1 = (x - 1)^2$ has exactly one root. In fact, x is a real number such that

$$(x - 1)^2 = 0$$
$$\Leftrightarrow x - 1 = 0$$
$$\Leftrightarrow x = 1$$

The polynomial $x^2 + 1$ has no roots, for if x were a real number satisfying

$$x^2 + 1 = 0$$

then

$$x^2 = -1$$

which is impossible since the square of a real number cannot be negative.

Example 5

To find the roots of the polynomial given by $(x - 2)^2 - 9$, we solve the equation $(x - 2)^2 - 9 = 0$.

$$(x - 2)^2 - 9 = 0$$
$$\Leftrightarrow (x - 2)^2 = 9$$
$$\Leftrightarrow x - 2 = -3 \quad \text{or} \quad x - 2 = 3$$
$$\Leftrightarrow x = 2 - 3 = -1 \quad \text{or} \quad x = 2 + 3 = 5$$

The polynomial given by $(x + 3)^2 + 9$ has no roots for if

$$(x + 3)^2 + 9 = 0$$

then

$$(x + 3)^2 = -9$$

which is impossible since for each real number x, $(x + 3)^2 \geq 0$.

Example 5 illustrates how to find roots of quadratic polynomials that can be written in the form

$$(x - d)^2 - e \tag{1}$$

In general, x is a root of this polynomial if and only if

$$(x - d)^2 = e \tag{2}$$

If $e < 0$, then our polynomial has no roots, for if x is a real number $(x - d)^2 \geq 0$.

If $e = 0$, then (2) becomes

$$(x - d)^2 = 0$$

and we can see that $x = d$ is the only root of our polynomial.

If $e > 0$, then $(x - d)^2 = e$ if and only if $x - d = \sqrt{e}$ or $x - d = -\sqrt{e}$. In this case, our polynomial has two roots, $x = d + \sqrt{e}$ and $x = d - \sqrt{e}$.

The problem of determining if an arbitrary second-degree polynomial has roots and, if it does, the value of the roots can always be solved by examining a polynomial of the form given in (1). Before we can show how this is done, we will need to learn a technique known as **completing the square.**

Completing the square is a method used to determine what constant, c, must be added to a polynomial of the form $x^2 + bx$ to make $x^2 + bx + c$ a perfect square. This means we want

$$x^2 + bx + c = (x + d)^2$$

for some constant d.

Example 6

> A.   If we add 9 to $x^2 + 6x$, we will have completed the square of $x^2 + 6x$. In particular,
>
> $$x^2 + 6x + 9 = (x + 3)^2$$
>
> B.   To complete the square of $x^2 - 8x$, we want to find a number c such that
>
> $$x^2 - 8x + c = (x + d)^2 \qquad\qquad (3)$$
>
> for some number d. Now (3) is equivalent to
>
> $$x^2 - 8x + c = x^2 + 2dx + d^2$$
>
> So to complete the square, we must choose
>
> $$c = d^2$$
>
> where
>
> $$2d = -8$$
>
> Hence,
>
> $$d = \frac{-8}{2} = -4 \quad \text{and} \quad c = d^2 = (-4)^2 = 16$$
>
> Therefore,
>
> $$x^2 - 8x + 16 = (x - 4)^2$$

Now in general,

$$x^2 + bx + c = (x + d)^2$$

if and only if

$$x^2 + bx + c = x^2 + 2dx + d^2$$

which is equivalent to requiring that

$$b = 2d \quad \text{and} \quad c = d^2$$

That is,

$$d = \tfrac{1}{2}b \quad \text{and} \quad c = (\tfrac{1}{2}b)^2$$

In general, to complete the square of $x^2 + bx$, we add on $(\tfrac{1}{2}b)^2$, the square of one-half the coefficient of x. Then

$$x^2 + bx + (\tfrac{1}{2}b)^2 = \left(x + \frac{b}{2}\right)^2$$

Example 7

To complete the square for the polynomial

$$x^2 + 6x$$

we add on $(\tfrac{1}{2}6)^2 = 9$ since, in this case, $b = 6$. Note that

$$x^2 + 6x + 9 = (x + 3)^2$$

To complete the square for the polynomial

$$x^2 - \pi x$$

we add on

$$[\tfrac{1}{2}(-\pi)]^2 = \frac{\pi^2}{4}$$

since $b = -\pi$. Note that

$$x^2 - \pi x + \frac{\pi^2}{4} = \left(x - \frac{\pi}{2}\right)^2$$

Exercises 2.4.2

In Exercises 1–10, find the roots of the given linear polynomial.

1. $2x - 3$ 2. $-7x + 2$ 3. $-5x - 3$
4. $\pi x + 4$ 5. $\tfrac{1}{4}y - 3$ 6. $-\tfrac{1}{8}t + \tfrac{1}{2}$
7. $0.05x - 3$ 8. $\pi x + \pi^2$ 9. $\sqrt{2}x + 5$
10. $\sqrt{3}x - 3$

Problems 11–16 can be reduced to finding the root of a linear polynomial.

11. For what real number does $2x + 3$ assume the value 7?
12. For what real number do $x + 5$ and $2x + 1$ have the same value?
13. For what real number does $3x + 5$ assume a value 4 greater than the value assumed by $8x - 3$?
14. For what real number is the value of $8x - 2$ twice the value of $2x + 3$?
15. Show that there is no real number for which the value of $8x - 2$ is twice the value of $4x + 1$.
16. Show that for each real number the value of $8x - 2$ is twice the value of $4x - 1$.

In Problems 17–23, find the roots, if any, of the given polynomials.

17. $(x + 3)^2 - 16$ 18. $(x - 7)^2 + 25$ 19. $(x + 8)^2$
20. $(x + 4)^2 - \pi$ 21. $(x + 3)^2 + 5$ 22. $(x - 4)^2$
23. $(x - 12)^2 + \sqrt{2}$

In Exercises 24–32, complete the square of the given polynomial, and write your result in the form $(x - c)^2$.

24. $x^2 + 10x$ 25. $x^2 - 9x$ 26. $x^2 + 5x$
27. $x^2 - \pi x$ 28. $x^2 - x$ 29. $x^2 + x$
30. $x^2 + 12x$ 31. $x^2 - 12x$ 32. x^2

33. A company has purchased a machine to manufacture bolts. The cost of the machine is $5,000. The cost of producing 1000 bolts (above the cost of the machine) is $5. Find a linear polynomial whose value at a nonnegative real number r gives the total cost of producing r thousand bolts. What is the cost of producing 10,000 bolts? 20,000 bolts? How many bolts can be produced for $6000? $7000?

34. a. Complete the square of $4x^2 + 20x$, and show your answer can be expressed in the form $(rx + s)^2$.
    b. Repeat (a) for the polynomial $3x^2 - 4x$.
    c. Assuming $a > 0$, what constant c must be added to $ax^2 + bx$ to create a perfect square? (*Hint:* Write $ax^2 + bx + c = (rx + s)^2$ and expand the right-hand side.)
    d. Explain why you cannot complete the square of $-x^2 + 2x$.

35. Find the roots of $6x^2 - 5x + 1$ using the procedure outlined in (a)–(e) below.

    a. Reduce $6x^2 - 5x + 1 = 0$ to an equation of the form

    $$x^2 + bx - d = 0 \tag{4}$$

    b. Rewrite equation (4) as

    $$x^2 + bx = d \tag{5}$$

c. Complete the square: Construct equation (6) equivalent to (5) by adding to both sides of (5) that number c which completes the square of $x^2 + bx$.

$$x^2 + bx + c = d + c \qquad (6)$$

d. Use the fact that you have completed the square on $x^2 + bx$ to write (6) in the form (2) of the text.

e. Find the roots by the methods of this section.

2.4.3

Roots of Quadratic Polynomials, Continued

Completing the square has many applications in mathematics. In Example 1, we will illustrate how to use the method of completing the square to find the roots of quadratic polynomials. Later in this section we will use the technique outlined in Example 1 to derive a formula, called *the quadratic formula*, which can be used to find the roots of quadratic polynomials.

Example 1

In this example, we will determine the roots of $2x^2 - 3x + 1$ and $3x^2 - 3x + 1$ by completing the square.

$$2x^2 - 3x + 1 = 0$$
$$\Leftrightarrow 2x^2 - 3x = -1$$
$$\Leftrightarrow x^2 - \frac{3}{2}x = -\frac{1}{2}$$
$$\Leftrightarrow x^2 - \frac{3}{2}x + \frac{9}{16} = -\frac{1}{2} + \frac{9}{16} = \frac{1}{16} \qquad \text{(Here we are completing the square)}$$
$$\Leftrightarrow \left(x - \frac{3}{4}\right)^2 = \frac{1}{16}$$
$$\Leftrightarrow \left(x - \frac{3}{4}\right) = \pm\frac{1}{4}$$
$$\Leftrightarrow x = \frac{3}{4} \pm \frac{1}{4}$$
$$\Leftrightarrow x = 1 \quad \text{or} \quad x = \frac{1}{2}$$

Similarly,

$$3x^2 - 3x + 1 = 0$$
$$\Leftrightarrow 3x^2 - 3x = -1$$
$$\Leftrightarrow x^2 - x = -\frac{1}{3}$$

$$\Leftrightarrow x^2 - x + \frac{1}{4} = -\frac{1}{3} + \frac{1}{4} = -\frac{1}{12} \quad \text{(Here we are completing the square)}$$

$$\Leftrightarrow \left(x - \frac{1}{2}\right)^2 = -\frac{1}{12}$$

This polynomial has no roots since it is impossible to find a real number x such that $(x - \frac{1}{2})^2$ is negative.

Suppose we are given an arbitrary quadratic polynomial

$$ax^2 + bx + c, \, a \neq 0 \tag{1}$$

and are asked to find its roots. We can use the same method we used in Example 1.

A real number x satisfies

$$ax^2 + bx + c = 0 \tag{2}$$

if and only if it satisfies each of the following equalities:

$$ax^2 + bx = -c \tag{3}$$

$$x^2 + \frac{b}{a}x = -\frac{c}{a} \tag{4}$$

By adding $[(1/2)(b/a)]^2$ to both sides of the equality (4), we complete the square of the left-hand side and obtain the equivalent equations:

$$x^2 + \frac{b}{a}x + \left[\left(\frac{1}{2}\right)\left(\frac{b}{a}\right)\right]^2 = \left[\left(\frac{1}{2}\right)\left(\frac{b}{a}\right)\right]^2 - \frac{c}{a} \tag{5}$$

$$\left(x + \frac{b}{2a}\right)^2 = \frac{b^2}{4a^2} - \frac{c}{a} \tag{6}$$

$$\left(x + \frac{b}{2a}\right)^2 = \frac{b^2 - 4ac}{4a^2} \tag{7}$$

Equation (7) has the form $(x - d)^2 = e$, which we learned how to solve in Section 2.4.2. Since $4a^2 > 0$, the sign of $(b^2 - 4ac)/4a^2$ is determined by $b^2 - 4ac$. We can draw the following conclusions about the roots of the polynomial given in (1).

A. If $b^2 - 4ac < 0$, then (1) has no roots.

B. If $b^2 - 4ac = 0$, then (1) has exactly one root, namely

$$x = -\frac{b}{2a}$$

C.   If $b^2 - 4ac > 0$, then (1) has two roots. In fact

$$x + \frac{b}{2a} = \pm \sqrt{\frac{b^2 - 4ac}{4a^2}}$$

$$x = -\frac{b}{2a} \pm \frac{\sqrt{b^2 - 4ac}}{2a}$$

$$x = \frac{-b \pm \sqrt{b^2 - 4ac}}{2a} \tag{8}$$

Formula (8) is called the **quadratic formula** and should be memorized. It can always be used to calculate the roots of a quadratic, whether there are one or two. Note that, if the expression under the square root symbol is negative, the polynomial has no roots.

Example 2

To use the quadratic formula to find the roots of

$$3x^2 - 2x - 4$$

note that $a = 3$, $b = -2$, and $c = -4$.
   The roots are given by

$$x = \frac{-(-2) \pm \sqrt{(-2)^2 - 4(3)(-4)}}{2(3)}$$

$$= \frac{2 \pm \sqrt{52}}{6}$$

$$= \frac{2 \pm 2\sqrt{13}}{6} = \frac{1 \pm \sqrt{13}}{3}$$

Example 3

To use the quadratic formula to find the roots of

$$x^2 + 6x + 9$$

note that $a = 1$, $b = 6$, and $c = 9$. The roots are given by

$$x = \frac{-6 \pm \sqrt{6^2 - 4(1)(9)}}{2(1)} = \frac{-6 \pm \sqrt{0}}{2} = -3$$

This polynomial has only one root.

Example 4

> To use the quadratic formula to find the roots of
>
> $$x^2 + x + 1$$
>
> note that $a = b = c = 1$. The roots are given by
>
> $$x = \frac{-1 \pm \sqrt{(1)^2 - 4(1)(1)}}{2(1)} = \frac{-1 \pm \sqrt{-3}}{2}$$
>
> Since the number under the square root symbol (i.e., $b^2 - 4ac$) is negative, we conclude that this polynomial has no roots.

We now have two methods for finding roots of quadratic polynomials. One method uses completing the square and the other uses the quadratic formula. We are now going to consider a third method. In the course of describing this method, we will study an important relationship between the roots of a quadratic polynomial and representations of that polynomial as a product.

Recall that, in Example 4 of Section 2.4.2, we showed that 1 and -1 were roots of $x^2 - 1$ and that $x^2 - 1$ could be expressed as a product of $(x - 1)$ and $(x + 1)$:

$$x^2 - 1 = (x - 1)(x + 1)$$

The polynomial $2x^2 - 3x + 1$ has 1 and $\frac{1}{2}$ as roots. (See Example 1.) Note that

$$2(x - \tfrac{1}{2})(x - 1) = 2(x^2 - \tfrac{3}{2}x + \tfrac{1}{2}) = 2x^2 - 3x + 1$$

The polynomial $x^2 - 2x + 1$ has one root, namely 1, and can be written as the product

$$(x - 1)^2 = (x - 1)(x - 1) = x^2 - 2x + 1$$

In Example 4, we showed that the polynomial $x^2 + x + 1$ has no roots. It is impossible to find linear polynomials $ax + b$ and $cx + d$ such that

$$x^2 + x + 1 = (ax + b)(cx + d)$$

If there did exist such real numbers a, b, c, and d, then

$$x^2 + x + 1 = 0$$
$$\Leftrightarrow (ax + b)(cx + d) = 0$$
$$\Leftrightarrow ax + b = 0 \quad \text{or} \quad cx + d = 0$$
$$\Leftrightarrow x = -\frac{b}{a} \quad \text{or} \quad x = -\frac{d}{c}$$

We see that if $x^2 + x + 1 = (ax + b)(cx + d)$ then the polynomial $x^2 + x + 1$ has roots, namely, $-(b/a)$ and $-(d/c)$. Since we know that $x^2 + x + 1$ has no roots, it must follow that $x^2 + x + 1$ cannot be written as a product of linear polynomials.

In general, if the quadratic polynomial $ax^2 + bx + c$ has roots r_1 and r_2 (where r_1 may equal r_2), we claim

$$ax^2 + bx + c = a(x - r_1)(x - r_2) \qquad (9)$$

Since we can use the quadratic formula to find the roots, we can check to see if (9) is correct. The roots of the polynomial $ax^2 + bx + c$ are

$$r_1 = \frac{-b + \sqrt{b^2 - 4ac}}{2a}$$

and

$$r_2 = \frac{-b - \sqrt{b^2 - 4ac}}{2a}$$

Now,

$$a(x - r_1)(x - r_2)$$

$$= a\left(x - \frac{-b + \sqrt{b^2 - 4ac}}{2a}\right)\left(x - \frac{-b - \sqrt{b^2 - 4ac}}{2a}\right)$$

$$= a\left[x^2 - \left(\frac{-b + \sqrt{b^2 - 4ac}}{2a}\right)x\right.$$

$$\left. - \left(\frac{-b - \sqrt{b^2 - 4ac}}{2a}\right)x + \frac{b^2 - (b^2 - 4ac)}{4a^2}\right]$$

$$= a\left(x^2 + \frac{b}{a}x + \frac{c}{a}\right) = ax^2 + bx + c$$

Conversely if $ax^2 + bx + c = a(x - r_1)(x - r_2)$, then the roots of $ax^2 + bx + c$ are r_1 and r_2.

Note that, if $ax^2 + bx + c$ has no roots, then it cannot be written as a product of linear polynomials, and if it cannot be written as a product of linear polynomials it has no roots.

Example 5

We can express $x^2 - 7x + 3$ as a product of linear polynomials $(x - r_1)$ $(x - r_2)$. (*Note:* $a = 1$.) The roots of $x^2 - 7x + 3$ can be found using the quadratic formula.

$$x = \frac{7 \pm \sqrt{7^2 - 4(3)}}{2} = \frac{7 \pm \sqrt{37}}{2}$$

The factorization is

$$x^2 - 7x + 3 = \left[x - \left(\frac{7 + \sqrt{37}}{2}\right)\right]\left[x - \left(\frac{7 - \sqrt{37}}{2}\right)\right]$$

Our discussions prior to Example 5 suggest a third method—the **factorization method**—for finding roots of quadratic polynomials. If $Q(x)$ is a quadratic polynomial, and we can find linear polynomials $ax + b$, $cx + d$ such that $Q(x) = (ax + b)(cx + d)$, we say $Q(x)$ has a **linear factorization.** The polynomials $ax + b$ and $cx + d$ are called **(linear) factors** of $Q(x)$. Now,

$$0 = Q(x) = (ax + b)(cx + d)$$

if and only if

$$ax + b = 0 \quad \text{or} \quad cx + d = 0$$

Hence, the roots of $Q(x)$ are the roots of the linear factors.

Of course, the factorization method is only useful if we can find a linear factorization for a quadratic polynomial. Recall that

$$(ax + b)(cx + d) = acx^2 + adx + bcx + db$$

The four terms acx^2, adx, acx, and bd represent the product of first, outer, inner, and last terms of the factors $ax + b$ and $cx + d$. Factorization of quadratics, which often involves trial and error, tries to reverse this process.

Suppose we are asked to find a linear factorization for $x^2 + 11x + 30$. We start by writing

$$x^2 + 11x + 30 = (x \quad)(x \quad)$$

since we want the product of the first terms to be x^2. (This is only one possible choice, since we could have written, for example, $(2x \quad)(\frac{1}{2}x \quad)$. However, because no fractions appear in any other coefficients the factorization $(x \quad)(x \quad)$ seems like a reasonable choice.) We know that the product of outer terms should be 30. This gives us a number of choices: 6 and 5, 10 and 3, -10 and -3, 30 and 1, and others. However, only one choice, 6 and 5, will give us $11x$ for the middle term:

$$(x + 5)(x + 6) = x^2 + 6x + 5x + 30 = x^2 + 11x + 30$$

Notice that the coefficient of x, i.e., 11, is just the sum of 5 and 6. This will always be the case if the coefficients of x in the linear factors are both 1.

Example 6

To find the roots of $x^2 + 3x - 4$, note that

$$x^2 + 3x - 4 = (x + 4)(x - 1)$$

so

$$0 = x^2 + 3x - 4 = (x + 4)(x - 1)$$
$$\Leftrightarrow x + 4 = 0 \quad \text{or} \quad x - 1 = 0$$
$$\Leftrightarrow x = -4 \quad \text{or} \quad x = 1$$

If you can easily factor a quadratic polynomial, the factorization technique is more convenient to use than the quadratic formula or completing the square. The method, however, has some drawbacks. First, factors may be difficult to find. (For example, we wonder how many students realize that $4x^2 + 7x + \frac{3}{2} = (2x + 3)(2x + \frac{1}{2})$.) Second, some quadratic polynomials have no linear factors and any search is futile.

We suggest that you first try the factorization method. However, if the factors are not easy to find, rely on the quadratic formula to determine the roots, if any, of a quadratic polynomial.

Exercises 2.4.3

In Exercises 1–8, determine the roots of the given polynomials using the quadratic formula.

1. $x^2 + 7x + 12$
2. $3x^2 - 4x + 1$
3. $2x^2 - 7x + 7$
4. $2x^2 + 8x + 8$
5. $-8x^2 + 2x + 3$
6. $25x^2 + 20x + 3$
7. $x^2 - 4x + 5$
8. $-4x^2 + 4x - 37$

In Exercises 9–16, determine the roots of the given polynomials by completing the square.

9. $x^2 - 4x + 4$
10. $x^2 + 8x - 9$
11. $x^2 - 3x + 3$
12. $3x^2 + 15x - 42$
13. $2x^2 + 4x + 3$
14. $-7x^2 + 6x - 1$
15. $\frac{1}{2}x^2 - \sqrt{6}x + 3$
16. $x^2 + \pi x - \pi$

In Exercises 17–24, determine the roots of the given polynomials by factoring. (See Example 6.)

17. $x^2 - x - 6$
18. $2x^2 - 5x + 3$
19. $4x^2 + 3x - 1$
20. $2x^2 - x - 21$
21. $2x^2 - 8$
22. $x^2 - 3$
23. $x^2 + (\pi^2 - \pi)x - \pi^3$
24. $2x^2 - 20x + 42$

25. Factor the following quadratic polynomials into products of linear polynomials. (*Hint:* Use the quadratic formula to find the roots of the

given polynomial and then use the roots to find the factors.)

    a. $x^2 + 3x - 5$ **b.** $2x^2 + \pi x + 1$ **c.** $3x^2 + 7x - 5$

26. Find a quadratic polynomial whose roots are 7 and -4.

27. Find a quadratic polynomial whose roots are -3 and -5.

28. Find a quadratic polynomial of the form $2x^2 + bx + c$ having roots $2 + \sqrt{3}$ and $2 - \sqrt{3}$.

29. Find a quadratic polynomial of the form $x^2 + bx + c$ having roots $-\pi$ and $\sqrt{7}$.

30. This problem is a review of ideas that can aid in the factorization of polynomials.

    a. Find formulas relating b and c to s_1 and s_2 if

$$x^2 + bx + c = (x - s_1)(x - s_2)$$

    and illustrate your results using $x^2 + 2x - 35$.

    b. Find formulas relating a, b, and c to r_1, s_1, r_2 and s_2 if

$$ax^2 + bx + c = (r_1 x + s_1)(r_2 x + s_2)$$

    and illustrate your results using $6x^2 + 7x - 5$.

31. A farmer has 1000 feet of fencing to make a rectangular field. One side of the field is bounded by a river and will require no fencing. Give a polynomial $P(x)$ whose value at a number r is the area of the rectangular field enclosed by the fencing if the width of the field is r feet. What should be the width if the area of the field is to be 80,000 ft²? 100,000 ft²?

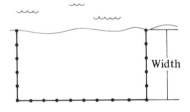

32. The printed matter on a page is to occupy a rectangle 5×7 inches. The margins at the top and bottom of the page are to be twice the margins on the sides of the page. Give a polynomial whose value at a number r is the area of a page with a side margin of r inches. What are the dimensions of the page if the area of the page is 75 in.²?

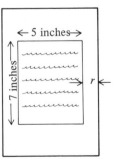

33. If air resistance is neglected, the height of an object dropped from an initial height of 1600 feet x seconds into flight is given by the polynomial

$$P(x) = -16x^2 + 1600$$

Find the time it takes the object to hit the ground. Find the time it takes the object to fall halfway to the ground.

SECTION 2.5   FACTORIZATION AND DIVISION OF POLYNOMIALS

Frequently it is useful in mathematics to represent an integer as a product of other integers. Such a representation is called a **factorization** and the integers that are multiplied together are called **factors.** When we write $48 = 2 \cdot 3 \cdot 8$ we have written a factorization of 48, and 2, 3, and 8 are the factors. We have exactly the same conventions for polynomials.

If $P(x)$, $P_1(x)$, . . . , $P_n(x)$ are polynomials such that $P(x) = P_1(x)P_2(x) \cdots P_n(x)$, then the product $P_1(x)P_2(x) \cdots P_n(x)$ is called a **factorization** of $P(x)$, and $P_1(x), \ldots, P_n(x)$ are called **factors** of $P(x)$. We have seen many examples of factorizations of quadratic polynomials in Section 2.4. For example, we saw

$$x^2 - 1 = (x - 1)(x + 1)$$
$$2x^2 - 3x + 1 = 2(x - \tfrac{1}{2})(x - 1)$$

Other examples of factorizations are

$$x^3 - 27 = (x - 3)(x^2 + 3x + 9)$$
$$x^4 - 16 = (x - 2)(x + 2)(x^2 + 4)$$
$$x^4 + 2x^3 - 7x^2 - 8x + 12 = (x - 1)(x - 2)(x + 2)(x + 3)$$

These factorizations may not be apparent, but they can easily be verified by taking the products of the terms in the factorizations on the right-hand sides.

You will recall that there are some polynomials $P(x)$ that cannot be factored into a product of polynomials whose degrees are less than the degree of $P(x)$ itself. For example, $x^2 + x + 1$ cannot be factored into a product of linear factors since $x^2 + x + 1$ has no roots.

Surprisingly, this is not the case for polynomials of degree greater than 2. An important result, which we shall not prove, concerning polynomials and their factorizations is:

> If $P(x)$ is a polynomial of degree greater than 2, then $P(x)$ can be factored into a product
>
> $$P(x) = P_1(x)P_2(x)$$
>
> where the degrees of $P_1(x)$ and $P_2(x)$ are at least 1.

This result does not tell us what the factors are or how to find them. It just assures us that there are factors.

Example 1

The polynomial $x^4 + 1$ has a factor with degree greater than or equal to 1. In fact,

$$x^4 + 1 = \left(x^2 - \sqrt{2}x + 1\right)\left(x^2 + \sqrt{2}x + 1\right)$$

The polynomial $x^2 - \sqrt{2}x + 1$ will not factor into a product of linear factors. If it did, it would have roots. Using the quadratic formula the roots would be

$$\frac{-\left(-\sqrt{2}\right) \pm \sqrt{\left(-\sqrt{2}\right)^2 - 4(1)(1)}}{2(1)} = \frac{\sqrt{2} \pm \sqrt{2 - 4}}{2}$$

$$= \frac{\sqrt{2} \pm \sqrt{-2}}{2}$$

Since the number under the second square root sign is negative, we can conclude $x^2 - \sqrt{2}x + 1$ has no roots. An analogous argument shows $x^2 + \sqrt{2}x + 1$ will not factor into a product of linear factors.

Example 2

The polynomial $x^4 - 1$ can be factored since its degree is $4 > 2$.

$$x^4 - 1 = (x^2 - 1)(x^2 + 1) = (x - 1)(x + 1)(x^2 + 1)$$

Recall that $x^2 + 1$ has no roots and, hence, no linear factors.

Suppose $P(x)$ is a polynomial that has been factored into a product $P_1(x)P_2(x)$ where the degrees of $P_1(x)$ and $P_2(x)$ are at least 1. If the degree of $P_1(x)$ is greater than 2, it can be factored into a product $Q_1(x)Q_2(x)$ where the degrees of $Q_1(x)$ and $Q_2(x)$ are at least 1. Then

$$P(x) = Q_1(x)Q_2(x)P_2(x)$$

Continuing in this way, we see that $P(x)$ could be factored into a product of linear and/or quadratic factors. We know that a quadratic factor can be factored into linear factors if it has roots; otherwise, it is not factorable. We call such nonfactorable quadratics **irreducible.** Then:

> If $P(x)$ is a polynomial whose degree is at least 1 then $P(x)$ can be factored into a product of linear and/or irreducible quadratic factors.

Each of the factorizations that we have seen in this section illustrate this type of factorization.

Example 3

The polynomial $x^6 - 1$ has the factorization

$$x^6 - 1 = (x^3 - 1)(x^3 + 1)$$

Since the degrees of $x^3 - 1$ and $x^3 + 1$ are greater than 2, these polynomials can also be factored:

$$x^3 - 1 = (x - 1)(x^2 + x + 1)$$
$$x^3 + 1 = (x + 1)(x^2 - x + 1)$$

So,

$$x^6 - 1 = (x - 1)(x + 1)(x^2 + x + 1)(x^2 - x + 1)$$

Both $x^2 + x + 1$ and $x^2 - x + 1$ are irreducible.

One reason we are interested in factoring polynomials is that the factorizations help us find roots, as the next example illustrates.

Example 4

We saw $x^4 - 16 = (x - 2)(x + 2)(x^2 + 4)$. Now x is a root of this polynomial if and only if

$$(x - 2)(x + 2)(x^2 + 4) = 0$$

which is equivalent to

$$x - 2 = 0 \qquad x + 2 = 0 \quad \text{or} \quad x^2 + 4 = 0$$

Since $x^2 + 4$ is not factorable, it is impossible for $x^2 + 4 = 0$. Thus, the roots of $x^4 - 16$ satisfy $x - 2 = 0$ or $x + 2 = 0$, and so the roots are 2 and -2.

In Example 4, the roots of $x^4 - 16$ were precisely the roots of the linear factors of $x^4 - 16$. This illustrates an important property about polynomials and their roots:

> If $P(x)$ is a polynomial, then r is a root of $P(x)$, if and only if r is the root of a linear factor of $P(x)$.

To prove this statement, suppose $P(x)$ is a polynomial having r as a root. We know that every polynomial is the product of linear and irreducible quadratic factors. To simplify our discussion, we assume that

$$P(x) = L_1(x)L_2(x)Q_1(x)Q_2(x)$$

where $L_1(x)$ and $L_2(x)$ are the linear factors and $Q_1(x)$ and $Q_2(x)$ are the irreducible quadratics. (Our work can be easily generalized to an arbitrary number of linear and quadratic factors.) Now,

$$P(r) = 0$$

if and only if

$$L_1(r) = 0 \qquad L_2(r) = 0 \qquad Q_1(r) = 0 \quad \text{or} \quad Q_2(r) = 0$$

$Q_1(x)$ and $Q_2(x)$ are irreducible and have no roots, so $Q_1(r) \neq 0$ and $Q_2(r) \neq 0$. Therefore, $P(r) = 0$ if and only if $L_1(r) = 0$ or $L_2(r) = 0$. Hence, r is a root of $P(x)$ if and only if it is a root of a linear factor.

Example 5

$$5x^4 + 12x^3 - 29x^2 - 48x + 36$$
$$= (5x - 3)(x - 2)(x + 2)(x + 3)$$
$$= 5(x - \tfrac{3}{5})(x - 2)(x + 2)(x + 3)$$

The roots of this polynomial are $\tfrac{3}{5}$, 2, -2, and -3.

In Example 5, we first factored the polynomial as $(5x - 3)(x - 2)(x + 2)$ $(x + 3)$. The root of the first linear factor is $\tfrac{3}{5}$. In the second line, we factored the first linear factor into $5(x - \tfrac{3}{5})$. This illustrates an important fact about a root, r, of a polynomial and the possible linear factors of $P(x)$ having r as a root:

> r is a root of a polynomial, $P(x)$, if and only if $x - r$ is a factor of $P(x)$.

To prove this fact, suppose

$$P(x) = (x - r)Q(x)$$

(That is, $x - r$ is a factor of $P(x)$.) Then

$$P(r) = (r - r)Q(r) = 0 \cdot Q(r) = 0$$

so r is a root of $P(x)$.

Now suppose r is a root of $P(x)$. There is a linear factor $ax + b$ of $P(x)$ having r as a root. Recall that $r = -(b/a)$. If

$$P(x) = (ax + b)Q(x)$$

then

$$P(x) = a\left[x - \left(-\frac{b}{a}\right)\right]Q(x) = a(x - r)Q(x)$$

so $x - r$ is a factor of $P(x)$.

Example 6

The polynomial $x^4 + 1$ has no roots. We saw

$$x^4 + 1 = (x^2 - \sqrt{2}x + 1)(x^2 + \sqrt{2}x + 1)$$

and both factors are unfactorable (i.e., have no roots).

Example 7

The polynomial $x^3 - 27$ has one root since

$$x^3 - 27 = (x - 3)(x^2 + 3x + 9)$$

and $x^2 + 3x + 9$ is unfactorable. The root, of course, is 3.

We have emphasized factorization in order to use factorization of polynomials to determine roots. Our discussions also show how roots may be used to determine factorizations of a polynomial. In particular, if we know that r is a root of $P(x)$ then

$$P(x) = (x - r)Q(x)$$

In addition, if s is a root of $P(x)$ different from r, then $0 = P(s) = (s - r)Q(r)$. Since $s - r \neq 0$, $Q(s) = 0$. Conversely, if $Q(s) = 0$, then s is a root of $P(x)$. Hence, the roots of $P(x)$ different from r are the roots of $Q(x)$. The following examples illustrate how these ideas can be useful in finding all the roots of a polynomial.

Example 8

By inspection (i.e., trial and error) we see that 1 is a root of $x^3 + x - 2$. Indeed $(1)^3 + (1) - 2 = 0$, so $x - 1$ is a factor of $x^3 + x - 2$. By taking the product on the right-hand side of the following equality, we see

$$x^3 + x - 2 = (x - 1)(x^2 + x + 2)$$

Any roots, other than 1, of $x^3 + x - 2$ must be roots of $x^2 + x + 2$. The roots of $x^2 + x + 2$ can be found using the quadratic formula,

$$x^2 + x + 2 = 0$$
$$x = \frac{-1 \pm \sqrt{(1)^2 - 4(1)(2)}}{2} = \frac{-1 \pm \sqrt{-7}}{2}$$

Therefore, $x^2 + x + 2$ has no roots, and 1 is the only root of $x^3 + x - 2$.

Many students feel uncomfortable seeing "trial and error" suggested as a method for solving problems in mathematics. Surely mathematicians could have thought of something better! In fact, trial and error methods are often used to solve problems (and with good results). Facility in using this method to find roots of polynomials comes with practice (which you will get doing the exercises).

We can, however, suggest one fact that can help you find integer roots of polynomials with integer coefficients. If $P(x) = a_n x^n + a_{n-1} x^{n-1} + \cdots + a_1 x + a_0$, and $a_n, a_{n-1}, \ldots, a_1, a_0$ **are integers and r is an integer root of $P(x)$, then r divides the constant coefficient a_0 (i.e., a_0/r is an integer).** For the polynomial of Example 8, the only possible integer roots are ± 1 and ± 2 since these are the only integers that divide the constant coefficient, -2.

Example 9

By inspection, -1 is a root of

$$x^4 + 3x^3 + x^2 - 3x - 2$$
$$(\text{because } (-1)^4 + 3(-1)^3 + (-1)^2 - 3(-1) - 2 = 0)$$

So, $[x - (-1)] = x + 1$ is a factor of this polynomial. In fact

$$x^4 + 3x^3 + x^2 - 3x - 2 = (x + 1)(x^3 + 2x^2 - x - 2)$$

By inspection, -2 is a root of

$$x^3 + 2x^2 - x - 2$$
$$(\text{because } (-2)^3 + 2(-2)^2 - (-2) - 2 = 0)$$

and we can easily see that

$$x^3 + 2x^2 - x - 2 = [x - (-2)](x^2 - 1)$$

So

$$x^4 + 3x^3 + x^2 - 3x - 2 = (x + 1)(x + 2)(x^2 - 1)$$

Finally

$$x^2 - 1 = (x - 1)(x + 1)$$

so

$$x^4 + 3x^3 + x^2 - 3x - 2 = (x + 1)(x + 2)(x - 1)(x + 1)$$

and the roots of our polynomial are -1, -2, and 1.

Examples 8 and 9 illustrate an important method of finding roots of a polynomial $P(x)$: If r is known to be a root of $P(x)$, then $P(x) = (x - r)Q(x)$; the remaining roots of $P(x)$ are the roots of $Q(x)$ and, since the degree of $Q(x)$ is less than the degree of $P(x)$, it is often easier to use $Q(x)$ rather than $P(x)$ itself to find these roots.

The next obvious question is: how do you find $Q(x)$? In Example 9

$$P(x) = x^4 + 3x^3 + x^2 - 3x - 2$$

$$r = -1$$

$$Q(x) = x^3 + 2x^2 - x - 2$$

since $x^4 + 3x^3 + x^2 - 3x - 2 = [x - (-1)](x^3 + 2x^2 - x - 2)$. To most people, such factorizations are not obvious. Fortunately, there is a division algorithm similar to the long division algorithm for real numbers that can be used to find the factor $Q(x)$. Since this technique of dividing polynomials is useful in many mathematical contexts, we will discuss how, in general, to divide one polynomial by another.

Before we begin to describe this division algorithm, notice that, if $P(x)$ and $Q(x)$ are polynomials, then $Q(x)$ will not necessarily divide $P(x)$ "evenly"; that is, there may not be any polynomial $S(x)$ such that $P(x) = Q(x)S(x)$ (i.e., $P(x)/Q(x) = S(x)$). Recall that the quotient of two integers is not necessarily an integer (for example $17/7$ is not an integer) but it is possible to express the ratio of two integers p/q as an integer plus a non-negative fraction less than 1. For example, $17/7 = 2 + (3/7)$ and, in general, $p/q = s + (r/q)$ where $0 \le r < q$. Equivalently, $p = q \cdot s + r$ where $0 \le r < q$. A similar result is true for polynomials.

> If $P(x)$ and $Q(x) \ne 0$ are polynomials, then there are (unique) polynomials $S(x)$ and $R(x)$ such that
>
> $$P(x) = S(x)Q(x) + R(x) \qquad (1)$$
>
> and $R(x) = 0$ or the degree of $R(x)$ is less than the degree of $Q(x)$. $S(x)$ is called the **quotient** and $R(x)$ is the remainder of dividing $P(x)$ by $Q(x)$.

Frequently, equation (1) is written using the following notation:

$$\frac{P(x)}{Q(x)} = S(x) + \frac{R(x)}{Q(x)} \qquad (2)$$

Although we will not formally discuss the meaning of ratios of polynomials, such as $(x + 1)/(x^2 - 3)$, until Chapter 3, we will use the fractional notation in (2) since it suggests that we are attempting to divide $P(x)$ by $Q(x)$ and that $S(x)$ is the quotient and $R(x)$ the remainder of division.

Rather than describing this algorithm in complete generality, we will illustrate the algorithm when $P(x) = 2x^3 + 1$ and $Q(x) = x - 1$ hoping that you will be able to easily generalize the algorithm for arbitrary $P(x)$ and $Q(x)$.

We begin by selecting ax^k such that the degree of

$$P(x) - ax^k Q(x) = 2x^3 + 1 - ax^k(x - 1)$$

is less than the degree of $P(x)$. A little reflection shows that ax^k is found by dividing x (the leading term in $x - 1$) into $2x^3$ (the leading term in $2x^3 + 1$). In fact, $2x^3/x = 2x^2$. Note

$$(2x^3 + 1) - 2x^2(x - 1) = 2x^2 + 1 \tag{3}$$

Repeat the above process with $2x^2 + 1$ replacing $2x^3 + 1$. Dividing $2x^2$ by x, we get $2x^2/x = 2x$ so

$$(2x^2 + 1) - 2x(x - 1) = 2x + 1 \tag{4}$$

Combining (3) and (4) we get

$$(2x^3 + 1) - 2x^2(x - 1) - 2x(x - 1) = 2x + 1$$

i.e.,

$$(2x^3 + 1) - (2x^2 + 2x)(x - 1) = 2x + 1 \tag{5}$$

We repeat our process once more with $2x + 1$ replacing $2x^3 + 1$. Dividing $2x$ by x we get $2x/x = 2$ so

$$2x + 1 - 2(x - 1) = 3 \tag{6}$$

Combining (5) and (6) we get

$$(2x^3 + 1) - (2x^2 + 2x)(x - 1) - 2(x - 1) = 3$$

i.e.,

$$(2x^3 + 1) - (2x^2 + 2x + 2)(x - 1) = 3 \tag{7}$$

The polynomial 3, on the right-hand side of (7), has a degree less than our divisor $Q(x) = x - 1$. We stop at this point. Rearranging (7) we get

$$(2x^3 + 1) = (2x^2 + 2x + 2)(x - 1) + 3 \tag{8}$$

For our particular example, $S(x) = 2x^2 + 2x + 2$ and $R(x) = 3$. Equation (8) corresponds to equation (1) above. Equation (8) can also be written in the form

$$\frac{2x^3 + 1}{x - 1} = (2x^2 + 2x + 2) + \frac{3}{x - 1} \tag{9}$$

Equation (9) corresponds to equation (2) above.

It is not necessary to carry out the calculations as we did above. The following method, which resembles "long division," will help you divide one polynomial by another to find quotient and remainder.

Step 1 Write $P(x)$ and $Q(x)$ in descending powers of x as follows: $Q(x) \overline{)\ P(x)}$. For our example,

$$x - 1 \overline{)\ 2x^3 + 1}$$

Step 2 Divide the leading term of $P(x)$ by the leading term of $Q(x)$ and enter the result above $P(x) = 2x^3 + 1$.

$$\begin{array}{r} 2x^2 \\ x - 1 \overline{)\ 2x^3 + 1} \end{array} \qquad \left(\frac{2x^3}{x} = 2x^2 \right)$$

Step 3 Multiply $Q(x) = x - 1$ by $2x^2$ and enter the result below $2x^3 + 1$.

$$\begin{array}{r} 2x^2 \\ x - 1 \overline{)\ 2x^3 + 1} \\ 2x^3 - 2x^2 \end{array}$$

Step 4 Subtract the new line from the one above it.

$$\begin{array}{r} 2x^2 \\ x - 1 \overline{)\ 2x^3 + 1} \\ \underline{2x^3 - 2x^2 } \\ 2x^2 + 1 \end{array}$$

Steps 5–? Repeat steps 2–4 using the last line written to replace $P(x)$.

$$\begin{array}{r} 2x^2 + 2x \\ x - 1 \overline{)\ 2x^3 + 1} \\ \underline{2x^3 - 2x^2 } \\ 2x^2 + 1 \end{array} \qquad \left(\frac{2x^2}{x} = 2x \right)$$

$$\begin{array}{r} 2x^2 + 2x \\ x - 1 \overline{)\ 2x^3 + 1} \\ \underline{2x^3 - 2x^2 } \\ 2x^2 + 1 \\ 2x^2 - 2x \end{array}$$

$$\begin{array}{r} 2x^2 + 2x \\ x - 1 \overline{)\ 2x^3 + 1} \\ \underline{2x^3 - 2x^2 } \\ 2x^2 + 1 \\ \underline{2x^2 - 2x} \\ 2x + 1 \end{array}$$

$$\begin{array}{r}
2x^2 + 2x + 2 \\
x - 1 \overline{\smash{\big)}\ 2x^3 + 1} \\
\underline{2x^3 - 2x^2} \\
2x^2 + 1 \\
\underline{2x^2 - 2x} \\
2x + 1
\end{array}$$

$$\left(\frac{2x}{x} = 2\right)$$

$$\begin{array}{r}
2x^2 + 2x + 2 \\
x - 1 \overline{\smash{\big)}\ 2x^3 + 1} \\
\underline{2x^3 - 2x^2} \\
2x^2 + 1 \\
\underline{2x^2 - 2x} \\
2x + 1 \\
\underline{2x - 2} \\
3
\end{array}$$

Last Step Stop when the current remainder is 0 or has a degree less than your divisor. In our example, we stop when we get a remainder of 3, which has a degree less than the degree of divisor, $x - 1$. Your quotient, $(2x^2 + 2x + 2)$, is found on the first line and your remainder, 3, on the last.

Example 10

> If $P(x) = x^4 + 5x^2 + 2x - 1$ and $Q(x) = x^2 + 3$, the polynomials $S(x)$ and $R(x)$ such that
>
> $$P(x) = S(x)Q(x) + R(x)$$
>
> and $R(x) = 0$ or the degree of $R(x)$ is less than the degree of $S(x)$ are found as follows:
>
> $$\begin{array}{r}
> x^2 + 2 \\
> x^2 + 3 \overline{\smash{\big)}\ x^4 + 5x^2 + 2x - 1} \\
> \underline{x^4 + 3x^2} \\
> 2x^2 + 2x - 1 \\
> \underline{2x^2 + 6} \\
> 2x - 7
> \end{array}$$
>
> $$S(x) = x^2 + 2 \qquad R(x) = 2x - 7$$

Example 11

> When the degree of $Q(x)$ is greater than the degree of $P(x)$, then $S(x)$ and $R(x)$ of equation (1) above are 0 and $Q(x)$, respectively. For example, if $P(x) = x^2 + 1$ and $Q(x) = x^3$, then
>
> $$(x^2 + 1) = 0x^3 + (x^2 + 1)$$

Suppose $P(x)$ is a polynomial and $Q(x) = x - r$. Since the degree of $Q(x)$ is 1 and polynomials of degree 0 are real numbers, we can express $P(x)$ as

$$P(x) = Q(x)S(x) + R$$

where R is a real number. If we use the notation $P(r)$ to indicate the value of $P(x)$ when $x = r$, then r is a root of $P(x)$ if and only if $P(r) = 0$.

Since $P(x) = Q(x)S(x) + R = (x - r)S(x) + R$,

$$P(r) = (r - r)S(r) + R$$

Therefore, $P(r) = 0$ if and only if $0[S(r)] + R = 0$, which is equivalent to $R = 0$. We see again that r is a root of $P(x)$ if and only if $x - r$ divides $P(x)$ evenly. So, when r is a root of $P(x)$, to find $Q(x)$ such that $P(x) = (x - r)Q(x)$, just divide $P(x)$ by $x - r$ according to our algorithm. The remainder will always be zero.

Example 12

By inspection, a root of $x^4 + x^3 + 2x^2 + 4x - 8$ is seen to be 1. $(1^4 + 1^3 + 2(1)^2 + 4(1) - 8 = 0.)$

$$
\begin{array}{r}
x^3 + 2x^2 + 4x + 8 \\
x - 1 \overline{\smash{)}\, x^4 + x^3 + 2x^2 + 4x - 8} \\
\underline{x^4 - x^3} \\
2x^3 + 2x^2 + 4x - 8 \\
\underline{2x^3 - 2x^2} \\
4x^2 + 4x - 8 \\
\underline{4x^2 - 4x} \\
8x - 8 \\
\underline{8x - 8} \\
0
\end{array}
$$

So

$$x^4 + x^3 + 2x^2 + 4x - 8 = (x - 1)(x^3 + 2x^2 + 4x + 8)$$

By inspection, we can see -2 is a root of $x^3 + 2x^2 + 4x + 8$.

$$
\begin{array}{r}
x^2 + 4 \\
x + 2 \overline{\smash{)}\, x^3 + 2x^2 + 4x + 8} \\
\underline{x^3 + 2x^2} \\
4x + 8 \\
\underline{4x + 8} \\
0
\end{array}
$$

So

$$x^4 + x^3 + 2x^2 + 4x - 8 = (x - 1)(x + 2)(x^2 + 4)$$

The polynomial $x^2 + 4$ cannot be factored.

Exercises 2.5

In each of the Exercises 1–10, two polynomials $P(x)$ and $Q(x)$ are given. Use the division algorithm to find polynomials $S(x)$ and $R(x)$ such that $P(x)$, $Q(x)$, $R(x)$, and $S(x)$ satisfy formula (2) of the text.

1. $P(x) = x^5 + 1 \qquad Q(x) = x^2 - 2$
2. $P(x) = x^4 + 3x^2 - 7x + 3 \qquad Q(x) = 2x^2 + x - 4$
3. $P(x) = 3x^3 + 2x^2 + x \qquad Q(x) = x - 1$
4. $P(x) = x^5 + 3x^4 + 4x^2 + x - 2 \qquad Q(x) = x^2 + 1$
5. $P(x) = x^2 + 2x - 3 \qquad Q(x) = x^3 + 5x - 1$
6. $P(x) = x^7 + x^5 - 3x^4 + x^3 - 2x^2 - 3x - 1 \qquad Q(x) = x^3 - 3$
7. $P(x) = x^3 + 1 \qquad Q(x) = x - 1$
8. $P(x) = x^4 + 1 \qquad Q(x) = x - 1$
9. $P(x) = 3x^5 + 5x^4 + 7x^3 + 3x^2 + 3x - 5 \qquad Q(x) = x^2 + 3x$
10. $P(x) = x^5 + 7x^3 - 3x^2 + 10x - 15 \qquad Q(x) = x^2 - 15$

In Exercises 11–20, factor the given polynomial into a product of linear and/or irreducible quadratic factors, and find all roots of the polynomial.

11. $x^4 - x^2$
12. $x^3 + 6x^2 + 3x - 10$
13. $x^4 + 3x^3 + 18x^2 + 48x + 32$
14. $x^3 - 8$
15. $x^4 + 10x^2 + 24$
16. $x^4 - 13x^2 + 36$
17. $x^4 - 36$
18. $x^3 - 6x^2 + 12x - 8$
19. $x^4 - 4bx^3 + 6b^2x^2 - 4b^3x + b^4$ (b, a constant)
20. $x^4 + 4bx^3 + 6b^2x^2 + 4b^3x + b^4$ (b, a constant)

21. If $P(x)$ is a polynomial of degree n, what is the maximum number of roots $P(x)$ may have? What is the minimum number?

In Exercises 22–28, give a polynomial, $P(x)$, with the specified properties.

22. $P(x)$ has 1, 0, 2, and -3 as roots.
23. $P(x)$ has degree 3, has -1 as a root, and has an irreducible quadratic factor.
24. $P(x)$ has degree 3, has -1 as its only root, and has no irreducible quadratic factors.
25. $P(x)$ has degree 4 and no roots.
26. $P(x)$ has degree 3, and has 2 and 3 as its only roots.
27. $P(x)$ has degree 4, and has -1, 0, and 1 as its only roots.
28. $P(x)$ has degree 5, and has π and $\sqrt{2}$ as its only roots.

29. If $P(x) = a x^n + a_{n-1}x^{n-1} + \cdots + a_1 x + a_0$, and 0 is a root of $P(x)$, what is the value of a_0?

30. A box with a rectangular base has a volume of 160 in.^{3}. Its length is twice its width, and its height is 1 inch longer than its width. What are the dimensions of the box?

31. The strength, S, of a rectangular beam is given by $S = 2d^2w$ where d is the depth and w the width of the beam. Find the depth and width of a beam for which $S = 108$ and the depth is one-half the width.

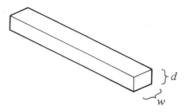

32. Solve Problem 31 if the depth is one unit larger than the width and $S = 96$.

33. An oil drum in the shape of a right circular cylinder is to be constructed to have a volume of 80π ft^3. If the height of the can is to be 1 foot longer than the radius of the base, what are the dimensions of the can? (*Hint:* The volume of a right circular cylinder with height h and base radius r is $V = \pi r^2 h$.)

APPENDIX CHAPTER 2

COMPLEX NUMBERS

Suppose the only numbers we had any knowledge of were the positive integers. (This should not be difficult to suppose. There was a time, admittedly long ago, when this was the case.) If we wanted to solve the equation $x + 1 = 0$, we could not. We now know that the solution is $x = -1$. However, if we were unaware of negative numbers, we would have to say that the equation is unsolvable. Similarly, we would say that the equations $x + 2 = 0$ and $x + 3 = 0$ were also unsolvable. It is only when we enlarge our set of numbers to include the symbols -1, -2, and -3, symbols we now call negative numbers, that we can solve these equations.

We are in a similar position when we work with real numbers. We saw that there are quadratic equations with no real numbers as solutions. Examples are $x^2 + 1 = 0$, $x^2 + x + 2 = 0$, and $x^2 + 2x + 4 = 0$. In each case, when we try to solve the equations using the quadratic formula, we get a negative value for $b^2 - 4ac$. Since there is no real number whose square is negative, we say that the equations are unsolvable.

It is possible, and in more advanced mathematics very helpful, to enlarge our collection of numbers in such a way that these equations also have solutions. This can be accomplished by introducing a new symbol, i, defined by

$$i^2 = -1$$

The symbol, i, does not represent any real number, since squaring real numbers gives non-negative results. When the symbol was introduced in the sixteenth century, it was called an imaginary number, a name we still use.

We can use the imaginary number, i, to represent roots of quadratic polynomials that have no real roots. Suppose we wanted to find the roots of

$$P(x) = x^2 - 2x + 2$$

We solve the equation

$$x^2 - 2x + 2 = 0$$

using the quadratic formula with $a = 1$, $b = -2$, and $c = 2$. As the solution, we get

$$x = \frac{2 \pm \sqrt{4 - 4 \cdot 1 \cdot 2}}{2} = \frac{2 \pm \sqrt{-4}}{2}$$

that is,

$$x = \frac{2 + \sqrt{-4}}{2} \quad \text{or} \quad x = \frac{2 - \sqrt{-4}}{2}$$

Since there is a negative number under the square root sign, we conclude that there are no real roots. However, we can represent the roots using i. We will perform arithmetic with the symbol i by assuming that it obeys all the usual rules of arithmetic obeyed by real numbers, but with the additional condition that $i^2 = -1$. Since $-4 = 4(-1)$, $\sqrt{-4} = \sqrt{4} \cdot \sqrt{-1} = 2i$. Substituting this into our solution,

$$x = \frac{2 + 2i}{2} \quad \text{or} \quad x = \frac{2 - 2i}{2}$$

Dividing,

$$x = 1 + i \quad \text{or} \quad x = 1 - i$$

The roots of this polynomial have the form $a + bi$ where a and b are real numbers. (We can think of $1 + i$ as $a + bi$ where $a = 1$ and $b = 1$. Similarly, $1 - i = a + bi$ where $a = 1$ and $b = -1$.) Symbols of the form $a + bi$ are called **complex numbers.** A complex number, $a + bi$, consists of a **real part**, a, and an **imaginary part**, bi.

One reason that we call these symbols numbers is that there are rules for adding, subtracting, multiplying, and dividing them and deciding when they are equal, just as there are for real numbers. When we perform arithmetic with complex numbers, we treat the imaginary symbol i the same way we would treat a variable, x, in the expression $a + bx$, keeping in mind that $i^2 = -1$. The next example illustrates arithmetic with complex numbers.

Example 1

A. $(3 + i) + (4 + 2i) = 7 + 3i$ (Compare this with $(3 + x) + (4 + 2x)$.)

B. $6(7 + 4i) = 42 + 24i$

C. $(3 + i)(2 - 3i) = 6 + 2i - 9i - 3i^2$

$\qquad\qquad\qquad\quad = 6 - 7i - 3i^2$

$\qquad\qquad\qquad\quad = 6 - 7i + 3 \qquad$ (Since $i^2 = -1$.)

$\qquad\qquad\qquad\quad = 9 - 7i$

Alternatively,

$$
\begin{array}{r}
3 + i \\
2 - 3i \\
\hline
6 + 2i \\
-9i - 3i^2 \\
\hline
6 - 7i - 3i^2 = 9 - 7i
\end{array}
$$

Before we give an example of division of complex numbers, it will be helpful to introduce a new definition. **The conjugate** of the complex number $a + bi$ is the complex number $a - bi$. (An identical terminology is used for the real numbers $a + \sqrt{b}$ and $a - \sqrt{b}$.) Notice that if we multiply complex conjugates together, the result is a real number, not a complex number.

$$(a + bi)(a - bi) = a^2 + abi - abi - b^2i^2$$

$$= a^2 - b^2i^2$$

$$= a^2 + b^2 \qquad \text{(Since } i^2 = -1.)$$

We use conjugates to divide complex numbers as illustrated in the next example.

Example 2

We will use conjugates to find $(7 - i) \div (3 + i)$. Write the quotient as a fraction, $(7 - i)/(3 + i)$, and multiply numerator and denominator by the conjugate of the denominator.

$$\frac{7 - i}{3 + i} \cdot \frac{3 - i}{3 - i} = \frac{21 - 3i - 7i + i^2}{9 + 3i - 3i - i^2} = \frac{20 - 10i}{10} = 2 - i$$

This method of dividing complex numbers is very similar to a method you may have learned in algebra for rationalizing denominators.

We define equality for complex numbers by saying that $a + bi = c + di$ if and only if $a = c$ and $b = d$. Equivalently, complex numbers are equal if and only if their real parts are equal and their imaginary parts are equal.

Early in this section we said that the complex numbers $1 + i$ and $1 - i$ were solutions of the quadratic equation $x^2 - 2x + 2 = 0$. Now that we can do arithmetic with complex numbers, we can check this result. We will check that $1 + i$ is a solution and leave a similar check of $1 - i$ as an exercise.

$$(1 + i)^2 - 2(1 + i) + 2 = 1 + 2i + i^2 - 2 - 2i + 2$$
$$= 2i - 2 - 2i + 2$$
$$= 0$$

Example 3

To factor $x^2 + x + 2$, we solve the equation $x^2 + x + 2 = 0$. Using the quadratic formula,

$$x = \frac{-1 + \sqrt{-3}}{2} = \frac{-1 + i\sqrt{3}}{2} \quad \text{or} \quad x = \frac{-1 - i\sqrt{3}}{2}$$

Using the relationship between roots and factors,

$$x^2 + x + 2 = \left[x - \left(\frac{-1 + i\sqrt{3}}{2}\right)\right]\left[x + \left(\frac{-1 - i\sqrt{3}}{2}\right)\right]$$

In Section 2.5, we stated that every polynomial can be written as a product of linear and/or irreducible quadratic factors. As Example 3 shows, quadratics that are irreducible when we work with real numbers have linear factors when we work with complex numbers. As a result, when we work with complex numbers every polynomial can be written as a product of linear factors.

Exercises
Appendix

1. Check that $1 - i$ is also a solution of $x^2 - 2x + 2 = 0$.

In Exercises 2–10, find complex roots for each polynomial.

2. $x^2 - x - 3$	**3.** $2x^2 - 4x + 4$	**4.** $x^2 + x + 2$
5. $x^2 - 2x + 5$	**6.** $x^2 - 4x + 13$	**7.** $4x^2 - 4x + 5$
8. $x^2 - 4x + 8$	**9.** $2x^2 + x + 1$	**10.** $x^2 + 12$

11. Find a polynomial whose roots are $2 + i$ and $2 - i$.
12. Find a polynomial whose roots are $3 + 2i$ and $3 - 2i$.
13. **a.** Show that $-1 + i\sqrt{3}$ is a root of $x^3 - 8$.
    b. Show that $-1 - i\sqrt{3}$ is a root of $x^3 - 8$.
    c. By inspection, find a real root of $x^3 - 8$.
14. Find three cube roots of -1.   (*Hint:* This is equivalent to finding three roots of $x^3 + 1$. You should be able to find one real root by inspection. Use results we obtained about the relationship between roots and factors.)

Review Exercises
Chapter 2

In Exercises 1–5, find $P(x) + Q(x)$, $P(x) - Q(x)$ and $P(x)Q(x)$.

1. $P(x) = 3x^2 + 2x - 5$ $Q(x) = x^3 - 3x^2 + 2x - 5$
2. $P(x) = x^2 - 2$ $Q(x) = 3x + 5$
3. $P(x) = x^2 + 2\sqrt{2}x - 3$ $Q(x) = \pi x^3 - \sqrt{2}x + 1$
4. $P(x) = -x^4 + x^3 + 3x - 1$ $Q(x) = 2x^2 + 3x - 1$
5. $P(x) = x^3 + 3x^2 + 9x + 27$ $Q(x) = x - 3$

In Exercises 6–10, complete the squares of the given polynomials.

6. $x^2 + 10x$ **7.** $x^2 - 10x$ **8.** $x^2 + \pi x$
9. $x^2 - \sqrt{2}\,x$ **10.** $x^2 + \sqrt{3}\,x$

In Exercises 11–15, find the roots of the given quadratic polynomial by completing the square.

11. $2x^2 - 7x + 3$ **12.** $x^2 + 9x + 14$ **13.** $6x^2 - x - 1$
14. $15x^2 + x - 2$ **15.** $x^2 + 2x + 2$

In Exercises 16–20, find the roots of the given quadratic polynomial by factoring.

16. $x^2 - x - 12$ **17.** $2x^2 + 5x - 12$ **18.** $6x^2 + 17x + 5$
19. $x^2 - 5$ **20.** $x^2 + 25x + 156$

In Exercises 21–30, find the roots of the given quadratic polynomial by using the quadratic formula.

21. $25x^2 - 5x + 6$ **22.** $x^2 + x - 1$ **23.** $x^2 + x + 1$
24. $x^2 - 9x + 20$ **25.** $x^2 + 7x - 3$ **26.** $2x^2 - 3x - 4$
27. $15x^2 - x - 6$ **28.** $7x^2 - 15x + 2$ **29.** $3x^2 + 5x + 6$
30. $x^2 - 98x - 200$

In Exercises 31–40, find all roots of the given polynomial and factor the polynomial into a product of linear and irreducible quadratic factors.

31. $x^3 - 3x + 2$ **32.** $x^3 + 3x^2 + 5x + 15$
33. $x^4 - 8x^2 + 16$ **34.** $x^4 - 5x^2 + 6$
35. $x^4 + x^3 - 8x^2 - 9x - 9$
36. $x^5 - 3x^4 + 6x^3 - 18x^2 + 5x - 15$
37. $x^3 - 27$ **38.** $x^4 - 81$
39. $x^6 + 3x^5 + 7x^4 + 5x^3$ **40.** $x^7 + 2x^6 + x^5$

In Exercises 41–45, find polynomials $S(x)$ and $R(x)$ such that

$$P(x) = Q(x)S(x) + R(x)$$

and $R(x) = 0$ or the degree of $R(x)$ is less than the degree of $Q(x)$.

41. $P(x) = x^4 + 3x^2 + 2x - 3$ $Q(x) = x^2 + 2$
42. $P(x) = 3x^3 + 2x^2 + 4x + 1$ $Q(x) = x^2 + 1$
43. $P(x) = x^2 + 5$ $Q(x) = x^3 + 2x - 3$
44. $P(x) = x^3 + 2x - 3$ $Q(x) = x^2 + 5$
45. $P(x) = x^5 + 3x^3 + 2x^2 + x$ $Q(x) = -x^2 - 3x + 1$

46. Let $P(x) = x^2 + 2x + 2$.
  a. Find $P(2)$, $P(-2)$, $P(3)$, and $P(0)$.
  b. Find the real numbers r for which $P(r) = 5$.
  c. Find the real numbers r for which $P(r) = 7$.
  d. Find the real number r for which $P(r) = 1$.

47. Let $P(x) = x^3 + 7x^2 + 7x - 10$.
  a. Find $P(-2)$, $P(-1)$, $P(0)$, $P(1)$, and $P(2)$.
  b. Find the real numbers r for which $P(r) = 5$.
  c. Find the real numbers r for which $P(r) = 40$.

48. When a tire manufacturer uses r hours of labor he can produce $r^2 + r$ tires. The cost of labor is $5 per hour, the cost of the raw material used to make one tire is $4, and the fixed costs needed to set up the plant to begin production are $5000. Give a polynomial whose value at a number r gives the total cost in producing tires when r hours of labor are used. What is the cost of production when 100 hours of labor are used for production? If the company can spend $15,450 in tire production, how many hours of labor can be used? How many tires will be produced if $15,450 is spent for production?

49. A piece of wire 12 feet long is cut into two pieces. One piece is to be bent into the shape of a square. The other piece is to be bent into the shape of a rectangle that is twice as long as it is wide. Find a polynomial whose value at a number r is the sum of the areas of the two figures if a side of the square has length r. What is the sum of the areas if the piece of wire used to make the square is 4 feet long? What is the length of the wire used to make the square if the sum of the areas of the figures is 44/9 ft²?

50. An open box is to have a square base and a height which is 1 inch longer than twice its width. Find a polynomial whose value at a real number r gives the surface area of the box if the length of a side of the base is r inches. What is the surface area of the box if the base measures 4×4 inches? What is the surface area of the box if the height of the box is 11 inches? What are the dimensions of the box if its surface area is 518 in.³?

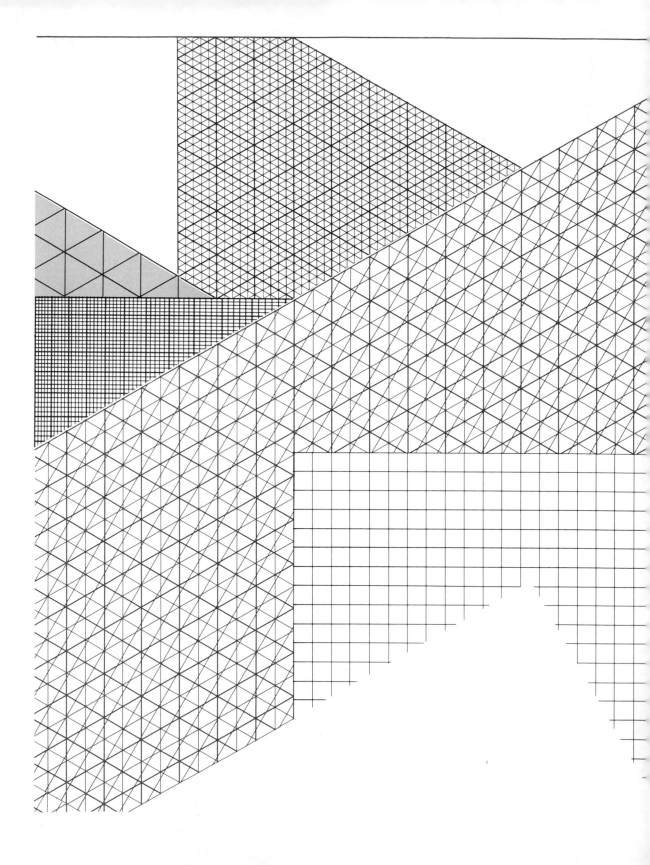

3 Rational Expressions

SECTION 3.1

INTRODUCTION

In Chapter 2, we saw that polynomials and real numbers possess many similar properties. Actually, the set of polynomials bears more resemblance to the set of integers than it does to the set of real numbers. In particular, we saw that the sum, difference, or product of two polynomials is also a polynomial. However, the quotient or ratio of two polynomials is not necessarily a polynomial. Note that analogous statements hold for the integers. The sum, difference, or product of two integers is an integer, but the ratio of two integers, for example, $\frac{7}{5}$, is not necessarily an integer.

Example 1

> To illustrate that the ratio of two polynomials may not be a polynomial, consider the ratio $(x^2 + 1)/x$. If there were a polynomial $P(x)$ such that $P(x) = (x^2 + 1)/x$, then we would require
>
> $$xP(x) = x^2 + 1$$
>
> The degree of $x^2 + 1$ is 2, and in order for $xP(x)$ to have degree 2, $P(x)$ must be of the form $ax + b$. But
>
> $$xP(x) = x(ax + b)$$
> $$= ax^2 + bx$$
>
> and the constant term of this polynomial is 0. The constant term of $x^2 + 1$ is 1, so it is impossible for $ax^2 + bx$ to equal $x^2 + 1$.

A little reflection on everyday calculations suggests that the set of integers is an inadequate collection of numbers. At the very least, we need the set of rational numbers to talk about fractional parts of quantities, such as inches or acres. Although very useful, the polynomials are also inadequate for scientific investigations. A collection that is more useful, and contains the set of polynomials, is the set of rational expressions, or ratios of polynomials.

Keep in mind as you read this chapter that rational expressions bear many similarities to rational numbers, and many arithmetic properties of rational expressions and rational numbers are identical.

SECTION 3.2   **DEFINITIONS AND PROPERTIES OF RATIONAL EXPRESSIONS**

> If $P(x)$ and $Q(x)$ are polynomials and $Q(x) \neq 0$, then $P(x)/Q(x)$ is called a **rational expression.**

Example 1

The following are rational expressions

$$\frac{x^2 + 1}{x} \qquad \frac{x^3 + x^2 + x + 1}{x^2 - 1}$$

$$\frac{x^{50} + \sqrt{2}\, x^{31} - \pi x + \sqrt{7}}{x + 2} \qquad \frac{0}{x - 1}$$

The following are *not* rational expressions

$$\frac{x^2 + 1}{0} \qquad \frac{0}{0} \qquad \frac{\sqrt{x + 1}}{2x - 5} \qquad \sqrt{\frac{x^2 + 1}{x - 1}}$$

Two rational expressions

$$\frac{P_1(x)}{Q_1(x)} \quad \text{and} \quad \frac{P_2(x)}{Q_2(x)}$$

are equal (we will sometimes say equivalent) if and only if

$$P_1(x)Q_2(x) = Q_1(x)P_2(x)$$

In this case we write

$$\frac{P_1(x)}{Q_1(x)} = \frac{P_2(x)}{Q_2(x)}$$

Note the similarity of equality of rational expressions and equality of rational numbers. For example, $4/5 = 8/10$ since $4 \cdot 10 = 5 \cdot 8$.

Example 2

A. $\dfrac{x^2 - 1}{x^2 - 2x + 1} = \dfrac{x + 1}{x - 1}$

since $(x^2 - 1)(x - 1) = (x^2 - 2x + 1)(x + 1)$.
  In fact

$$(x^2 - 1)(x - 1) = x^3 - x^2 - x + 1$$

and

$$(x^2 - 2x + 1)(x + 1) = x^3 - 2x^2 + x + x^2 - 2x + 1$$
$$= x^3 - x^2 - x + 1.$$

B. $$\frac{x-1}{x} \neq \frac{x+1}{x^2}$$

since

$$x^2(x-1) \neq x(x+1)$$

Suppose $P(x)$ and $Q(x)$ are polynomials with $Q(x) \neq 0$ and $S(x)$ is a nonzero polynomial. Then

$$\frac{P(x)}{Q(x)} = \frac{P(x)S(x)}{Q(x)S(x)}$$

since $P(x)[Q(x)S(x)] = Q(x)[P(x)S(x)]$. This means that, if $P(x), Q(x)$ are polynomials with $Q(x) \neq 0$ and $P(x) = P_1(x)S(x)$, $Q(x) = Q_1(x)S(x)$, then

$$\frac{P(x)}{Q(x)} = \frac{P_1(x)}{Q_1(x)}$$

Here we are saying that if both numerator and denominator of a rational expression have the same nonzero factor, then dividing by this factor in both numerator and denominator gives an equivalent rational expression.

Example 3

A. $$\frac{x^2-1}{x^2-2x+1} = \frac{(x+1)(x-1)}{(x-1)(x-1)} = \frac{x+1}{x-1}$$

B. $$\frac{x^2+3x+2}{x^2+4x+3} = \frac{(x+2)(x+1)}{(x+3)(x+1)} = \frac{x+2}{x+3}$$

C. $$\frac{x^4+x^3+x^2}{x^3+2x} = \frac{x^2(x^2+x+1)}{x(x^2+2)} = \frac{x(x^2+x+1)}{x^2+2}$$

$$= \frac{x^3+x^2+x}{x^2+2}$$

A common error in simplifying rational expressions by dividing is to divide a factor of either numerator or denominator with part of a sum of denominator or numerator, respectively. For example, students often incorrectly write

$$\frac{x^2+x}{x}$$

as x^2 or $x^2 + 1$ (trying to cancel the x terms). Actually

$$\frac{x^2+x}{x} = \frac{x(x+1)}{x} = \frac{x+1}{1} = x+1$$

The division technique described above can be used (using the terminology of rational numbers) to put rational expressions into simplest form. To be precise we will say

> A rational expression $P(x)/Q(x)$ is in **simplest form** if $P(x)$ and $Q(x)$ have no common factor with degree at least 1.

Example 4

The rational expression

$$\frac{x^4 + x^3 - x^2 + x - 2}{x^4 + 3x^3 + 3x^2 + 3x + 2}$$

is not in simplest form. But

$$\frac{x^4 + x^3 - x^2 + x - 2}{x^4 + 3x^3 + 3x^2 + 3x + 2} = \frac{(x^2 + 1)(x - 1)(x + 2)}{(x^2 + 1)(x + 2)(x + 1)} = \frac{x - 1}{x + 1}$$

and

$$\frac{x - 1}{x + 1}$$

is in simplest form. Note that

$$\frac{x - 1}{x + 1} = \frac{\frac{1}{2}(2x - 2)}{\frac{1}{2}(2x + 2)} = \frac{2x - 2}{2x + 2}$$

so $x - 1$ and $x + 1$ do have common factors of degree 0, but they do not have common factors of degree 1 or greater.

We have, throughout our discussions of rational expressions, tacitly assumed that each polynomial is a rational expression. In general, if $P(x)$ is a polynomial, we can think of it as the rational expression $P(x)/1$, but we usually write $P(x)$.

Finally note that, if $Q(x)$ is any nonzero polynomial, then

$$\frac{0}{Q(x)} = 0 = \frac{0}{1}$$

since $0 = 1 \cdot 0 = Q(x) \cdot 0 = 0$. Furthermore, if

$$\frac{P(x)}{Q(x)} = 0 = \frac{0}{1}$$

then $1 \cdot P(x) = P(x) = Q(x) \cdot 0 = 0$, so $P(x) = 0$.

Exercises 3.2   In Exercises 1–8, determine if the given expression is a rational expression.

1. $\dfrac{2x + \pi}{x^2 + 4}$

2. $\dfrac{\sqrt{2}x + x^2}{x^3 + x + 1}$

3. $\dfrac{\sqrt{2}x + x^2}{\sqrt{x}}$

4. $x^3 + 3x^2 - 4x + 7$

5. $\dfrac{0}{x^2 + 4}$

6. $\dfrac{x^2 + 4}{0}$

7. $\dfrac{0}{0}$

8. $\dfrac{x^2 + \sqrt{7}}{x^2 + 2}$

In Exercises 9–16, express the given rational expression in simplest form.

9. $\dfrac{x^2 - 1}{x^2 + 2x + 1}$

10. $\dfrac{x^3 + 1}{x^2 - x + 1}$

11. $\dfrac{x^5 + x^3}{x^3 - x^2}$

12. $\dfrac{x^4 - 9x^3 + 20x^2}{x^3 - 25x}$

13. $\dfrac{x + 1}{x^2 - 1}$

14. $\dfrac{x^2 + 1}{x^4 - 1}$

15. $\dfrac{x^3 - 3x^2 - 4x + 12}{x^4 - 3x^3 + x - 3}$

16. $\dfrac{x^4 + x^2 - 20}{x^3 - 2x^2 + 5x - 10}$

In Exercises 17 and 18, determine if the two given rational expressions are equal, using the given definition of equality of rational expressions.

17. $\dfrac{x - 5}{x + 5}$ $\dfrac{x^2 - 10x + 25}{x^2 - 25}$

18. $\dfrac{x^2 + 2x - 6}{x^3 + 3x - 4}$ $\dfrac{x - 3}{x - 2}$

19. Express each of the rational expressions given below in the form $P(x)/Q(x)$ where *both* $P(x)$ and $Q(x)$ have the factor $x - 1$.

a. $\dfrac{x^2 + 2}{x + 3}$

b. $\dfrac{x^5 + x - 1}{x^3 - 3}$

c. $\dfrac{x^2 - 1}{x - 1}$

20. In the two following problems, you will find two rational expressions

$$R_1(x) = \dfrac{P_1(x)}{Q_1(x)} \quad \text{and} \quad R_2(x) = \dfrac{P_2(x)}{Q_2(x)}$$

Find equivalent forms of $R_1(x)$ and $R_2(x)$ with the same denominators.

a. $R_1(x) = \dfrac{x + 1}{x - 1}$ $R_2(x) = \dfrac{x^3 + x^2 + 1}{x + 3}$

b. $R_1(x) = \dfrac{x^3 + 5}{x^2 - 1}$ $R_2(x) = \dfrac{x^4 + 3x + 1}{x^2 + 2x + 1}$

21. Find equivalent rational expressions for $R_1(x)$ and $R_2(x)$ in 20(b) having denominators that are equal and of degree 3.

SECTION 3.3

THE ARITHMETIC OF RATIONAL EXPRESSIONS

In this section, we will learn how to add, subtract, multiply, and divide rational expressions. In particular, if

$$\frac{P_1(x)}{Q_1(x)} \quad \text{and} \quad \frac{P_2(x)}{Q_2(x)}$$

are rational expressions, then

$$\frac{P_1(x)}{Q_1(x)} + \frac{P_2(x)}{Q_2(x)} = \frac{P_1(x)Q_2(x) + P_2(x)Q_1(x)}{Q_1(x)Q_2(x)} \tag{1}$$

and

$$\frac{P_1(x)}{Q_1(x)} \cdot \frac{P_2(x)}{Q_2(x)} = \frac{P_1(x)P_2(x)}{Q_1(x)Q_2(x)} \tag{2}$$

Example 1

A. $\dfrac{2x + 1}{x^2} + \dfrac{x^2 + 2}{x - 1} = \dfrac{(2x + 1)(x - 1) + (x^2 + 2)x^2}{x^2(x - 1)}$

$$= \frac{2x^2 - x - 1 + x^4 + 2x^2}{x^3 - x^2}$$

$$= \frac{x^4 + 4x^2 - x - 1}{x^3 - x^2}$$

B. $\dfrac{2x + 1}{x^2} \cdot \dfrac{x^2 + 2}{x - 1} = \dfrac{(2x + 1)(x^2 + 2)}{x^2(x - 1)} = \dfrac{2x^3 + x^2 + 4x + 2}{x^3 - x^2}$

Example 2

A. $\dfrac{x}{x - 3} + \dfrac{7}{2x - 5} = \dfrac{x(2x - 5) + 7(x - 3)}{(x - 3)(2x - 5)}$

$$= \frac{2x^2 - 5x + 7x - 21}{2x^2 - 11x + 15} = \frac{2x^2 + 2x - 21}{2x^2 - 11x + 15}$$

B. $\dfrac{x}{x - 3} \cdot \dfrac{7}{2x - 5} = \dfrac{7x}{(x - 3)(2x - 5)} = \dfrac{7x}{2x^2 - 11x + 15}$

Note that the rules used to add and multiply rational expressions (i.e., (1) and (2)) are identical to the rules used to add and multiply rational numbers;

and in general, arithmetic properties of the rational numbers are valid for rational expressions. After we have used and illustrated them in our examples, we will list the most basic and fundamental of these properties.

Subtraction can be defined by

$$\frac{P_1(x)}{Q_1(x)} - \frac{P_2(x)}{Q_2(x)} = \frac{P_1(x)Q_2(x) - P_2(x)Q_1(x)}{Q_1(x)Q_2(x)} \tag{3}$$

and if

$$\frac{P_2(x)}{Q_2(x)} \neq 0 \quad \text{so} \quad P_2(x) \neq 0$$

then division can be defined by

$$\frac{P_1(x)}{Q_1(x)} \div \frac{P_2(x)}{Q_2(x)} = \frac{P_1(x)}{Q_1(x)} \cdot \frac{Q_2(x)}{P_2(x)} = \frac{P_1(x)Q_2(x)}{Q_1(x)P_2(x)} \tag{4}$$

Division is accomplished by the rule "invert and multiply," and frequently, instead of writing

$$\frac{P_1(x)}{Q_1(x)} \div \frac{P_2(x)}{Q_2(x)}$$

we write

$$\frac{[P_1(x)/Q_1(x)]}{[P_2(x)/Q_2(x)]}$$

Note that

$$(-1)\frac{P(x)}{Q(x)} = \frac{-P(x)}{Q(x)} = \frac{P(x)}{-Q(x)}$$

and subtraction can be viewed as follows:

$$\frac{P_1(x)}{Q_1(x)} - \frac{P_2(x)}{Q_2(x)} = \frac{P_1(x)}{Q_1(x)} + (-1)\frac{P_2(x)}{Q_2(x)} = \frac{P_1(x)}{Q_1(x)} + \frac{-P_2(x)}{Q_2(x)}$$

$$= \frac{P_1(x)Q_2(x) - P_2(x)Q_1(x)}{Q_1(x)Q_2(x)}$$

We frequently write

$$-\frac{P_1(x)}{Q_1(x)} \quad \text{for} \quad (-1)\frac{P_1(x)}{Q_1(x)}$$

Example 3

A. $\dfrac{x+1}{x^2+2} - \dfrac{x}{x+3} = \dfrac{(x+1)(x+3) - x(x^2+2)}{(x^2+2)(x+3)}$

$\qquad\qquad\qquad = \dfrac{(x^2+4x+3) - (x^3+2x)}{x^3+3x^2+2x+6}$

$$= \frac{-x^3 + x^2 + 2x + 3}{x^3 + 3x^2 + 2x + 6}$$

B. $\dfrac{x + 1}{x^2 + 2} \div \dfrac{x}{x + 3} = \dfrac{[(x + 1)/(x^2 + 2)]}{[x/(x + 3)]} = \dfrac{x + 1}{x^2 + 2} \cdot \dfrac{x + 3}{x}$

$$= \frac{(x + 1)(x + 3)}{(x^2 + 2)x} = \frac{x^2 + 4x + 3}{x^3 + 2x}$$

In the next example, we show how the distributive law, $a(b + c) = ab + ac$, can be used with rational expressions.

Example 4

$$\frac{x^2 + x}{x + 2}\left(\frac{x + 3}{x} + \frac{x - 5}{x + 1}\right)$$

$$= \frac{x^2 + x}{x + 2}\left(\frac{(x + 3)(x + 1) + (x - 5)x}{x(x + 1)}\right)$$

$$= \frac{x^2 + x}{x + 2}\left(\frac{x^2 + 4x + 3 + x^2 - 5x}{x^2 + x}\right) = \frac{x^2 + x}{x + 2}\left(\frac{2x^2 - x + 3}{x^2 + x}\right)$$

$$= \frac{(x^2 + x)(2x^2 - x + 3)}{(x + 2)(x^2 + x)} = \frac{2x^2 - x + 3}{x + 2}$$

We now evaluate this product using the distributive law.

$$\frac{x^2 + x}{x + 2}\left(\frac{x + 3}{x} + \frac{x - 5}{x + 1}\right)$$

$$= \frac{x^2 + x}{x + 2} \cdot \frac{x + 3}{x} + \frac{x^2 + x}{x + 2} \cdot \frac{x - 5}{x + 1}$$

$$= \frac{x(x + 1)(x + 3)}{(x + 2)x} + \frac{x(x + 1)(x - 5)}{(x + 2)(x + 1)}$$

$$= \frac{(x + 1)(x + 3)}{x + 2} + \frac{x(x - 5)}{x + 2}$$

$$= \frac{x^2 + 4x + 3}{x + 2} + \frac{x^2 - 5x}{x + 2}$$

$$= \frac{(x + 2)(x^2 + 4x + 3) + (x^2 - 5x)(x + 2)}{(x + 2)(x + 2)}$$

$$= \frac{x^3 + 6x^2 + 11x + 6 + x^3 - 3x^2 - 10x}{x^2 + 4x + 4}$$

$$= \frac{2x^3 + 3x^2 + x + 6}{x^2 + 4x + 4}$$

Note that

$$\frac{2x^3 + 3x^2 + x + 6}{x^2 + 4x + 4} = \frac{(x + 2)(2x^2 - x + 3)}{(x + 2)(x + 2)} = \frac{2x^2 - x + 3}{x + 2}$$

and so the two answers obtained when we computed

$$\frac{x^2 + x}{x + 2}\left(\frac{x + 3}{x} + \frac{x - 5}{x + 1}\right)$$

are equivalent.

When two rational expressions $R_1(x)$ and $R_2(x)$ are to be added, subtracted, multiplied, or divided, it is beneficial to reduce both $R_1(x)$ and $R_2(x)$ to their simplest forms before combining them.

Example 5

To subtract

$$\frac{x - 1}{x^2 - 1} \quad \text{from} \quad \frac{x^2 + 2x}{3x^2 + x}$$

first note that

$$\frac{x - 1}{x^2 - 1} = \frac{x - 1}{(x - 1)(x + 1)} = \frac{1}{x + 1}$$

and

$$\frac{x^2 + 2x}{3x^2 + x} = \frac{x(x + 2)}{x(3x + 1)} = \frac{x + 2}{3x + 1}$$

Thus,

$$\frac{x^2 + 2x}{3x^2 + x} - \frac{x - 1}{x^2 - 1} = \frac{x + 2}{3x + 1} - \frac{1}{x + 1}$$

$$= \frac{(x + 2)(x + 1) - 1(3x + 1)}{(3x + 1)(x + 1)}$$

$$= \frac{x^2 + 3x + 2 - (3x + 1)}{3x^2 + 4x + 1}$$

$$= \frac{x^2 + 1}{3x^2 + 4x + 1}$$

When the rules given by (1)–(4) are used to add, multiply, subtract, or divide rational expressions, the resulting polynomials in numerators and denominators often have common factors. As a result, they are cumbersome to use since they have relatively large degrees. There are some arithmetic techniques that can help us obtain sums or products of rational expressions in forms simpler than what might be obtained if we relied on (1)–(4) alone.

If two rational expressions have the same denominator, addition can be accomplished as follows:

$$\frac{P_1(x)}{Q(x)} + \frac{P_2(x)}{Q(x)} = \frac{P_1(x) + P_2(x)}{Q(x)} \tag{5}$$

We can verify (5), using (1), as follows:

$$\frac{P_1(x)}{Q(x)} + \frac{P_2(x)}{Q(x)} = \frac{P_1(x)Q(x) + P_2(x)Q(x)}{Q(x)Q(x)}$$

$$= \frac{Q(x)[P_1(x) + P_2(x)]}{Q(x)Q(x)} = \frac{P_1(x) + P_2(x)}{Q(x)}$$

Example 6

$$\frac{x^5 + 1}{x^2 + 2} + \frac{x - 3}{x^2 + 2} = \frac{(x^5 + 1) + (x - 3)}{x^2 + 2} = \frac{x^5 + x - 2}{x^2 + 2}$$

Our verification of (5) shows that, if the sum

$$\frac{P_1(x)}{Q(x)} + \frac{P_2(x)}{Q(x)}$$

is found using the rule of addition given by (1), then numerator and denominator of the sum

$$\frac{P_1(x)Q(x) + P_2(x)Q(x)}{Q(x)Q(x)}$$

have the common factor $Q(x)$ and, thus, the sum is not in simplest form.

Even when rational expressions with different denominators are added using (1) the result may not always be in simplest form. In fact, if the denominators of two rational expressions have a common factor, then the sum of these rational expressions, given by (1), *will not* be in simplest form. The next example illustrates this phenomenon and suggests why it occurs.

Example 7

The two polynomials $x^2 - 4 = (x + 2)(x - 2)$ and $x^2 + 4x + 4 = (x + 2)^2$ have the factor $x + 2$ in common. Using (1) to find the sum of

$$\frac{3}{x^2 - 4} \quad \text{and} \quad \frac{x}{x^2 + 4x + 4}$$

we have

$$\frac{3}{x^2 - 4} + \frac{x}{x^2 + 4x + 4} = \frac{3}{(x + 2)(x - 2)} + \frac{x}{(x + 2)^2}$$

$$= \frac{3(x + 2)^2 + x(x + 2)(x - 2)}{(x + 2)(x - 2)(x + 2)^2}$$

$$= \frac{(x + 2)[3(x + 2) + x(x - 2)]}{(x + 2)(x - 2)(x + 2)^2}$$

Note that $x + 2$ is a factor of both numerator and denominator of the rational expression giving the sum of

$$\frac{3}{x^2 - 4} \quad \text{and} \quad \frac{x}{x^2 + 4x + 4}$$

We can avoid the difficulty discussed in Example 7 by using the following method of addition, called the **method of least common multiples.** To add

$$\frac{P_1(x)}{Q_1(x)} \quad \text{and} \quad \frac{P_2(x)}{Q_2(x)}$$

we proceed as follows:

A.   Find a polynomial $Q(x)$ of smallest degree such that $Q_1(x)$ and $Q_2(x)$ are factors of $Q(x)$. $Q(x)$ is called a **least common multiple** of $Q_1(x)$ and $Q_2(x)$.

B.   Write

$$\frac{P_1(x)}{Q_1(x)} \quad \text{and} \quad \frac{P_2(x)}{Q_2(x)}$$

as equivalent rational expressions with denominators $Q(x)$. Suppose

$$\frac{P_1(x)}{Q_1(x)} = \frac{T_1(x)}{Q(x)}$$

$$\frac{P_2(x)}{Q_2(x)} = \frac{T_2(x)}{Q(x)}$$

C.   Add

$$\frac{P_1(x)}{Q_1(x)} \quad \text{and} \quad \frac{P_2(x)}{Q_2(x)}$$

using the equivalent rational expressions of (B) and the rule of addition given by (5):

$$\frac{P_1(x)}{Q_1(x)} + \frac{P_2(x)}{Q_2(x)} = \frac{T_1(x)}{Q(x)} + \frac{T_2(x)}{Q(x)}$$

$$= \frac{T_1(x) + T_2(x)}{Q(x)}$$

Example 8

We will apply the above method to find the sum

$$\frac{3}{x^2 - 4} + \frac{x}{x^2 + 4x + 4}$$

(which was computed in Example 7 using formula (1)).

A least common multiple of $x^2 - 4 = (x - 2)(x + 2)$ and $x^2 + 4x + 4 = (x + 2)^2$ must have $(x - 2)(x + 2)$ and $(x + 2)^2$ as factors. A polynomial of least degree satisfying these conditions is $(x + 2)^2(x - 2)$.

Now,

$$\frac{3}{x^2 - 4} = \frac{3}{(x + 2)(x - 2)} = \frac{3(x + 2)}{(x + 2)^2(x - 2)}$$

and

$$\frac{x}{x^2 + 4x + 4} = \frac{x}{(x + 2)^2} = \frac{x(x - 2)}{(x + 2)^2(x - 2)}$$

Hence,

$$\frac{3}{x^2 - 4} + \frac{x}{x^2 + 4x + 4} = \frac{3(x + 2)}{(x + 2)^2(x - 2)} + \frac{x(x - 2)}{(x + 2)^2(x - 2)}$$

$$= \frac{3(x + 2) + x(x - 2)}{(x + 2)^2(x - 2)}$$

$$= \frac{x^2 + x + 6}{x^3 + 2x^2 - 4x - 8}$$

We would like to develop a method for finding least common multiples. Before we describe this method, it will be helpful to introduce the concept of **multiplicity of factors.** Sometimes when we factor a polynomial, a factor occurs more than once. For example, the factor $x + 2$ occurs twice in the factorization $x^2 + 4x + 4 = (x + 2)^2$.

The number of times a factor appears in a factorization of a polynomial is called the **multiplicity** of the factor.

For example, if $P(x)$ has the factorization $(x + 2)^2(x + 1)(x - 3)^4$, then the factor $x + 2$ has multiplicity 2, the factor $x + 1$ has multiplicity 1, and the factor $x - 3$ multiplicity 4.

Suppose we want to find the least common multiple of two polynomials $Q_1(x)$ and $Q_2(x)$. Each of the polynomials can be written as a product of linear factors of the form $x - r$, irreducible quadratic factors of the form $x^2 + ax + b$ and a constant term. (Remember that the multiplicity of any of the factors can be greater than 1.)

> To construct $Q(x)$, a least common multiple of $Q_1(x)$ and $Q_2(x)$, take the product of all polynomials of the form $[S(x)]^n$ where $S(x)$ is a linear or irreducible quadratic factor of $Q_1(x)$ or $Q_2(x)$ and n is the larger of this factor's multiplicities in the two factorizations.

The next two examples illustrate how to use this method to find least common multiples.

Example 9

If
$$Q_1(x) = 2(x - 1)^2(x + 1)(x^2 + 4)$$
and
$$Q_2(x) = (x - 1)(x + 2)(x^2 + 4)^3$$
then a least common multiple of $Q_1(x)$ and $Q_2(x)$ is
$$Q(x) = (x - 1)^2(x + 1)(x + 2)(x^2 + 4)^3$$

Note that $2(x - 1)^2(x + 1)(x + 2)(x^2 + 4)^3$ is also a least common multiple of $Q_1(x)$ and $Q_2(x)$. In fact, any nonzero constant multiple of $Q(x)$ will be a least common multiple of $Q_1(x)$ and $Q_2(x)$.

Example 10

A least common multiple of
$$Q_1(x) = (x + 3)^3(x - 2)^2(x^2 + x + 1)(x^2 + 2x + 3)^3$$
and
$$Q_2(x) = (x + 3)^2(x + 2)(x - 2)(x^2 + x + 1)^2$$
is
$$Q(x) = (x + 3)^3(x + 2)(x - 2)^2(x^2 + x + 1)^2(x^2 + 2x + 3)^3$$

Example 11 illustrates how to use least common multiples to add rational expressions. Since the difference $R_1(x) - R_2(x)$ of two rational expressions is the sum $R_1(x) + [-R_2(x)]$, the least common multiple method can also be used to subtract rational expressions. This method is illustrated in (B) of Example 11.

Example 11

A. $\dfrac{x}{x^3 - 8} + \dfrac{2}{x^2 - 4} = \dfrac{x}{(x - 2)(x^2 + 2x + 4)} + \dfrac{2}{(x - 2)(x + 2)}$

$$= \dfrac{x(x + 2)}{(x - 2)(x^2 + 2x + 4)(x + 2)}$$

$$+ \dfrac{2(x^2 + 2x + 4)}{(x - 2)(x^2 + 2x + 4)(x + 2)}$$

$$= \dfrac{x(x + 2) + 2(x^2 + 2x + 4)}{(x - 2)(x^2 + 2x + 4)(x + 2)}$$

$$= \dfrac{3x^2 + 6x + 8}{x^4 + 2x^3 - 8x - 16}$$

B. $\dfrac{x + 1}{(x - 1)(x - 2)^2(x^2 + 1)^4(x^2 + x + 1)}$

$$- \dfrac{3x^2 - 4}{(x - 1)^2(x^2 + 1)^3(x^2 + x + 1)^2}$$

$$= \dfrac{(x + 1)(x - 1)(x^2 + x + 1)}{(x - 1)^2(x - 2)^2(x^2 + 1)^4(x^2 + x + 1)^2}$$

$$- \dfrac{(3x^2 - 4)(x - 2)^2(x + 1)}{(x - 1)^2(x - 2)^2(x + 1)^4(x^2 + x + 1)^2}$$

$$= \dfrac{(x + 1)(x - 1)(x^2 + x + 1) - (3x^2 - 4)(x - 2)^2(x + 1)}{(x - 1)^2(x - 2)^2(x + 1)^4(x^2 + x + 1)^2}$$

In Example 9, we noted that if $Q(x)$ is the least common multiple of the polynomials $Q_1(x)$ and $Q_2(x)$, then $cQ(x)$, where c is a nonzero real number, is also a least common multiple of $Q_1(x)$ and $Q_2(x)$.

Frequently when finding a sum

$$\dfrac{P_1(x)}{Q_1(x)} + \dfrac{P_2(x)}{Q_2(x)}$$

using the least common multiple method, a suitable choice of c will ensure that the coefficients we work with will be integers.

Example 12

$$\dfrac{1}{6x + 6} + \dfrac{1}{4x^2 - 4} = \dfrac{1}{6(x + 1)} + \dfrac{1}{4(x - 1)(x + 1)}$$

Using the method described in the text to find a least common multiple of the denominators, $6(x + 1)$ and $4(x - 1)(x + 1)$, we get a least common multiple of $(x + 1)(x - 1)$.

If we use this least common multiple to add the rational expressions, we proceed as follows:

$$\frac{1}{6(x + 1)} + \frac{1}{4(x + 1)(x - 1)}$$

$$= \frac{\frac{1}{6}(x - 1)}{(x + 1)(x - 1)} + \frac{\frac{1}{4}}{(x + 1)(x - 1)}$$

$$= \frac{\frac{1}{6}(x - 1) + \frac{1}{4}}{(x + 1)(x - 1)} = \frac{\frac{1}{6}x + \frac{1}{12}}{(x + 1)(x - 1)}$$

We can avoid using the rational coefficients $\frac{1}{6}$ and $\frac{1}{12}$ by choosing as our least common multiple $12(x + 1)(x - 1)$. We choose 12 because it is the smallest positive integer that is evenly divisible by both 4 and 6. (By evenly divisible we mean that the result of the division is an integer.)

With this choice of least common multiple, we proceed as follows:

$$\frac{1}{6(x + 1)} + \frac{1}{4(x + 1)(x - 1)}$$

$$= \frac{2(x - 1)}{12(x + 1)(x - 1)} + \frac{3}{12(x + 1)(x - 1)}$$

$$= \frac{2x + 1}{12(x + 1)(x - 1)}$$

We have seen examples where two rational expressions

$$\frac{P_1(x)}{Q_1(x)} \quad \text{and} \quad \frac{P_2(x)}{Q_2(x)}$$

were in simplest form but their sum found according to (1) was not. We developed the least common multiple method to help us avoid having common factors appear in the numerator and denominator of a sum of two rational expressions. If two rational expressions are given in simplest form and are added using the least common multiple method, is the sum always a rational expression in simplest form? Unfortunately, as the next example illustrates, the answer to this question is no.

Example 13

Both

$$\frac{3x + 1}{(x + 1)(x - 1)} \quad \text{and} \quad \frac{5x - 11}{(x + 2)(x - 1)}$$

are in simplest form. Using the least common multiple method to find the sum of these two rational expressions, we get

$$\frac{3x+1}{(x+1)(x-1)} + \frac{5x-11}{(x+2)(x-1)}$$

$$= \frac{(3x+1)(x+2)}{(x+1)(x+2)(x-1)} + \frac{(5x-11)(x+1)}{(x+1)(x+2)(x-1)}$$

$$= \frac{(3x+1)(x+2) + (5x-11)(x+1)}{(x+1)(x+2)(x-1)}$$

$$= \frac{8x^2+x-9}{(x+1)(x+2)(x-1)}$$

But,

$$\frac{8x^2+x-9}{(x+1)(x+2)(x-1)}$$

is not in simplest form since

$$\frac{8x^2+x-9}{(x+1)(x+2)(x-1)} = \frac{(8x+9)(x-1)}{(x+1)(x+2)(x-1)}$$

$$= \frac{8x+9}{(x+1)(x+2)}$$

$$= \frac{8x+9}{x^2+3x+2}$$

We can use the method of least common multiples to add more than two rational expressions as illustrated in Example 14.

Example 14

A. $$\frac{2}{(x-3)(x+1)} + \frac{3}{(x-3)^2(x+2)} + \frac{2x}{(x+2)(x-1)}$$

$$= \frac{2(x-3)(x+2)(x-1)}{(x-3)^2(x+1)(x+2)(x-1)}$$

$$+ \frac{3(x+1)(x-1)}{(x-3)^2(x+1)(x+2)(x-1)}$$

$$+ \frac{2x(x-3)^2(x+1)}{(x-3)^2(x+1)(x+2)(x-1)}$$

$$= \frac{2(x-3)(x+2)(x-1) + 3(x+1)(x-1) + 2x(x-3)^2(x+1)}{(x-3)^2(x+1)(x+2)(x-1)}$$

B. $$\frac{2}{x-3} + \frac{4x-1}{x^2+4} - \frac{5x}{(x-3)(x^2+2)}$$

$$= \frac{2(x^2+4)(x^2+2)}{(x-3)(x^2+4)(x^2+2)} + \frac{(4x-1)(x-3)(x^2+2)}{(x-3)(x^2+4)(x^2+2)}$$

$$-\frac{5x(x^2+4)}{(x-3)(x^2+4)(x^2+2)}$$

$$=\frac{2(x^2+4)(x^2+2)+(4x-1)(x-3)(x^2+2)-5x(x^2+4)}{(x-3)(x^2+4)(x^2+2)}$$

Finally, we will list the general arithmetic properties possessed by rational expressions. They are not listed to be memorized but to indicate the extent to which the arithmetic properties of rational expressions are similar to the arithmetic properties of the real numbers.

If $R_1(x)$, $R_2(x)$, $R_3(x)$ are rational expressions, then

A. $R_1(x) + R_2(x) = R_2(x) + R_1(x)$
    $R_1(x) \cdot R_2(x) = R_2(x) \cdot R_1(x)$

B. $R_1(x) + [R_2(x) + R_3(x)] = [R_1(x) + R_2(x)] + R_3(x)$
    $R_1(x) \cdot [R_2(x) \cdot R_3(x)] = [R_1(x) \cdot R_2(x)] \cdot R_3(x)$

C. $R_1(x) + 0 = R_1(x)$
    $R_1(x) \cdot 1 = R_1(x)$

D. There is a rational expression, $-R_1(x)$, such that
    $R_1(x) + [-R_1(x)] = 0$. If $R_1(x) \neq 0$, there is a rational
    expression, namely $1 \div R_1(x)$, such that $R_1(x)[1 \div R_1(x)] = 1$.

E. $R_1(x)[R_2(x) + R_3(x)] = R_1(x)R_2(x) + R_1(x)R_3(x)$

F. If $R_1(x) = P(x)/Q(x) \neq 0$, then:

$$1 \div R_1(x) = \frac{1}{R_1(x)} = \frac{1}{1} \div \frac{P(x)}{Q(x)}$$

$$= \frac{1}{1} \cdot \frac{Q(x)}{P(x)} = \frac{Q(x)}{P(x)}$$

For example,

$$\frac{1}{\dfrac{x^2+4}{x-3}} = \frac{x-3}{x^2+4}$$

Exercises 3.3

In Exercises 1–8, find a least common multiple of the given polynomials.

1. $x^2 - 1, x^2 + 2x + 1$
2. $x^3 - 2x^2 + x - 2, x^4 + 2x^2 + 1$
3. $4x^2 + 4x - 24, 6x^2 + 12x - 18$

4. $x^3 - 5x^2 + 8x - 4, x^3 - 4x^2 + 5x - 2$

5. $x^3 - 1, x^4 - 1$

6. $x^3 - 1, (x^4 - 1)^2$

7. $6(x - 1)^2(x + 3)^3(x^2 + 4)(x^2 + x + 10)^3,$
   $15(x - 1)^3(x + 4)(x^2 + 5)(x^2 + x + 10)^2$

8. $(x + \pi)^2(x - 2)(x^2 + x + 1)^4, (x - \pi)(x - 2)^2(x^2 + x + 1)^4$

In Exercises 9–41, perform the indicated arithmetic operation. Express all answers in simplest form.

9. $\dfrac{x + 1}{x - 3} + \dfrac{2x + 4}{x}$

10. $\dfrac{x}{x - 1} + \dfrac{x^2 + x - 2}{x^2 - 1}$

11. $\dfrac{x^3 - 7x + 6}{x^2 + x - 6} + \dfrac{-x^3 + x^2 - 5x + 5}{x^2 + 5}$

12. $\dfrac{5}{x^2 + 1} + \dfrac{1}{x - 3}$

13. $\dfrac{-3}{2x^2 + 2x - 4} + \dfrac{4}{x^2 + 2x - 3}$

14. $\dfrac{-x - 2}{x^3 - x^2 + 5x - 5} + \dfrac{2x + 1}{x^3 - 3x^2 + 5x - 15}$

15. $\dfrac{x^2 - 1}{4x^3 - 4} + \dfrac{3x}{6x^2 - 6x}$

16. $\dfrac{x^2 + 1}{x^3 - 4x^2 + x - 4} + \dfrac{3}{x^2 - 6x + 8}$

17. $\dfrac{x + 4}{x^3 - 5x^2 + 8x - 4} + \dfrac{x^2 + 1}{x^2 - 3x - 2}$

18. $\dfrac{x^2 - 2}{x^3 + x^2 - 2x - 2} + \dfrac{3}{x + 1}$

19. $\dfrac{x + 1}{x^2 - 2x + 1} + \dfrac{3}{x^3 - 5x^2 + 6x - 3}$

20. $\dfrac{1}{x - 3} - \dfrac{2}{x + 5}$

21. $\dfrac{x}{x^2 + 4} - \dfrac{x + 1}{x - 1}$

22. $\dfrac{x^2 - 1}{x^2 + 2x + 1} - \dfrac{3x}{x + 2}$

23. $\dfrac{2x - 1}{3x^2 - 9x + 6} - \dfrac{x + 2}{2x^2 + 2x - 4}$

24. $\dfrac{6x + 31}{x^3 + 2x^2 + x + 2} - \dfrac{3x - 7}{x^3 - 2x^2 + x - 2}$

25. $\dfrac{x+2}{(x-1)^3(x^2+x+1)^2(x-2)} - \dfrac{3}{(x-1)^2(x^2+x+1)^2}$

26. $\dfrac{x-3}{x+5} \cdot \dfrac{x+1}{x-4}$

27. $\dfrac{x^2+1}{x-1} \cdot \dfrac{x^2+1}{x+1}$

28. $\dfrac{x^2+2x+1}{x-3} \cdot \dfrac{x^2-6x+9}{x^2-1}$

29. $\dfrac{x^3+1}{x-1} \cdot \dfrac{x^2+4}{x^2+3x+2}$

30. $\dfrac{x^2-4}{x+2} \cdot \dfrac{(x+2)^2}{x^3+x^2-2x-2}$

31. $\dfrac{3x-1}{x+2} \div \dfrac{1}{x+1}$

32. $\dfrac{x^2-1}{x^3-27} \div \dfrac{x-3}{x+1}$

33. $\dfrac{x^2-1}{x^3-27} \div \dfrac{x+1}{x-3}$

34. $\dfrac{(x^2+5)^3(x+2)(x-1)}{(x^2+5)(x^2+4)(x+4)} \div \dfrac{(x^2+5)^2(x+2)^2}{(x^2+4)^2(x-1)}$

35. $\dfrac{(x^2+2x+2)^2(x^2-9)}{(x^2-x-6)(x+1)^4} \div \dfrac{(x^2+2x+2)(x+3)}{(x+1)^3(x-3)}$

36. $\dfrac{1}{x-1} + \dfrac{1}{x+2} + \dfrac{x}{x-5}$

37. $\dfrac{2x}{2(x-1)^2} + \dfrac{x+1}{3(x-1)(x+1)} + \dfrac{x-3}{x+1}$

38. $\dfrac{x+3}{(x^2+1)^2} + \dfrac{2}{x^2+1} + \dfrac{5x^2-1}{x-1}$

39. $\dfrac{2}{x-1} + \dfrac{3}{(x-1)^2} + \dfrac{4}{(x-1)^3}$

40. $\dfrac{2}{x^2+5} + \dfrac{3}{(x^2+5)^2} + \dfrac{4}{(x^2+5)^3}$

41. $\dfrac{-1}{x-3} + \dfrac{2}{x+1} - \dfrac{3}{(x+1)^2} - \dfrac{2x-3}{x^2-2x-3}$

42. The value of a rational expression $P(x)/Q(x)$ at a real number r is the numerical value obtained by replacing x in the expression by r provided the value of the polynomial $Q(x)$ at r is not 0. If the value of $Q(x)$ at r is 0, we say $P(x)/Q(x)$ has no value at r. For example, the value of $x/(x^2+1)$ at 2 is

$$\frac{2}{2^2+1} = \frac{2}{5}$$

Find the values of the following rational expressions at the indicated values of r.

a. $\dfrac{x^2+2}{x^3+1}$ $r = -2, 1$

b. $\dfrac{x}{x^2-3x+2}$ $r = \sqrt{2}, -1, 7$

c. $\dfrac{x^3-27}{x^2-9}$ $r = -3, 0, 2$ **d.** $\dfrac{x^2+8}{x^3+1}$ $r = -1, 0, 1$

43. When computing values of rational expressions, we do not make the usual identification of rational expressions. For example,

$$\frac{x}{x-1} = \frac{x^2}{x^2-x}$$

but we say $x/(x-1)$ has value 0 at $r = 0$ since $0/(0-1) = 0$, while $x^2/(x^2-x)$ has *no value* at $r = 0$ since the denominator $x^2 - x$ has value 0 at $r = 0$. However, if

$$\frac{P_1(x)}{Q_1(x)} = \frac{P_2(x)}{Q_2(x)}$$

then the values of

$$\frac{P_1(x)}{Q_1(x)} \quad \text{and} \quad \frac{P_2(x)}{Q_2(x)}$$

are identical at r if the values of both $Q_1(x)$ and $Q_2(x)$ are not zero. This means that the simplest form of $P(x)/Q(x)$ can be used to compute the value of $P(x)/Q(x)$ at r provided $P(x)/Q(x)$ has a value at r (i.e., $Q(x)$ at r is not zero). For example,

$$\frac{x^2-1}{x^2+x-2} = \frac{(x-1)(x+1)}{(x-1)(x+2)} = \frac{x+1}{x+2}$$

The roots of $x^2 + x - 2 = (x-1)(x+2)$ are 1 and -2, so $(x^2-1)/(x^2+x-2)$ has no values at 1 and -2. At all other numbers, the value of $(x^2-1)/(x^2+x-2)$ is equal to the value of $(x+1)/(x+2)$.

Reduce each of the following to its simplest form and find values at the indicated real numbers.

a. $\dfrac{x^2 + 5x + 6}{x^2 - 2x - 8}$ $r = -3, -2, -1, 0, 1, 4$

b. $\dfrac{x^3 - 2x^2 - x + 2}{x^3 + x^2 + 4x + 4}$ $r = -2, -1, 0, 1, 2$

44. The **zeros** of a rational expression $P(x)/Q(x)$ are those numbers at which the value of $P(x)/Q(x)$ is zero. Finding the zeros of $P(x)/Q(x)$ is equivalent to solving the equation

$$\frac{P(x)}{Q(x)} = 0$$

If $Q(x) \neq 0$, $P(x)/Q(x) = 0$ is equivalent to $P(x) = 0$. So to find the zeros of $P(x)/Q(x)$, we find the roots of polynomial $P(x)$ and remove the roots of $Q(x)$ from this list.

Find the zeros of the following:

a. $\dfrac{x^2 + 2x + 1}{x^2 + 4}$ b. $\dfrac{x^2 + 2x + 1}{x^2 + 3x + 2}$

45. Find the numbers at which the value of

$$\frac{x^2 + 2x + 1}{x^2 - 3}$$

is 7. For what number is the value -7?

46. The cost, in dollars, a company incurs when producing r light switches is given by the value of the polynomial $P(x) = 1000 + 0.1x$ at r. Give a rational expression whose value at r is the average cost of producing a switch when r switches are produced.

What is the average cost of a switch if 1000 switches are produced? What is the average cost if 10,000 switches are produced?

How many switches must be produced for the average cost of a switch to be $0.3?

47. A company produces two types of airplanes, T_1 and T_2. Because the same machinery and labor is used to produce both types of plane, if x represents the number of T_1 planes produced per year and y the number of T_2 planes produced per year, then

$$y = \frac{100 - 10x}{5 + x}$$

How many T_2-type planes can be produced if no T_1 planes are produced? 5 T_1 planes? 10 T_1 planes?

How many T_1-type planes can be produced if 15 T_2 planes are produced?

Review Exercises
Chapter 3

In Exercises 1–5, determine if the given expression is a rational expression.

1. $\dfrac{x + \pi}{\sqrt{2}x - 1}$ **2.** $x^{50} + x^{30} + x^{10} + 2$

3. $\dfrac{\sqrt{x} + 1}{x + 5}$ **4.** $\dfrac{\sqrt{x^2 + 1}}{\sqrt{x^6 + 2}}$

5. $\dfrac{x^3 - 2x - 7}{\sqrt[3]{8x^2} - 3x + 2}$

In Exercises 6–11, determine if the given rational expressions are equal.

6. $\dfrac{2x-4}{6x+12}$ $\dfrac{x-2}{3x+4}$

7. $\dfrac{x-1}{x+4}$ $\dfrac{x^3-2x^2-5x+6}{x^3+3x^2-10x-24}$

8. $\dfrac{x^3-39x^2+141x-35}{x^3-34x^2-33x-70}$ $\dfrac{x^2-4x+1}{x^2+x+1}$

9. $x+1-\dfrac{2}{x-1}$ $\dfrac{x^2+2x-3}{x-1}$

10. $\dfrac{x^4+8x^2+2x+17}{x^2+5}$ $x^2+3+\dfrac{2x+2}{x^2+5}$

11. $\dfrac{1}{(x-1)^2}+\dfrac{2}{(x-1)(x+2)}$ $\dfrac{3x^2-3x}{(x-1)(x+2)}$

In Exercises 12–20, reduce the given rational expression to its simplest form.

12. $\dfrac{x^2-16}{x^2+8x+16}$

13. $\dfrac{x^3+7x^2+16x+12}{x^2+5x+6}$

14. $\dfrac{x^3-x^2+5x-5}{x^3+x^2+5x+5}$

15. $\dfrac{x^3-5x^2+10x-12}{x^3+8}$

16. $\dfrac{x^4+14x^2+40}{x^4+7x^2+12}$

17. $\dfrac{x^3-1}{x^4-1}$

18. $\dfrac{x^4-16}{x^5-32}$

19. $\dfrac{x^3+9x^2-x-105}{x^3+17x^2+95x+175}$

20. $\dfrac{x^4+2x^3+3x^2+2x+1}{x^3+4x^2+4x+3}$

In Exercises 21–40, perform the indicated operation and reduce your answer to its simplest form.

21. $\dfrac{x}{x-\pi}+\dfrac{3x+2}{x-\pi}$

22. $\dfrac{2x}{(x-1)^2(x+4)}+\dfrac{x-4}{(x-1)(x+4)}$

23. $\dfrac{3x-7}{(x-2)^2}-\dfrac{x+1}{x^2-x-2}$

24. $\dfrac{x^4+8x}{x+3}+\dfrac{40x+63}{x+3}$

25. $\dfrac{12x-17}{x^2-5x+6}-\dfrac{8x+35}{x^2-x-6}$

26. $\dfrac{1}{6x-4}+\dfrac{3x^2+4}{x-2}$

27. $\dfrac{2x+3}{6x^2+17x+5}+\dfrac{3x}{9x^2+6x+1}$

28. $\dfrac{x^5+1}{x^2-4}+\dfrac{x-2}{x^2+4x+4}$

29. $\dfrac{x^5 - 1}{x^2 - 1} + \dfrac{x}{x + 1}$

30. $\dfrac{1}{x - 3} + \dfrac{3x}{(x - 3)^2} + \dfrac{3x - 2}{x + 1}$

31. $\dfrac{2x + 3}{x^2 + 4} + \dfrac{3}{(x^2 + 4)^2} - \dfrac{1}{x - 4}$

32. $\dfrac{3}{(2x^2 + 10)(x - 3)^2(x + 2)} + \dfrac{3x - 5}{(x^2 + 5)(x - 3)(x + 2)}$

$$- \dfrac{2x - 3}{(x^2 + 5)(x + 2)}$$

33. $\dfrac{x^3 - 8}{2x^2 + 5x - 3} \cdot \dfrac{2x^2 - 7x + 3}{3x^2 - 7x + 2}$

34. $\dfrac{(x + 2)^2(4x - 8)^3}{(x - 2)(x^2 + 4x + 4)} \cdot \dfrac{(x + 3)}{(6x - 12)(x + 1)}$

35. $\dfrac{x^4 - 16}{x^3 - 3x^2 - 4x + 12} \div \dfrac{x^2 - 6x + 9}{2x^2 + 9x + 10}$

36. $\dfrac{x^4 - 16}{x^3 - 3x^2 - 4x + 2} \div \dfrac{2x^2 + 9x + 10}{x^2 - 6x + 9}$

37. $\dfrac{2x^2 + 3x - 1}{x + 5} \cdot \dfrac{2x^2 + 5}{x + 1}$

38. $\dfrac{(2x + 1)^3(x + 3)(x - 1)}{(3x + 5)(x - 4)^2} \cdot \dfrac{x^2 - 16}{(x + \frac{1}{2})^3(x + 3)(1 - x)}$

39. $\dfrac{x + 1}{x - 5} \cdot \dfrac{2x^2 - 20x + 50}{x + 3} \cdot \dfrac{x + 1}{x^2 + 2x + 1}$

40. $\dfrac{2x + 3}{x^2 - 5} \cdot \dfrac{x^2 - 3\sqrt{5}x + 10}{4x + 6}$

Exercises 41–46 deal with the concepts discussed in Exercises 42–44 of Section 3.3.

41. Evaluate $\dfrac{x - 3}{x^2 + 3x + 2}$ at $-3, -1, 0, 2,$ and 3.

42. Evaluate $\dfrac{x^2 - 7x + 12}{x^2 - 2x - 3}$ at $-1, 0, 1,$ and 2.

43. Evaluate $\dfrac{2x + 7}{x^2 + 4}$ at $-2, 0,$ and 2.

44. Evaluate $\dfrac{(x - 1)(x^5 - 32)}{x^2 - 2x + 1}$ at 1 and 2.

45. At what number is the value of $\dfrac{x - 3}{x + 7}$ equal to 4?

46. At what numbers is the value of $\dfrac{x^2 + 3x - 4}{x^2 - 6x + 9}$ equal to 4?

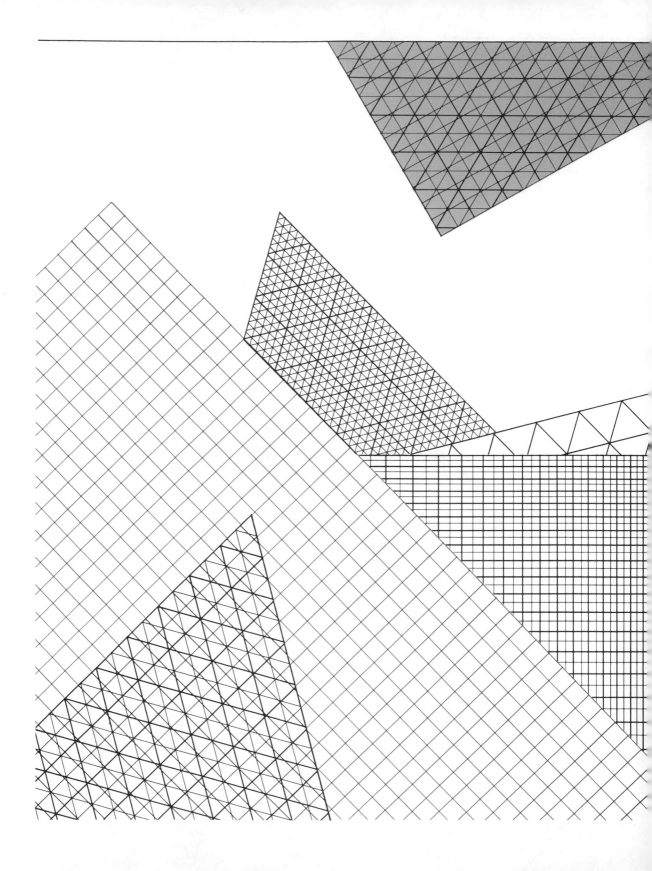

4 Functions

INTRODUCTION

In the scientific world much work is done in establishing the existence of relationships between two collections of objects and studying the nature of these relationships. For example, extensive work was involved in trying to establish a connection between cigarette smokers and people who contract lung cancer, in physicists' discovery that the pressure exerted by a gas with a fixed volume is closely related to the temperature of that gas, and in economists' studies of demand relationships that describe the total amounts of products purchased by consumers in terms of the unit prices of those products. Special types of relationships, called **functions,** have become so prevalent in scientific, social scientific, and mathematical investigations that functions themselves have become a topic of study.

The remainder of this text is primarily devoted to the study of functions. We will be concerned with abstractly characterizing functions, studying their common properties and developing general mathematical techniques that can be used to investigate functional relationships. Finally, we will study some special functions (linear, quadratic, polynomial, rational, trigonometric, exponential, and logarithmic) that have proven to be very important in a wide variety of scientific studies.

SECTION 4.2

DEFINITIONS AND PROPERTIES OF FUNCTIONS

In this section, we present a preliminary definition of function and give several examples of functions. Although this first definition is not the most rigorous (and, therefore, not very satisfying to a mathematician), it is fairly intuitive and allows us to discuss some of the general properties of functions. Later, we will give a more mathematically sound definition of function, providing further insight into this concept.

Definition

> Let D and R be sets. A **function f from D to R** is a rule that associates to each element x in D one and only one element y in R. If y in R is related to x in D by f, we denote y by $f(x)$ (read: f of x). Furthermore, $f(x)$ is called the **value of f at x.** The set D is called the **domain** of f and is denoted by D_f. The set of all values $f(x)$ where x is in D is called the **range** of f and is denoted by R_f.

Before proceeding to examples of functional and nonfunctional relationships, we would like to make some observations about the definition given above.

First, in this definition we have not defined the word *function* but the phrase *function from D to R*. The rule, symbolized by *f*, is referred to as a function, and strictly speaking, whenever a function is defined the domain *D* and a set *R* containing the range of that function should be specified. However, there are instances that will be discussed throughout this chapter when functions are defined with no reference to *D* or *R*. In such cases, either some universally accepted convention or the context of one's work implies the precise nature of *D* and *R*.

Second, the use of the word *rule* to define function is rather unsatisfactory to mathematicians since this word has not been given a well-defined meaning. This is perhaps the major objection to our definition, and the reason we will replace it with a more rigorous definition later. However, the connotations brought to mind by this word help us to understand what is meant by a function.

Third, the symbol *f* is used to denote the rule (function) associating points in *D* to points in *R*, while the symbol $f(x)$ is used to denote that particular element of *R* associated to the element *x*. There is nothing sacred about using *f* to represent a function, *x* to represent an element in *D*, or *y* to represent an element in *R*. Be prepared to see all types of symbols used to represent functions and elements of *D* and *R*.

Fourth, note that although R_f is always contained in *R*, it is not necessarily identical to *R*. Throughout this text you will find many examples of functions for which the actual determination of their ranges is virtually impossible. However, it will be easy to specify sets containing the ranges. Of course, given a function we would like to know its range, but frequently we will find it sufficient to know *R* rather than R_f.

Finally, note that, if *x* is an element of *D*, then *f* associates exactly one element of *R* to *x*. This stipulation is made to avoid ambiguities when discussing the values *f* associates to *x*. For example, if we let *D* be the set of all mothers; *R*, the set of all children; and *f*, the rule associating *x* in *D* to *y* in *R* whenever *x* is the mother of *y*, then *f* is not a function from *D* to *R* since there is some mother with more than one child. Hence, if *x* were a mother with more than one child, $f(x)$ would not have a well-defined meaning because we would not know which child $f(x)$ represents.

Example 1

Let *D* and *R* be the set of real numbers. The rule *f*, which associates to a number *x* in *D* the number $2x - 3$, is a function from *D* to *R*. Note that $f(3) = 2(3) - 3 = 3, f(-5) = 2(-5) - 3 = -13$ and, in general, if *x* is in *D*, $f(x) = 2x - 3$.

Example 2

An oil company has estimated that there are 100,000,000 gallons of oil in an oil field. The cost of erecting an oil well and drilling down to oil will be $500,000. Once oil is reached, the cost of removing each 100,000 gallons of oil is $5000.

A rule C associating the cost of removing oil from the ground to the number of gallons removed is given the following formula:

$$C(x) = 500{,}000 + 5000x \tag{1}$$

where x represents the amount of oil in hundreds of thousands of gallons and $C(x)$ represents the total cost of removing x units of oil from the ground.

The rule C described by (1) gives a function. However, to this point, we have not specified the domain and range for C. The domain and range of C are dictated by the physical characteristics of the problem. If we assume there are 100,000,000 gallons of oil to be removed, we can take D_C to be the interval [0, 1000]. (Recall that x represents the amount of oil in units of 100,000 gallons.) The set R can be taken to be the set of real numbers since R_C, the set of possible costs, is some subcollection of the set of real numbers. (Actually R_C is the set of numbers greater than or equal to 500,000 and less than or equal to 500,500,000.)

Note that equation (1) gives a formula that can be used to determine the value of C at any point in D_C. This is a typical method of defining functions.

Example 3

To manufacture buses a company must make an initial investment of $500,000—basically for the equipment necessary to begin production. The cost of producing each bus once the equipment has been obtained is $5000. A rule C associating the cost of producing buses to the number of buses produced is

$$C(x) = 500{,}000 + 5000x \tag{2}$$

when x represents the number of buses produced, and $C(x)$ represents the total cost of producing x buses.

Note that formulas (1) and (2) are identical. However, we have not defined the same functions in Examples 2 and 3 because the domains of the functions defined in these examples are different. The domain of the cost function of Example 3 cannot feasibly be the set of real numbers in [0, 1000]. Indeed, we would not consider the cost of manufacturing $\frac{1}{2}$, $\frac{1}{3}$, or $\sqrt{2}$ buses. The domain of the function defined in Example 3 should realistically be the set of non-negative integers, 0, 1, 2, 3 . . . .

The next two examples give more illustrations of functions. They also give a formula that can be used to compute the distance between two points in a plane. Since we will use this distance formula frequently throughout the text, you should pay careful attention to the results of these examples and memorize the formula.

Example 4

Let D be the set of all ordered pairs, (x, y), of real numbers and R the set of real numbers. For each (x, y) in D, let f be the rule associating the distance between x and y on the number line to the ordered pair (x, y). (See Figure 4.2.1.)

Figure 4.2.1

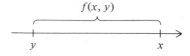

For example,

$$f(2, 3) = 1 \qquad f(3, 2) = 1 \qquad \text{(for this particular function, } f, \\ f(x, y) = f(y, x))$$

$$f(0, -4) = 4 \quad \text{and} \quad f(2, -1) = 3$$

For f to be a function from D to R, it must have the property that it associates to each element of D exactly one element of R. This is a trivial fact since the distance between any two points on the number line is unique.

If (x, y) is in D, then $f(x, y)$, the value f associates to (x, y), is $x - y$ if $x \geq y$, and $y - x$ if $x < y$. (See Figures 4.2.2 and 4.2.3.)

Figure 4.2.2

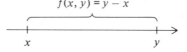

Figure 4.2.3

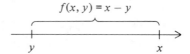

Since the distance function we have defined here is so widely used and the distance between x and y on the number line is either $x - y$ or $-(x - y) = y - x$, the special symbol $|x - y|$ is used to denote $f(x, y)$.

For example,

$$f(1, 4) = |1 - 4| = 3 \quad \text{and} \quad f(-1, -3) = |-1 - (-3)| = 2$$

Example 5

In this example, we will extend the idea of the distance function discussed in Example 4 to a distance function giving the distance between points in a plane. To do so, let D be the set of ordered pairs (P, Q) of points in a plane and R the set of real numbers. The rule, d, which associates to each ordered pair (P, Q) in D the distance between P and Q, is a function from D to R.

If this plane has a cartesian (rectangular) coordinate system, then each point in the plane can be identified with an ordered pair of numbers. Let us suppose, as shown in Figure 4.2.4, that P and Q are points represented by (x_1, y_1) and (x_2, y_2), respectively.

Figure 4.2.4

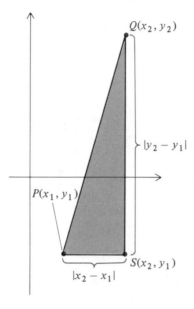

The perpendiculars drawn from P to the vertical axis and from Q to the horizontal axis meet at the point S with coordinates (x_2, y_1). The distance from Q to S is $|y_2 - y_1|$ and the distance from P to S is $|x_2 - x_1|$.

Using the Pythagorean Theorem we see

$$d(P, Q) = \sqrt{|x_2 - x_1|^2 + |y_2 - y_1|^2}$$

Now $|x_2 - x_1|^2 = [\pm(x_2 - x_1)]^2 = (x_2 - x_1)^2$, and similarly, $|y_2 - y_1|^2 = (y_2 - y_1)^2$. Hence,

$$d(P, Q) = \sqrt{(x_2 - x_1)^2 + (y_2 - y_1)^2}$$

For example, if P is $(1, -2)$ and Q is $(-3, -5)$, then

$$d(P, Q) = \sqrt{(-3 - 1)^2 + (-5 - (-2))^2} = \sqrt{16 + 9} = 5$$

Finally, note that, for convenience, the sets corresponding to R for the distance functions described in Examples 4 and 5 were defined to be the set of real numbers. Since distances are not negative, the range of each function is actually the set of non-negative real numbers.

Example 6

Let D and R represent the set of real numbers. The rule, f, which associates to each number in D the square of that number, is a function from D to R. If x is in D, then

$$f(x) = x^2$$

The range of f is the set of non-negative numbers since, if y is a non-negative number, there is a real number x with $x^2 = f(x) = y$ and, if y is negative, it cannot be the square of any number.

Example 7

Let D be the set of non-negative real numbers and R the set of real numbers. Let g be the rule that associates x in D to y in R if and only if $y^2 = x$. The rule g is **not** a function from D to R because there is at least one x in D that is associated to two different numbers in R. For example, 4 is in D and g associates both 2 and -2 to 4.

Let D and R be the set of real numbers and f and g functions from D to R defined by the equations

$$f(x) = x^2$$
$$g(x) = \sqrt{x^4}$$

Elementary rules of algebra tell us that for each real number x, $\sqrt{x^4} = x^2$. This suggests that f and g represent the same function since at each point in D the values of f and g are equal.

Definition

Let f and g be functions from D to R. We say f is **equal** to g (and write $f = g$) if for each x in D, $f(x) = g(x)$.

This definition says that, when considering functions, we are not primarily concerned with the defining rules but with the actual correspondences between points in the domain and points in the range space. In particular, each function from D to R is completely determined by the values it associates to points in D. We are not suggesting that the form a particular rule takes is unimportant. In fact, knowing different rules that define a particular function may be quite beneficial to our understanding and use of the relationship. For example, knowing $d(P, Q) = \sqrt{(x_2 - x_1)^2 + (y_2 - y_1)^2}$ and knowing $d(P, Q)$ gives the distance between $P(x_1, y_1)$ and $Q(x_2, y_2)$ are both important when using the function d.

Example 8

> Let D be the set of ordered pairs (x, y) of real numbers and R the set of real numbers. Let g be the function from D to R defined by
>
> $$g(x, y) = \sqrt{(x - y)^2}$$
>
> and f the function defined in Example 4. Then, $f = g$. To show this let (x, y) be in D. Either $x \geq y$ and $g(x, y) = \sqrt{(x - y)^2} = x - y = |x - y| = f(x, y)$ or $x < y$ and $g(x, y) = \sqrt{(x - y)^2} = \sqrt{(y - x)^2} = y - x = |x - y| = f(x, y)$. In any case $g(x, y) = f(x, y)$ at **each** point (x, y) in D and so $f = g$.

Exercises 4.2

1. Let D be the set of all children, R the set of all mothers, and f the rule associating x in D to y in R if and only if y is the mother of x. Is f a function from D to R?

2. Two sides of a cube are labeled with the number 1; the remaining four sides are labeled 2, 3, 4, and 5, respectively. The cube is then rolled as a die. Let D be the set of numbers that can appear on the face of the cube after it is rolled. Let P be the rule that associates to each number in D the probability of that number appearing when the cube is rolled.

    a. List the numbers in D.   b. Find $P(x)$ for each x in D.

    P is an example of a **probability density function.**

3. Let D be the set of numbers in $[0, 2\pi]$ and R the set of points on the circle of radius 1 centered at the origin. (See the figure below.)

    (a)   **(b)**

    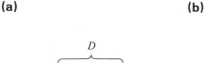

    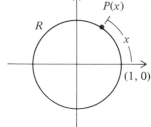

    Let P be the rule that associates to each x in D the point $P(x)$ on R, which has the property that the arc length measured along R from $(1,0)$ to $P(x)$ in a counterclockwise direction is x units. For example, $P(0) = (1, 0)$. Note that the circumference of the circle is 2π units and since $\pi/2$ represents $1/4$ the circumference of the circle, $P(\pi/2) = (0, 1)$.

a. Find $P(\pi)$, $P\left(\dfrac{3\pi}{2}\right)$, and $P(2\pi)$.

b. Find $P\left(\dfrac{\pi}{4}\right)$.

(*Hint:* Use the Pythagorean theorem and an appropriate right isosceles triangle.)

c. Find $P\left(\dfrac{3\pi}{4}\right)$.

4. Let D and R be the set of real numbers. Let f and g be the functions from D to R defined by

$$f(x) = x^2 + 1$$
$$g(x) = x^3 + 2$$

a. $f(2) = $ _____
b. $f(-3) = $ _____
c. If a is a real number:

$$f(a) = \text{_____}$$
$$f(a^2) = \text{_____}$$

d. $g(1) = $ _____
e. $g(-3) = $ _____
f. $f(g(1)) = $ _____
g. $f(f(1)) = $ _____

5. a. Let D and R be the set of real numbers and f, g functions from D to R defined by

$$f(x) = x^2$$
$$g(x) = x^4 + 2x^2 - 2x$$

Show $f \neq g$.

b. Let D be the set consisting of the numbers -1, 0, and 1, and R the set of real numbers. If f and g are the functions from D to R defined by the formulas in (a) above, show $f = g$.

6. Let D and R be the set of real numbers. In each part of this problem, you will be given two functions, f and g from D to R. Prove that either $f = g$ or $f \neq g$.

a. $f(x) = (x^3 + 2x + 1)^3$
   $g(x) = x^6 + 7x^5 + 14x^4 + 18x^3 + 14x^2 + 6x + 1$
b. $f(x) = x - 1$
   $g(x) = \dfrac{x^3 - x^2 + x - 1}{x^2 + 1}$
c. $f(x) = |x - 0|$
   $g(x) = \sqrt{x^2}$

7. The pressure, P, exerted by a gas at fixed temperature is a function of the volume, V, of the gas. Given a gas, there is a constant c such that

$$P(V) = \frac{c}{V}$$

What is a reasonable domain for the function P?

8. Each of the functions defined in this problem is to have its domain and range as subsets of the real numbers. What is the largest collection of real numbers that could act as the domain of each function?

    a. $f(x) = \dfrac{1}{x}$ **b.** $g(x) = \sqrt{x}$

SECTION 4.3

REAL VALUED FUNCTIONS OF A REAL VARIABLE

When we defined function, we made no restrictions as to the nature of possible domains and ranges. They could be sets of apples, numbers, letters, or kazoos. It made no difference as long as our rule associated elements in the domain to elements of the range in a unique fashion. However, functions most commonly used in applications are functions whose domains and ranges are sets of real numbers. Such functions are called **real valued functions of a real variable.** Understanding these functions is often prerequisite to understanding other functions, and much of the remainder of this text will be devoted to studying them. *For the remainder of this chapter you may assume that function means real valued function of a real variable.*

Let f be a function and suppose we can find a mathematical expression (for example $\sqrt{x^2 + 1}$ or $(3x + 2)/x^2$) involving a single variable (or unknown), x, such that for each element t in D_f the value $f(t)$ and the numerical value of the expression obtained when x is replaced by t are equal. This mathematical expression can be used to define f. We have seen illustrations of this idea in Example 3 of Section 4.2 where the cost function C was defined by the mathematical expression $500,000 + 5000x$ and in Example 6 where the squaring function was defined by writing $f(x) = x^2$.

Example 1

> Let D and R be the set of real numbers and f the function from D to R that associates to each real number the cube of that number. The expression x^3 defines f completely, for if t is in D, then $f(t) = t^3$. For example, $f(2) = 2^3 = 8$ and $f(-3) = (-3)^3 = -27$.

Usually, when defining a function f, we do not give the rule of association verbally if there is a mathematical expression available that can be used to define f. To define f, we write

$$f(x) = \text{mathematical expression in variable } x \tag{1}$$

Example 2

> Let D and R be the set of real numbers. The function f from D to R defined by
>
> $$f(x) = x^2 + 2x + 1$$
>
> could be described as follows: f is the rule that associates to each number the sum of the square of that number, 2 times that number, and 1.

You may already see the advantages of using mathematical expressions to define functions. Frequently, verbal descriptions of functions are unnecessarily complicated, contain words that are interpreted ambiguously, and in the final analysis, must be replaced by a mathematical expression when we try to evaluate the function at different points in the domain. If you are not convinced of the value of using a formula like (1) above to define a function, try to give an English sentence that describes the function defined by

$$f(x) = \sqrt{\left[\sqrt{x^2 + 1} + x^{5/3}\right]^2 + 1}$$

Note that (1) does not represent an equality of numbers; it is simply a symbolic way of defining the function f.

However, when x is replaced in (1) with a number from D_f, we have a true equality. For example,

$$f(x) = x^2 \tag{2}$$

is a definition of the squaring function. If π is in D_f, then $f(\pi)$ actually equals π^2. Furthermore, when defining a function by means of a formula like (1), it is not necessary to use the variable (or unknown) x to define f. Any symbol will do. For example, if we view t, α, or $\square$ as variables, the squaring function given in (2) can be defined equally well by $f(t) = t^2$, $f(\alpha) = \alpha^2$, or $f(\square) = \square^2$.

When working with functions, symbols such as x, t, or α are used in several ways. They are viewed as variables to define functions as in (1), or they are frequently viewed as numbers. Hence, if we have a function defined by, for example,

$$f(x) = x^2 + x + 1$$

and we are told that t is in the domain of f, then $f(t) = t^2 + t + 1$ must be viewed as an equality of numbers and not a formula defining f. There is obviously a certain amount of ambiguity in the meaning of $f(t) = t^2 + t + 1$, but the context of the work usually tells us the meaning we should attach to $f(t) = t^2 + t + 1$. Furthermore, the importance of using $t^2 + t + 1$ as an expression to define f at one time and as a value of f at another is perhaps not apparent. But as you read on in this text, and in your studies of calculus, you will see examples where the distinction is important. The next example will illustrate a situation where the distinction between t as a variable and as a number is used.

Example 3

> If a projectile is shot vertically upward from the surface of the earth with an initial velocity of 400 ft/sec, the height of the object can be thought of as a function, f, of time, measured in seconds after launch. Using calculus, it can be shown that f can be defined by $f(t) = -16t^2 + 400t$. (Here t has been used as a variable to define f.) The following question is typical when investigating such motion: At what time is the projectile at a height of 1600 feet? To solve this problem, let t represent a number giving the time (if such a time exists) when the projectile reaches 1600 feet. The height at this time is $f(t) = -16t^2 + 400t$. So we must have
>
> $$-16t^2 + 400t = 1600 \qquad \text{(Here } t \text{ is assumed to be a number in } D_f$$
> $$\text{where the value of } f \text{ is 1600.)}$$
> $$0 = 16t^2 - 400t + 1600$$
> $$0 = t^2 - 25t + 100$$
> $$0 = (t - 5)(t - 20)$$
>
> Hence, $t = 5$ or $t = 20$.

When a function f is defined by means of a formula like (1), then to find the value of f at a point t in D_f, we need only go back to the defining relation (1) and replace x (or whatever symbol is used there) by t. The following examples illustrate the computation of functional values using this replacement rule.

Example 4

> Let D be the set of non-negative real numbers, R the set of real numbers, and f the function from D to R defined by
>
> $$f(x) = \sqrt{x}$$
>
> $f(4) = \sqrt{4} = 2.$
>
> $f(9) = \sqrt{9} = 3.$
>
> $f(-4)$ is not defined since -4 is not in the domain of f.
>
> $f(t) = \sqrt{t}$ if t is in D.
>
> $f(t^2) = \sqrt{t^2}$ for any number t. (Note that $f(t^2)$ is not necessarily t. For example, if $t = -2$, then $f(t^2) = f((-2)^2) = f(4) = 2 \neq -2$.)
>
> $f(t^4) = t^2$ for any number t.

A symbol is frequently used both to define a function and to represent a number as these examples illustrate.

$$f(x^2) = \sqrt{x^2}$$
$$f(x^4) = \sqrt{x^4} = x^2$$
$$f(x^4 + 2x^2 + 1) = \sqrt{x^4 + 2x^2 + 1} = \sqrt{(x^2 + 1)^2} = x^2 + 1$$

Example 5

Let D be the set of all real numbers except 0, R the set of real numbers, and g the function defined by

$$g(z) = \frac{z + 1}{z}$$

$$g(1) = \frac{1 + 1}{1} = 2$$

$$g(-1) = \frac{-1 + 1}{-1} = 0$$

$$g(t^2) = \frac{t^2 + 1}{t^2}$$

$$g(z^2) = \frac{z^2 + 1}{z^2}$$

$$g\left(\frac{z + 1}{z}\right) = \frac{\left(\frac{z + 1}{z}\right) + 1}{\left(\frac{z + 1}{z}\right)} = \frac{\left(\frac{2z + 1}{z}\right)}{\left(\frac{z + 1}{z}\right)} = \frac{2z + 1}{z + 1}$$

Functions are often defined without specifying the domain of the function. (Since we are discussing real valued functions, the set R containing the range may always be taken to be the set of real numbers.) Suppose a function f is defined by a mathematical expression, as in (1), $f(x) =$ mathematical expression in variable x. If D_f is not specified and there are no physical circumstances dictating the domain of the function, we take D_f to be the set of all real numbers for which the mathematical expression in x "makes sense" (i.e., when a number in D_f replaces x, the mathematical expression should compute a real number using the usual arithmetic and algebraic operations).

Example 6

We would take the domain of the function f defined by $f(x) = \sqrt{x}$ to be the set of non-negative real numbers. If x is negative, then x has no square roots and so $\sqrt{x}$ is not computable. If x is non-negative, then we can compute $\sqrt{x}$ since there is exactly one non-negative number whose square is x.

The function g defined by $g(z) = (z + 1)/z$ would be given the domain consisting of all real numbers except 0 since $(0 + 1)/0$ is meaningless and $(z + 1)/z$ has a well-defined value if $z \neq 0$.

The ideas illustrated in Example 6 will present themselves frequently when we are trying to determine the domain of a function f given in the form of (1). In particular, a number, t, is ruled out of the domain if it forces us to compute the square root of a negative number or to divide by 0 when computing $f(t)$.

Example 7

Let G be the function defined by

$$G(t) = \frac{\sqrt{t + 1}}{t - 2}$$

The domain of G would consist of those numbers t for which $t + 1 \geq 0$ and $t - 2 \neq 0$, or all real numbers greater than or equal to -1 and different from 2.

Example 8
(Polynomial Functions)

When a function is defined by means of a formula of type (1), and the right-hand side of (1) is a polynomial, the function is called a **polynomial function.** Examples of polynomial functions are

$$f(x) = 2$$
$$g(x) = 2 + 3x$$
$$h(x) = 2 + 3x + 4x^2$$
$$k(x) = 2 + 3x + 4x^2 + 10x^5$$

If the domain of a polynomial function is not specified, we can take it to be the set of all real numbers.

Example 9
(Rational Functions)

When a function is defined by a formula of type (1) and the right-hand side of (1) is a rational expression, the function is called a **rational function.** Examples of rational functions are

$$f(x) = \frac{2}{x}$$

$$g(x) = \frac{2}{x(x - 3)}$$

$$h(x) = \frac{2(x-3)}{x(x-3)}$$

In general, a rational function, r, can be defined by

$$r(x) = \frac{P(x)}{Q(x)}$$

where $P(x)$ and $Q(x)$ are polynomials and $Q(x)$ is not the zero polynomial. If t is a number for which $Q(t) \neq 0$, then $P(t)/Q(t)$ or $r(t)$ is a well-defined number. If $Q(t) = 0$, then $r(t)$ does not exist. So if the domain of r is not specified, we take D_r to consist of the real numbers t for which $Q(t) \neq 0$.

For f, g, and h given above, D_f consists of all real numbers different from 0, and D_g and D_h consist of all real numbers different from 0 and 3.

Note that $h(x) = 2/x$ if $x \neq 3$, but $h \neq f$ since $D_f \neq D_h$. (In particular $f(3)$ is defined but $h(3)$ is not.)

Exercises 4.3

1. Describe each of the following functions in the form (1) of the text.

   a. $D = [0, +\infty)$, $R = (-\infty, +\infty)$, and f is the function that assigns to each number in D the sum of the cube root and the positive square root of that number.

   b. $D = [0, +\infty)$, $R = (-\infty, +\infty)$, and F is the function that assigns to each number in D the sum of the cube root and the negative square root of that number.

   c. $D = [0, +\infty)$, $R = (-\infty, +\infty)$, and S is the function that assigns to each number x in D the area of a square with side x.

2. A manufacturer of dining room chairs has estimated from his past records that the number of chairs he can produce in a month is a function of the number of hours of labor he uses. In fact, the function defined by

   $$C(x) = 100x - x^2$$

   where $C(x)$ represents the number of chairs produced using x hours of labor seems to describe this relationship. Give the function, A, whose value at a number r is the average number of chairs produced per hour of labor when r hours of labor are used in production.

   a. What are appropriate domains for C and A?
   b. Find $A(1)$, $A(50)$, and $A(75)$.

3. Give a function V whose value at a number t is the volume of a right circular cone with height t and for which the radius of the base is one-half the height of the cone. (Recall that the volume of a right circular cone

with height h and base radius r is $\frac{1}{3}\pi r^2 h$.)

a. What is a reasonable domain for this function?
b. What are $V(1)$, $V(10)$, and $V(20)$?
c. What are the dimensions of the cone with the above specifications if its volume is 18 cubic units?

4. Water is being pumped into a swimming pool at the rate of 10 ft³/min. The pool had 100 ft³ of water in it before the pumping began. Give a function f whose value at a number t gives the volume of water in the pool t minutes after pumping begins. If the pool can hold 1000 ft³ of water,

a. What is a reasonable domain for f?
b. What are $f(10)$ and $f(20)$?
c. How long does it take to fill the pool?

5. Let f be the function defined by

$$f(x) = x^3 - 3$$

Find:

a. $f(1)$ b. $f(-1)$ c. $f(4)$
d. $f(-3)$ e. $f(t)$ f. $f(t^2)$
g. $f(t^3)$ h. $f(-t^2)$ i. $f(x^2)$
j. $f(x^2 + 1)$ k. $f(\sqrt{x})$

6. Let g be the function defined by

$$g(u) = \frac{u^2}{u^2 - 1}$$

Find:

a. $g(3)$ b. $g(-\pi)$ c. $g(x)$
d. $g(x + 1)$ e. $g((x + 1)^2)$ f. $g(u^2)$

g. $g\left(\dfrac{u}{u + 1}\right)$ h. $g\left(\dfrac{u}{u - 1}\right)$ i. $g\left(\sqrt{\dfrac{u}{u - 1}}\right)$

j. $g(1)$

7. Let f and g be the functions defined by

$$f(x) = x^2 + 2 \qquad g(x) = \frac{1}{x^2 + 2}$$

Find:

a. $f(2)$ b. $g(6)$ c. $g(f(2))$

d. $g(f(x))$ e. $f(g(x))$ f. $f\left(\dfrac{1}{g(x)}\right)$

8. Let

$$H(x) = \frac{1}{\sqrt{x - 5}}$$

Find:

a. $H(6)$ b. $H(9)$ c. $H(x^2)$

d. $[H(x)]^2$ e. $H(\sqrt{x - 5})$ f. $H(0)$

g. $H(5)$ h. $H(x^2 + x)$ i. $H(x + 5)$

j. $H(x - 5)$

In Exercises 9–17, find the domain of the given function.

9. $f(x) = x^2 + 1$

10. $g(u) = \frac{1}{u^2 - 1}$

11. $h(v) = \sqrt{v - 2}$

12. $F(x) = \frac{\sqrt{x + 2}}{x^2 - 9}$

13. $G(x) = \frac{x^2 + 1}{x^2 + 9}$

14. $H(\alpha) = \frac{\alpha^2 - \alpha}{\alpha - 1}$

15. $T(x) = \sqrt{\frac{1}{x - 1}}$

16. $T(y) = y^2 + 1 + \frac{1}{\sqrt{y}}$

17. $Z(t) = \frac{t^2 + 1}{t^3 - t^2 - 10t - 8}$

SECTION 4.4

THE GRAPH OF A FUNCTION

Given a (real valued) function (of a real variable), we can frequently discover properties of the function by examining its graph. The word *graph* probably has some meaning to you already but we will use the following definition.

Definition

> If f is a function from D to R, the **graph** of f is the collection of ordered pairs (x, y) where x is in D and $y = f(x)$.

If we have a plane with a rectangular coordinate system, we think of points in the plane as ordered pairs of numbers and vice versa. The graph of f can be viewed as a curve in the plane consisting of the collection of points $P(x, y)$ where x is D and $y = f(x)$. (Frequently, we will say the graph consists of the points $(x, f(x))$.)

Example 1

Consider the function f having a domain consisting of the numbers -4, -3, -2, -1, 0, 1, 2, 3, 4 and defined by $f(x) = x^2$. The graph of f consists of the points $(-4, 16)$, $(-3, 9)$, $(-2, 4)$, $(-1, 1)$, $(0, 0)$, $(1, 1)$, $(2, 4)$, $(3, 9)$, $(4, 16)$. The graph can also be viewed as the collection of points marked off in Figure 4.4.1.

Figure 4.4.1

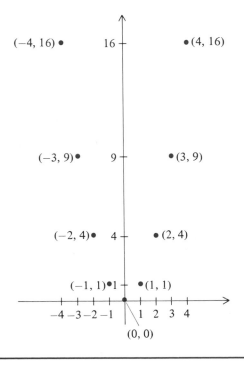

Example 2

Consider the function g whose domain is the set of real numbers and is defined by $g(x) = x^2$. Since the domain of g is infinite, it is impossible to list all the ordered pairs constituting the graph of g. We can, however, attempt to sketch the graph of g by plotting enough points on the graph of g to suggest the true nature of this curve.

The ordered pairs making up the graph of the function f described in Example 1 belong to the graph of g. The following points are also on the graph of g:

$$\left(\pm \frac{1}{2}, \frac{1}{4}\right) \quad \left(\pm \frac{3}{2}, \frac{9}{4}\right) \quad \left(\pm \frac{5}{2}, \frac{25}{4}\right) \quad \text{and} \quad \left(\pm \frac{7}{2}, \frac{49}{4}\right)$$

This collection of ordered pairs together with the collection from Example 1 is plotted in Figure 4.4.2.

Figure 4.4.2

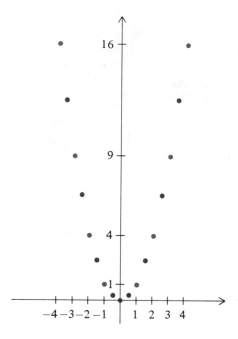

The entire graph of *g* can be viewed as a curve that passes through the points plotted in Figure 4.4.2. Figure 4.4.3 shows several possibilities for this curve.

Figure 4.4.3

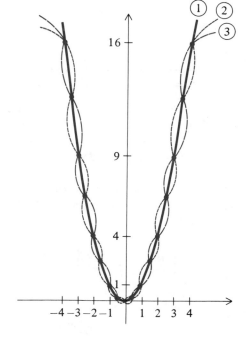

At this time each of the curves labeled 1, 2, and 3 qualifies as a possible graph of *g*. By plotting a few more (critically placed) points, we could immediately rule out curves 2 and 3. Actually, in trying to picture the graph of a function by plotting points, we plot what we feel is an appropriate number of points to suggest the true shape of the graph and then draw, through these points, a curve whose bends and turning points are dictated by the plotted points. Needless to say, the more points plotted, the more accurate the graph. From this point of view, the graph labeled 1 in Figure 4.4.3 would best approximate the graph of *g*.

Example 3

In this example, we will sketch the graph of the function *F* given by

$$F(u) = \frac{1}{u}$$

using the point-plotting method described above. First, note that the domain of *F* consists of all real numbers different from 0. Instead of listing ordered pairs in the graph of *F* in the form $(u, F(u))$ we will construct a table with two rows: The first row will list a sampling of numbers from the domain of *F*, and the second row will contain the values of *F* corresponding to the numbers listed in the first row.

u	-5	-4	-3	-2	-1	$-\frac{1}{2}$	$-\frac{1}{3}$	$-\frac{1}{4}$	$\frac{1}{4}$	$\frac{1}{3}$	$\frac{1}{2}$	1	2	3	4	5
$F(u)$	$-\frac{1}{5}$	$-\frac{1}{4}$	$-\frac{1}{3}$	$-\frac{1}{2}$	-1	-2	-3	-4	4	3	2	1	$\frac{1}{2}$	$\frac{1}{3}$	$\frac{1}{4}$	$\frac{1}{5}$

In Figure 4.4.4, we have plotted the points determined by the above table and have sketched the graph of *F*.

Figure 4.4.4   *F*(*u*) = 1 / *u*

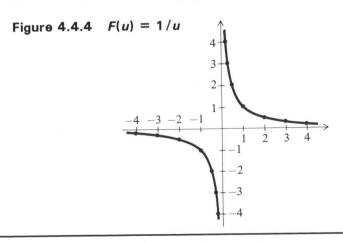

Example 4

In this example, we will sketch the graph of the cost function, C, described in Example 2 of Section 4.2. Recall that $D_C = [0, +\infty)$ and

$$C(x) = 500{,}000 + 5000x$$

Figure 4.4.5 shows both a table of the collection of points on the graph of C and a sketch of the graph of C. Note that the graph of f lies on and to the right of the vertical axis since D_C contains no negative numbers.

Figure 4.4.5 $C(x) = 500{,}000 + 5000x$

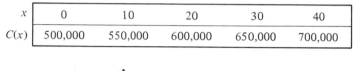

x	0	10	20	30	40
$C(x)$	500,000	550,000	600,000	650,000	700,000

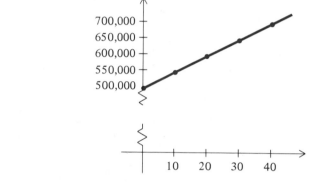

Given a function f and its graph, any vertical line (a line perpendicular to the horizontal axis) can meet the graph of f in at most one point. To see this, recall that (a, b) and (c, d) are on a line perpendicular to the horizontal axis if and only if $a = c$. If, as shown in Figure 4.4.6, we have a vertical line meeting the graph of f in two distinct points, (x, y_1) and (x, y_2) where $y_1 \neq y_2$, then $f(x) = y_1$ and $f(x) = y_2$ contradicting the fact that f associates exactly one value to x.

Figure 4.4.6

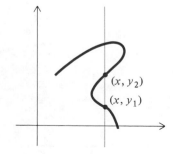

We can be more specific about the intersections of vertical lines and graphs of functions. If x is in D_f, then the vertical line through the point labeled x on the horizontal axis meets the graph of f at exactly one point, namely $(x, f(x))$. If x is not in the domain of f, then the vertical line through the point marked x does not meet the graph of f. (Notice in Example 3 above, the graph of f did not cross the vertical axis since 0 was not in the domain of F.)

Now suppose we are presented with a collection C of points which has the property that each vertical line meets this collection in at most one point. This implies that C is the graph of some function that we will denote by f. To define f we must specify D and R, and give a rule showing how f associates points in D to points in R (or equivalently give the value of f at each point in D). Since f is to be a real valued function (i.e., R_f will be contained in the set of real numbers), we can take R to be the set of real numbers. The domain D will consist of all real numbers x for which there is a point of the form (x, y) on C. D can be pictured by projecting the points of C down onto the horizontal axis. (See Figure 4.4.7.)

Figure 4.4.7

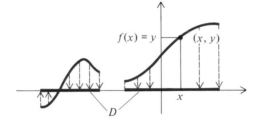

If x is a point in D, we will define $f(x)$ to be that number y for which (x, y) is a point on C. (See Figure 4.4.7.) Clearly f is a rule that associates points in D to points in R. If x is in D, then $f(x)$ is actually the second coordinate of the point of intersection of the curve C and the vertical line passing through x on the horizontal axis. Since there is exactly one such point of intersection, $f(x)$ is uniquely determined and f is a function from D to R. Clearly the graph of f is C.

Our discussion above tells us that there is a one-to-one correspondence between functions and graphs that intersect vertical lines in at most one point. Knowing the graph of a function f is equivalent to knowing $f(x)$ for each x in D_f (i.e., to knowing f). Thus, we can think of a function as either a rule or a graph. (Analogous types of identifications are not unusual. For example, when a plane has a rectangular coordinate system, we make no distinction between a point and its representation as an ordered pair of numbers because each point is uniquely identified with an ordered pair and vice versa.) This identification of a function with its graph permits us to give (as we promised) the following more mathematically rigorous definition of function.

Definition

> Let D and R be sets. A **function f from D to R** is a collection of ordered pairs (x, y) where x is in D and y is in R, for which the following properties hold:
>
> 1. If x is in D, there is a y in R such that (x, y) is in the collection.
> 2. There are no two points (x, y_1) and (x, y_2) in the collection with $y_1 \neq y_2$ (i.e., vertical lines meet the graph in at most one point).

Note:

A. Statement (1) in this definition guarantees that D is precisely the domain of f, and (2) says that, if x is in D, then f associates exactly one value to x.

B. The range of f consists of the second coordinates of the ordered pairs of f: those elements y for which there is an ordered pair (x, y) in the collection.

C. The domain, D, of the function f consists of the first coordinates of the ordered pairs of f: those elements x for which there is an ordered pair (x, y) in the collection.

D. A function f can now be unambiguously viewed as a rule or a graph.

As pointed out above, the range of a function f consists of the second coordinates of the ordered pairs comprising the function. Hence R_f may be obtained by projecting the graph of f onto the vertical axis as shown in Figure 4.4.8.

Figure 4.4.8

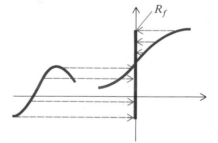

Example 5

> The collection, f, of ordered pairs $(1, 2)$, $(2, -1)$, $(3, 4)$, and $(4, 2)$ is a function. The graph of f is given in Figure 4.4.9. D_f consists of the points 1, 2, 3, and 4 and R_f consists of -1, 2, and 4.

Figure 4.4.9

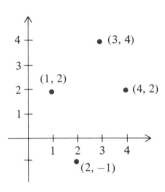

Example 6

The function g consisting of the ordered pairs (x, x^2) where x is a real number is nothing more than the squaring function given by

$$g(x) = x^2$$

If y is non-negative, there is a number, namely $\sqrt{y}$, whose square is y (i.e., $g(\sqrt{y}) = y$). If y is negative, there is no number whose square is y. Hence, R_g consists of the non-negative numbers. The graph of g is given in Figure 4.4.10. Note that the projection of the graph onto the vertical axis gives R_f as the set of non-negative numbers.

Figure 4.4.10 $g(x) = x^2$

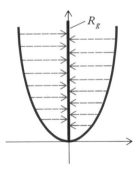

Example 7

Consider a graph consisting of the collection of points (x, y) where $x^2 + y^2 = 1$. You can sketch this graph easily if you recall the distance function defined in Section 4.2 and note that $x^2 + y^2 = (x - 0)^2 + (y - 0)^2$ is the square of the distance from (x, y) to $(0, 0)$. This graph consists of those points whose distances from $(0, 0)$ is 1 unit; that is, the graph is a circle of radius 1 centered at $(0, 0)$. This graph is *not* the graph of a function since, as illustrated in Figure 4.4.11, there are vertical lines meeting the graph at more than one point.

Figure 4.4.11

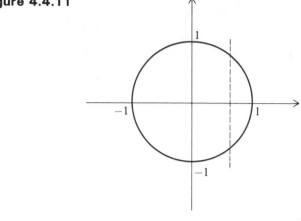

We have seen that functions can be compactly and informatively given in equation form. For example, we have investigated the functions defined by

$$f(x) = x^2 \tag{3}$$

and

$$C(x) = 500,000 + 5000x \tag{4}$$

To simplify matters and for historical reasons, the functional symbols on the left-hand sides of such equations are often replaced by a single variable.

$$y = x^2 \tag{3'}$$

$$C = 500,000 + 5000x \tag{4'}$$

The terminology we use to say that equations (3′) and (4′) define the functions f and C is "$y = x^2$ defines y as a function of x" and "$C = 500,000 + 5000x$ defines C as a function of x."

You probably know that the graph of an equation in two variables, for example, $y^3 = x^2$, consists of the collection of ordered pairs whose coordinates satisfy the equation. To be precise, we do the following to draw the graph of an equation in the variables x and y:

A. Construct a coordinate system.
B. Select the points on one of the axes to represent possible values of the variable, x. (This axis is then referred to as the x-axis.) The points on the other axis represent possible values of the other variable.
C. If the horizontal and vertical axes correspond to the x- and y-axes,

respectively, the graph consists of the ordered pairs (x_0, y_0) that satisfy the equation.

Definition

> We say **an equation in the variables x and y determines y as a function of x** if its graph, when the horizontal axis is the x-axis, is the graph of a function.

This definition is equivalent to each of the following statements:

A.   Each vertical line meets the graph of the equation (when the horizontal axis is chosen as the x-axis) in, at most, one point.
B.   There are no two distinct points of the form (x_0, y_1) and (x_0, y_2) with $y_1 \neq y_2$ satisfying the equation.

The following table lists a collection of ordered pairs that satisfy $y^3 = x^2$:

x	-27	-8	-1	0	1	8	27
y	9	4	1	0	1	4	9

Figure 4.4.12 shows the graph of $y^3 = x^2$ when the horizontal axis is made the x-axis and Figure 4.4.13 shows the graph when the vertical axis is made the x-axis.

Figure 4.4.12

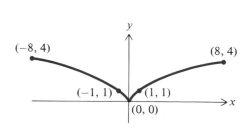

Figure 4.4.13

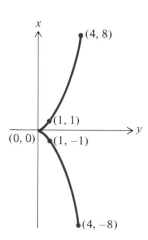

Figure 4.4.12 shows that $y^3 = x^2$ defines y as a function of x since each vertical line meets the graph in at most one point.

To determine if $y^3 = x^2$ defines x as a function of y, we can use the above definition with the roles of x and y reversed. In particular, we graph this equation selecting the horizontal axis to be the y-axis and check to see if vertical lines meet the graph in more than one point. Figure 4.4.13 shows us that $y^3 = x^2$ *does not* define x as a function of y. In particular, note that the line passing through 1 on the y-axis meets the graph at $(1, 1)$ and $(1, -1)$.

Some reflection on our arguments above shows us that the graph in Figure 4.4.12 was sufficient to show that $y^3 = x^2$ does not define x as a function of y. Indeed, drawing a vertical line through y_0 on the y-axis in Figure 4.4.13 is equivalent to drawing a horizontal line through y_0 on the vertical axis in Figure 4.4.12. We conclude that $y^3 = x^2$ would determine x as a function of y if each horizontal line in Figure 4.4.12 met the graph in, at most, one point.

In general,

> if the graph of an equation in the variables x and y is graphed with the horizontal axis as the x-axis, then the equation defines x as a function of y if each horizontal line meets the graph in, at most, one point (or, equivalently, if there are no two points of the form (x_1, y_0) and (x_2, y_0) with $x_1 \neq x_2$ on the graph).

It is often possible to solve an equation in two variables, x and y, for x in terms of y alone or for y in terms of x alone. For example, when we solve $y^3 = x^2$ for y in terms of x, we get

$$y^3 = x^2 \Longleftrightarrow y = \sqrt[3]{x^2}$$

and if we solve for x in terms of y, we get

$$y^3 = x^2 \Longleftrightarrow x = \pm \sqrt{y^3}$$

When such solutions are possible, they can tell us if our equation determines one of the variables as a function of the other. For example, $x = \pm \sqrt{y^3}$ shows us that, for each positive y value, y_0, there exist two different x values, $x_1 = +\sqrt{y_0^3}$ and $x_2 = -\sqrt{y_0^3}$, such that (x_1, y_0) and (x_2, y_0) both satisfy $x^3 = y^2$. On the other hand, $y = \sqrt[3]{x^2}$ shows that, for each value x_0 of x, there is exactly one value $y_0 = \sqrt[3]{x_0^2}$ such that $y_0^3 = x_0^2$, implying that $y^3 = x^2$ defines y as a function of x.

Example 8

> The equation $x^2 + y^2 = 1$ does not define x as a function of y. If we try to solve the equation for x in terms of y we get $x = \pm \sqrt{1 - y^2}$, indicating that there are certain values of y (for example, $y = 0$) to which the equation associates two values of x.

Example 9

The equation $z^2 = u$ does not define z as a function u. The graph of $z^2 = u$ is pictured in Figure 4.4.14.

Figure 4.4.14 $z^2 = u$

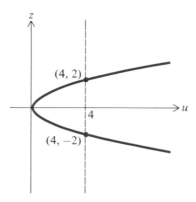

Notice that the vertical line through 4 on the horizontal axis meets the graph of $z^2 = u$ in two distinct points.

If we change our orientation and look at horizontal lines and their intersection with the graph of $z^2 = u$, we see that each horizontal line meets the graph in at most one point. We conclude that the equation $z^2 = u$ defines u as a function of z. We have investigated this function many times. It is the squaring function,

$$f(z) = z^2$$

Example 10

The equation $x^2 - y^2 = 0$ neither defines y as a function of x nor x as a function of y. For,

$$x^2 - y^2 = 0$$

is equivalent to

$$(x - y)(x + y) = 0$$

So, a point lies on the graph of this equation if and only if it satisfies $x - y = 0$ or $x + y = 0$. Hence, the graph of $x^2 - y^2 = 0$ consists of the graph of $x - y = 0$ together with the graph of $x + y = 0$, as shown in Figure 4.4.15. The vertical line through $(1, 0)$ on the x-axis meets the graph of $x^2 - y^2 = 0$ at the points $(1, 1)$ and $(1, -1)$, indicating y is not a function of x. The horizontal line through $(0, 1)$ meets the graph of $x^2 - y^2 = 0$ at $(1, 1)$ and $(-1, 1)$, indicating x is not a function of y.

Figure 4.4.15

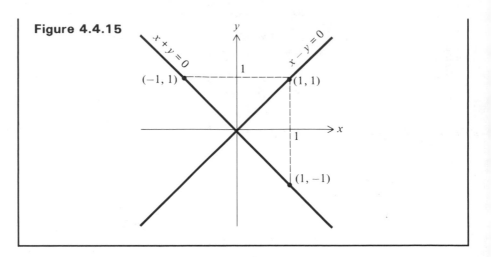

Example 11

The graph of the equation

$$(t^2 + s^2 + 1)(s - 2t) = 0 \tag{5}$$

consists of the graph of $t^2 + s^2 + 1 = 0$ together with the graph of $s - 2t = 0$. Since it is impossible for $t^2 + s^2 = -1 < 0$ for any pair of numbers, (t, s), the graph of (5) is the graph of $s - 2t = 0$ or $s = 2t$. The graph of $s = 2t$ is drawn in Figure 4.4.16. Notice that Equation (5) defines t as a function of s and s as a function of t.

Figure 4.4.16 $s = 2t$

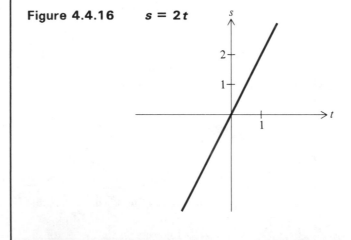

In mathematics texts, it is customary to use the points on the horizontal axis of a coordinate system to represent points in the domain of a function, f, and to use the symbol x to represent a point in the domain. For this reason, the horizontal axis is often referred to as the x-axis. The vertical axis is often referred to as the y-axis since y is most often used to represent a functional value, $f(x)$, and functional values are plotted along the vertical axis. These conventions are not universal, and in many instances it is convenient to deviate from these practices. Be prepared in your readings for such occurrences.

Exercises 4.4   In Exercises 1–11, give the domain and sketch the graph of the given function.

1. $f(x) = x^4$

2. $g(x) = \dfrac{1}{x - 1}$

3. $h(x) = x^2 + x + 1$

4. $F(u) = \begin{cases} u & u \le 1 \\ 3u + 1 & u > 1 \end{cases}$

5. $G(z) = \dfrac{z^2}{z + 1}$

6. $H(z) = \dfrac{z^2}{z^2 + 1}$

7. $T(x) = x^3 + x^2$

8. $f(x) = x$

9. $g(x) = -x$

10. $A(x) = \begin{cases} x & \text{if } x \ge 0 \\ -x & \text{if } x < 0 \end{cases}$

11. $h(x) = \dfrac{A(x)}{x}$ (A is the function in Exercise 10.)

12. If x is a real number, $[[x]]$ is used to denote the greatest integer less than or equal to x. For example, $[[2.5]] = 2$, $[[3]] = 3$, $[[-1.31]] = -2$, and $[[-\frac{5}{2}]] = -3$. Sketch the graph of the **greatest integer function**

$$f(x) = [[x]]$$

for values of x in $[-4, 4]$.

13. Let f be the function with domain D consisting of the numbers $-3, -1, 0, 2, 4$ and defined by

$$f(x) = \frac{x + 1}{x - 1}$$

a. List the ordered pairs comprising the graph of f.
b. Draw the graph of f.
c. What is the range of f?

14. Let f be the function with domain D consisting of the numbers $-1, 2, 3,$ and 5 and defined by

$$f(x) = \frac{x^2}{x - 1}$$

a. List the ordered pairs comprising the graph of f.
b. Draw the graph of f.
c. What is the range of f?

15. Consider the graph consisting of the points $(-2, 4)$, $(-1, 4)$, $(0, 3)$, $(1, 3)$, and $(2, 5)$. This is the graph of some function f.

    a. What is D_f?   **b.** What is R_f?   **c.** Find $f(x)$ for each x in D_f.

16. Consider the graph consisting of the points $(-2, 2)$, $(-1, -1)$, $(2, 5)$, $(3, 2)$ and $(7, 2)$. This is the graph of a function f.

    a. What is D_f?   **b.** What is R_f?   **c.** Find $f(x)$ for each x in D_f.

17. Sketch each of the following functions on the same coordinate system.

    a. $f(x) = x^3$ **b.** $g(x) = -x^3$
    c. $h(x) = x^3 + 1$ **d.** $F(x) = \frac{1}{2}x^3$
    e. $G(x) = (x - 1)^3$ **f.** $H(x) = (x + 1)^3$

18. Each curve in the four figures (a)–(d) is the graph of an equation in the variables x and y.

    a. For each curve determine if the curve is the graph of an equation that defines y as a function of x.

    b. For each curve determine if the curve is the graph of an equation that defines x as a function of y.

(a)

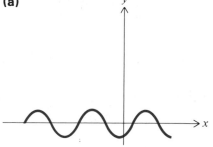

(b)

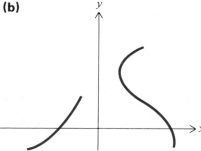

(c)

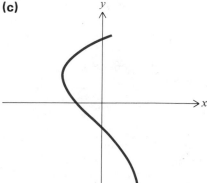

(d)

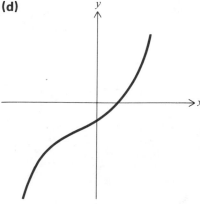

In Exercises 19–28, indicate whether the given equation determines y as a function of x. Also indicate if the equation determines x as a function of y.

19. $x^2 + 2y^2 = 1$ 20. $y = x^5$
21. $x^2 + 2xy + y^2 = 0$ 22. $x^2 + 2xy + y^2 = 1$
23. $y = x^2, x \geq 0$ 24. $y^3 = x^2 + 1, x \geq 0$
25. $y^3 = x^2 + 1, x \leq 1$ 26. $4y^2 + 9x^2 = 36, x \geq 0$
27. $4x^2 + 9y^2 = 36, y \geq 0$ 28. $4x^2 + 9y^2 = 36, x \geq 0, y \leq 0$

SECTION 4.5

INVERSE FUNCTIONS

If f is a function, we can think of f as a rule that will generate the number $f(x)$ in R_f when it is supplied with a number x in D_f. (See Figure 4.5.1.) We are frequently interested in the **inverse relationship** defined by f; i.e., the rule that associates to each y in R_f the number or numbers x in D_f such that $y = f(x)$. (See Figure 4.5.2.)

Figure 4.5.1   **Given x in D_f the function f determines f(x).**

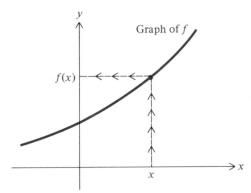

Figure 4.5.2   **The Inverse Relationship: Given y in R_f, the inverse relationship determines x in D_f such that y = f(x).**

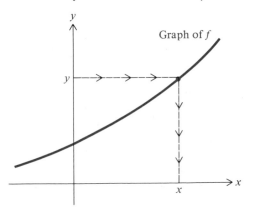

Example 1

Let f be the function defined by

$$f(x) = 2x + 1$$

The inverse relationship defined by f is the rule that, when supplied with a number y in R_f, generates the numbers x in D_f such that $y = f(x) = 2x + 1$. For example, 5 is in R_f. The number in D_f associated to 5 by f is the number x such that $5 = 2x + 1$. Solving this equation, we see $x = 2$. In general, given y in R_f

$$y = f(x)$$
$$\Leftrightarrow y = 2x + 1$$
$$\Leftrightarrow x = \frac{y - 1}{2}$$

Hence, the inverse relationship defined by f is the rule that associates to y in R_f the number $(y - 1)/2$ in D_f. Figure 4.5.3 illustrates this relationship graphically.

Figure 4.5.3

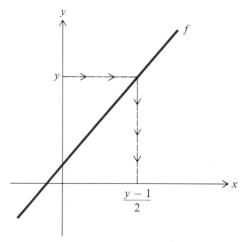

Notice that to each y in R_f there is precisely one x in D_f, namely $(y - 1)/2$, associated to y by this inverse relationship, and hence this relationship is a function. In fact, it is the function from R_f to D_f defined by

$$g(y) = \frac{y - 1}{2}$$

(It is customary to use x as the defining variable for g and write $g(x) = (x - 1)/2$.)

Example 2

Let g be the function defined by

$$g(x) = x^2$$

The inverse relationship defined by g is the rule that, when supplied with a number y in R_g, generates the numbers x in D_g such that $y = g(x) = x^2$. For example, 4 is in R_g, and the numbers in D_g associated to 4 by g are the numbers x for which $4 = x^2$. Solving this equation for x, we see $x = \pm 2$. In general, given y in R_g

$$y = g(x)$$
$$\Leftrightarrow y = x^2$$
$$\Leftrightarrow x = \pm \sqrt{y}$$

Hence, the inverse relationship defined by f is the rule that associates to y in R_f the numbers $\sqrt{y}$ and $-\sqrt{y}$ in D_g. Figure 4.5.4 illustrates this relationship graphically. Notice that the inverse relationship associated to g is not a function from R_g to D_g since there are numbers y in R_g that have two numbers, $\sqrt{y}$ and $-\sqrt{y}$, associated to y. Nevertheless, we could examine the inverse relationship by studying the two functions h_1 and h_2, where $D_{h_1} = R_g$ and $h_1(x) = \sqrt{x}$, and $D_{h_2} = R_g$ and $h_2(x) = -\sqrt{x}$.

Figure 4.5.4

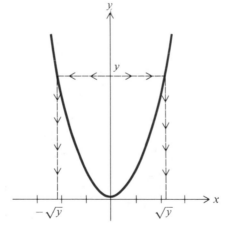

In this section, we will study inverse relationships defined by functions. We will, however, confine our attention to those functions whose inverse relationships are also functions. As indicated by Example 2 above, this is not always the case.

If f is a function, when is the inverse relationship defined by f also a function? It will be a function if and only if for each y in R_f there is precisely one x in D_f such that $y = f(x)$. There are several informative ways of restating this property. First, as indicated by Figures 4.5.3 and 4.5.4, the inverse relationship defined by f is a function if and only if each horizontal line meets the graph of f in at most one point. From our work in Section 4.4, it follows that the inverse relationship defined by f is a function if and only if the graph of $y = f(x)$ defines x as a function of y. (See Figure 4.5.5.)

Figure 4.5.5 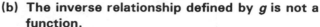 (a) The inverse relationship defined by *f* is a function.
(b) The inverse relationship defined by *g* is not a function.

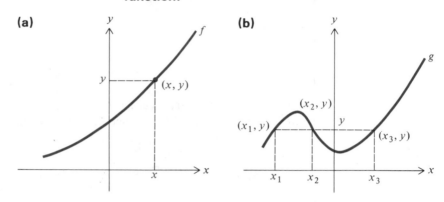

If the inverse relationship defined by a function f is also a function, we say that f is **one-to-one.** Indeed, in this case, we see that to each x in D_f, f associates precisely one number, $f(x)$, in R_f and to each of y in R_f there is precisely one number x in D_f such that $f(x) = y$. Hence, f establishes a one-to-one correspondence between the numbers in D_f and the numbers in R_f.

Since we will use each of the ideas presented above to determine if functions are one-to-one (and, hence, define inverse relationships that are functions), we restate below these equivalent definitions of a one-to-one function.

> A function f is one-to-one if it satisfies one of the following equivalent properties:
>
> A.   For each y in R_f there is precisely one x in D_f such that $y = f(x)$.
> B.   Each horizontal line meets the graph of f in at most one point.
> C.   The equation $y = f(x)$ defines x as a function of y.

Now, suppose f is a one-to-one function. The inverse relationship defined by f is a function denoted by f^{-1} and called the inverse of f. To be precise,

If f is a one-to-one function, the **inverse of** f is that function, f^{-1}, with domain

$$D_{f^{-1}} = R_f$$

range

$$R_{f^{-1}} = D_f$$

and defined by the rule

$$y = f^{-1}(x) \Leftrightarrow x = f(y)$$

Suppose f has an inverse. A point (a, b) is on the graph of f if and only if $b = f(a)$. But, $b = f(a)$ if and only if $a = f^{-1}(b)$ and $a = f^{-1}(b)$ if and only if (b, a) is on the graph of f^{-1}. Hence (a, b) is on the graph of f if and only if (b, a) is on the graph of f^{-1}. This fact can help us graph f^{-1}. If we have a table of values giving points $(x_1, y_1), (x_2, y_2), \ldots, (x_n, y_n)$ on the graph of f then the points $(y_1, x_1), (y_2, x_2), \ldots, (y_n, x_n)$ give a table of values for f^{-1}, and we can use these to sketch the graph. We will illustrate this idea in the following examples.

Example 3

We saw in Example 1 that the function f defined by $f(x) = 2x + 1$ is one-to-one and that $f^{-1}(x) = (x - 1)/2$. A table of values for f is

x	-4	-3	-2	-1	0	1	2	3
$f(x)$	-7	-5	-3	-1	1	3	5	7

Our discussion prior to this example shows that a table of values for f^{-1} is

x	-7	-5	-3	-1	1	3	5	7
$f^{-1}(x)$	-4	-3	-2	-1	0	1	2	3

These tables have been used to sketch the graphs of f and f^{-1} in Figure 4.5.6.

Figure 4.5.6

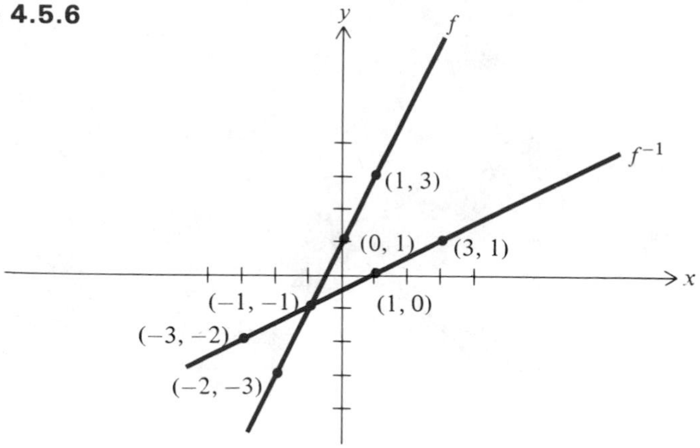

Example 4

Let f be the function with domain $[0, +\infty)$ and defined by

$$f(x) = x^2$$

As the graph of f (Figure 4.5.7) indicates, $R_f = [0, +\infty)$. Since horizontal lines meet the graph of f in at most one point, we see that f is one-to-one and f^{-1} exists.

Figure 4.5.7

x	0	1	2	3	4
$f(x)$	0	1	4	9	16

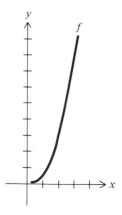

Now

$$x = f^{-1}(y)$$

$$\Leftrightarrow y = f(x)$$
$$\Leftrightarrow y = x^2 \quad \text{and} \quad x \ge 0 \qquad \text{(i.e., } x \text{ is in } D_f)$$
$$\Leftrightarrow x = \pm \sqrt{y} \quad \text{and} \quad x \ge 0$$
$$\Leftrightarrow x = \sqrt{y} \qquad (\text{i.e., } f^{-1}(y) = \sqrt{y})$$

Hence, we see that $D_{f^{-1}} = [0, +\infty)$, $R_{f^{-1}} = [0, +\infty)$, and, using x as the defining variable, $f^{-1}(x) = \sqrt{x}$.

Figure 4.5.8 gives a table of values (derived from the table of values for f given above) and the graph of f^{-1}.

Figure 4.5.8

x	0	1	4	9	16
$f^{-1}(x)$	0	1	2	3	4

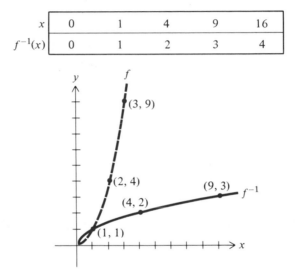

Example 5

Consider the function g defined by

$$g(x) = \sqrt{x - 1}$$

The domain of g consists of all real numbers x that have the property that $x - 1 \ge 0$. Hence, $D_g = [1, +\infty)$. The graph of g appears in Figure 4.5.9, which indicates that $R_g = [0, +\infty)$ and that g has an inverse.

$$x = f^{-1}(y)$$
$$\Leftrightarrow y = f(x)$$
$$\Leftrightarrow y = \sqrt{x - 1}$$
$$\Leftrightarrow y^2 = x - 1 \qquad y \ge 0$$
$$\Leftrightarrow y^2 + 1 = x \qquad y \ge 0 \qquad (\text{i.e., } f^{-1}(y) = y^2 + 1)$$

$D_{g^{-1}} = R_g = [0, +\infty)$, $R_{g^{-1}} = D_g = [1, +\infty)$, and using x as the defining variable, $g^{-1}(x) = x^2 + 1$.

A table of values for g^{-1} (derived from the table of values for g) and the graph of g^{-1} appear in Figure 4.5.10.

Figure 4.5.9

x	1	2	5	10	17
$g(x)$	0	1	2	3	4

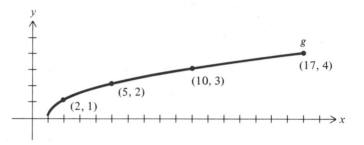

Figure 4.5.10

x	0	1	2	3	4
$g^{-1}(x)$	1	2	5	10	17

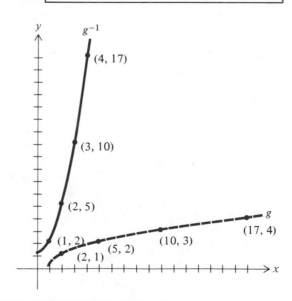

The above examples show that, if f is a function and we can solve the equation $y = f(x)$ for x in terms of y, this solution can tell us if f has an inverse and give us a defining relation for f^{-1}. Frequently, it is not possible to solve $y = f(x)$ for x in terms of y. Nevertheless, f^{-1} may exist and we may want to study it. To investigate f^{-1} in such cases, we will rely on graphical techniques and our definition of f^{-1}. (When you study calculus you will learn several concepts that can be used to determine if functions have inverses and, if they do, what properties the inverses have. Unfortunately, these concepts are beyond the scope of this text.) In the next example we investigate the inverse of a function f where we cannot solve the equation $y = f(x)$ for x in terms of y.

Example 6

Let h be the function defined by

$$h(x) = x^3 + x$$

We have not learned how to solve the equation

$$y = x^3 + x$$

for x in terms of y and so we will not be able to employ the procedure used in previous examples to find h^{-1}. However, we can still decide if h^{-1} exists and, if it does, what properties it has.

First, note that, since h is a polynomial function, $D_h = (-\infty, +\infty)$. The graph of h together with a table of values appears in Figure 4.5.11.

Figure 4.5.11

x	-4	-3	-2	-1	$-\frac{1}{2}$	0	$\frac{1}{2}$	1	2	3	4
$h(x)$	-68	-30	-10	-2	$-\frac{5}{8}$	0	$\frac{5}{8}$	2	10	30	68

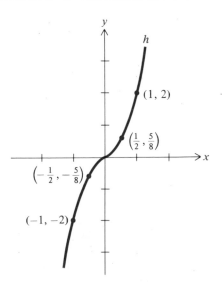

The graph of h indicates that $R_h = (-\infty, +\infty)$ and, since any horizontal line meets the graph of h in exactly one point, h has an inverse. Thus, we can conclude that h^{-1} exists and that

$$D_{h^{-1}} = R_h = (-\infty, +\infty) \quad \text{and} \quad R_{h^{-1}} = D_h = (-\infty, +\infty)$$

(See Figure 4.5.12.)

Figure 4.5.12

x	-68	-30	-10	-2	$-\frac{5}{8}$	0	$\frac{5}{8}$	2	10	30	68
$h^{-1}(x)$	-4	-3	-2	-1	$-\frac{1}{2}$	0	$\frac{1}{2}$	1	2	3	4

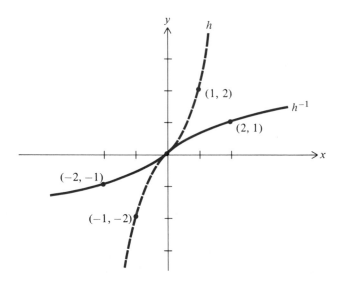

Unfortunately, all these conclusions are drawn from our belief that the sketch of the graph of h given in Figure 4.5.11 is accurate. We have plotted only a few points on the graph and sketched the remainder of the graph using the plotted points as guides. It is quite possible that the graph is not completely accurate and that there is a horizontal line meeting the graph of h in more than one point. When you study calculus, you will learn concepts that can be used to rigorously prove that $R_h = (-\infty, +\infty)$ and that h^{-1} exists.

Since we cannot solve $y = x^3 + x$ for x in terms of y, we cannot supply a mathematical expression for $h^{-1}(x)$ and we must rely on the statement $y = h^{-1}(x)$ if and only if $x = h(y) = y^3 + y$ as a rule of definition for h^{-1}. However, we can draw the graph of h^{-1}. A table of values for h^{-1} (constructed from the table of values for h) and a sketch of the graph of

h^{-1} are given in Figure 4.5.12.

In the remaining sections of this chapter, we will study additional properties of inverse functions. We will use these properties and the ideas presented in this section when we study inverse trigonometric functions in Chapter 6 and the exponential and logarithmic functions in Chapters 7 and 8.

Exercises 4.5

 In Exercises 1–15, determine if the given function has an inverse. If a function has an inverse, find a defining relation for the inverse and sketch the graphs of the function and its inverse.

1. $f(x) = 3x - 2$
2. $h(x) = \frac{1}{2}x + 3$
3. $g(x) = x^2$ $D_g = (-\infty, 0]$
4. $f(x) = x^2$ $D_f = [-1, 1]$
5. $g(x) = x^4$
6. $F(x) = x^4$ $D_F = [1, +\infty)$
7. $G(x) = x^4$ $D_G = (-\infty, -1]$
8. $H(x) = x^3$
9. $f(x) = x^3 - 1$
10. $g(x) = (x + 1)^3$
11. $f(x) = \dfrac{1}{x}$
12. $g(x) = \dfrac{1}{x - 1}$
13. $T(x) = x^2 + 4x + 4$
14. $U(x) = x^2 + 4x + 4$ $D_u = [-2, +\infty)$
15. $V(x) = x^2 + 4x + 4$ $D_v = [-3, +\infty)$

 In Exercises 16–20, determine by graphical methods if the given function has an inverse. If a function has an inverse, sketch the graph of the inverse.

16. $f(x) = \sqrt{x^3} + \sqrt{x}$
17. $g(x) = x^5 + x^3$
18. $h(x) = x^5 - x^3$
19. $F(x) = \begin{cases} x^2 \text{ if } x \geq 0 \\ -x^2 \text{ if } x < 0 \end{cases}$
20. $G(x) = x^3 + x^2$

Each function, f, given in Exercises 21–25 below has an inverse. Find a defining relation for f^{-1}. Then show $f(f^{-1}(x)) = x$ and $f^{-1}(f(x)) = x$.

21. $f(x) = -2x + 4$
22. $f(x) = \dfrac{3}{2x - 3}$
23. $f(x) = x^6$ $D_f = [0, +\infty)$
24. $f(x) = x^3$
25. $f(x) = x^2 + 6x - 4$ $D_f = (-\infty, -3]$

SECTION 4.6   GRAPH SKETCHING

The graph of a function is a picture of that function. If drawn with some care, it records considerable information in an easily recognizable form. Consequently, we suggest that, whenever you study a function, you sketch its graph.

Up to this point we have sketched the graphs of functions by plotting a few points that lie on the graph and using these points as a guide to sketch in the remainder of the graph. In this section, we will develop some concepts that can help in sketching graphs more accurately.

4.6.1   Intercepts

The intercepts of a graph—the points at which the graph crosses the axes of the coordinate system—frequently represent important information about functions used in applied studies. For example, the vertical axis intercept of the cost function, $C(x) = 500,000 + 5000x$, defined in Example 2 of Section 4.2 and graphed in Example 4 of Section 4.4, represents the "fixed" or "set up costs" in building and operating an oil well.

As another example, suppose an object is fired vertically upward with an initial velocity of 50 ft/sec. The velocity of the object t seconds after it is fired is given by the functional relationship

$$v(t) = -32t + 50$$

A sketch of the graph of v appears in Figure 4.6.1. Note that the graph intersects the v-axis at the point 50, which gives the initial velocity of the projectile. The graph intersects the t-axis at the point $\frac{50}{32}$, which represents the time at which the velocity is 0. (This also represents the time at which the projectile reaches its maximum height.)

Figure 4.6.1 $v(t) = -32t + 50$

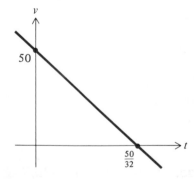

Because of the importance of intercepts, we suggest that you plot (if possible) all the intercepts when sketching the graph of a function.

We mentioned earlier that, when graphing a function, f, we usually call the vertical axis the *y*-axis, and the horizontal axis the *x*-axis. With this convention, the *y*-intercept of a function, *f*, can be found by evaluating $f(0)$. Actually, the point $(0, f(0))$ is the *y*-intercept, but we frequently refer to this intercept as $f(0)$. Note that, if 0 is not in the domain of *f*, then the graph of *f* has no *y*-intercept, and since the graph of *f* can cross the *y*-axis at most once, there can be no more than one *y*-intercept.

To find the intercepts of the graph of *f* with the *x*-axis, we find the values *x* where $f(x) = 0$. Although the *x*-intercepts are actually the points $(x, f(x))$ where $f(x) = 0$, we frequently refer to the *x*-coordinates of these points as the *x*-intercepts. (The *x*-intercepts are frequently called the **zeros** of *f*. Note also that, if *f* is a polynomial function, then the zeros of *f* are the roots of the polynomial defining *f*.)

Example 1

To find the intercepts of the graph of *f*, where $f(x) = x^3 + 3x^2 - x - 3$, we proceed with $f(0) = -3$. Thus, $(0, -3)$ or simply -3 is the *y*-intercept. To find the *x*-axis intercepts, we set $0 = x^3 + 3x^2 - x - 3$ and solve for *x*. Factoring the right-hand side of this equation, we get $0 = (x - 1)(x + 1)(x + 3)$. The solutions of our equation are $x = 1$, $x = -1$, and $x = -3$. The *x*-intercepts are $(1, 0)$, $(-1, 0)$, and $(-3, 0)$ (or simply $1, -1, -3$).

A list of points on the graph of *f* and the graph itself appear in Figure 4.6.2.

Figure 4.6.2

x	-4	-3	-2	-1	0	1	2
$f(x)$	-15	0	3	0	-3	0	15

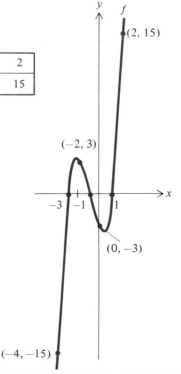

Example 2

Consider the function f defined by

$$f(x) = \frac{x^2 - 4}{\sqrt{x^2 - 9}}$$

Note that the domain of f consists of those real numbers x for which $x^2 - 9 > 0$, which is equivalent to requiring that $x > 3$ or $x < -3$. Since 0 is not in the domain of f, the graph of f has no y-intercepts.

To find horizontal axis intercepts, we solve

$$\frac{x^2 - 4}{\sqrt{x^2 - 9}} = 0 \tag{1}$$

Now x is a solution of (1) if and only if

$$x^2 - 4 = 0 \quad \text{and} \quad x^2 - 9 > 0$$
$$\Leftrightarrow x^2 = 4 \quad \text{and} \quad x^2 - 9 > 0$$
$$\Leftrightarrow x = \pm 2 \quad \text{and} \quad x^2 - 9 > 0$$

Neither 2 nor -2 is a solution of (1) since $(\pm 2)^2 - 9 < 0$. Thus, the graph of f has no x-intercepts.

The graph of f is sketched in Figure 4.6.3.

Figure 4.6.3

x	-6	-5	-4	-3.5	-3.1	-3.01	3.01	3.1	3.5	4	5	6
$f(x)$	6.16	5.25	4.54	4.58	7.18	20.64	20.64	7.18	4.58	4.54	5.25	6.16

(Accurate to two decimal places)

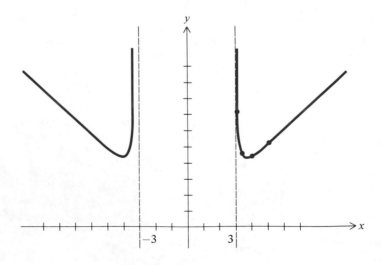

4.6.2   **Symmetry**

Frequently, the graph of a function possesses symmetries. These symmetries can be identified by studying a defining equation of the function; and as you will see in the examples below and in the exercises, recognizing symmetry can help greatly in graph sketching. Moreover, when functions are used to represent physical phenomena, symmetries in the graphs of these functions suggest symmetries in the actions of the physical phenomena. For example, when a stone is thrown into the air from ground level with an initial velocity of 160 ft/sec, the height of the stone at time t is given by the function s where

$$s(t) = -16t^2 + 160t$$

The graph of s is given in Figure 4.6.4. Notice the symmetry of the graph about the vertical line passing through the point (5, 400). The symmetry suggests, for example, that the height at time $5 - t$ is the same as height at the time $5 + t$ and that the time necessary for the projectile to reach its maximum height is equal to one-half of the time the projectile is in flight.

Figure 4.6.4 $s(t) = -16t^2 + 160t$

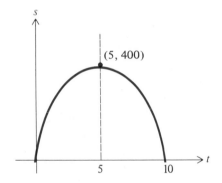

The above example relied on your intuition to understand the symmetry of a graph. Now let us be more specific about symmetry with respect to lines and points. Suppose we are given a point P and a line l in a plane α as pictured in Figure 4.6.5. A point Q in α is symmetric to P with respect to the line l if Q is the "mirror image" of P in l. The term "mirror image" means that, if a mirror were placed perpendicular to α along the line l, it would appear that the reflection of P in the mirror is the point Q. This is equivalent to saying that if the plane α were folded along the line l, the points P and Q would coincide.

Note that there is exactly one point that is symmetric to P with respect to the line l and if Q is symmetric to P with respect to l, then P must also be symmetric to Q with respect to l.

The definitions of symmetry with respect to a line given above are very intuitive but are hardly adequate from a (working) mathematical point of view; and so we give the following alternative definition that we will use in investigating the graphs of functions.

Figure 4.6.5

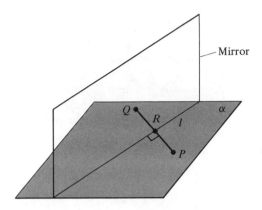

Definition

> Two points P and Q are **symmetric with respect to a line** l if l is the perpendicular bisector of the line segment PQ.

We define symmetry with respect to a point in a similar manner.

Definition

> Two points P and Q are **symmetric with respect to a point** R if R is the bisector of the line segment PQ. (See Figure 4.6.5.)

Example 3

Please refer to Figure 4.6.6. Let P be the point $(2, 3)$. If P and $Q_1(x_1, y_1)$ are symmetric with respect to the y-axis, then the line segment PQ_1 is perpendicular to the y-axis and so $y_1 = 3$. Since the distance from P to R is equal to the distance from Q_1 to R, $x_1 = -2$ and, hence, Q_1 is the point $(-2, 3)$.

Figure 4.6.6

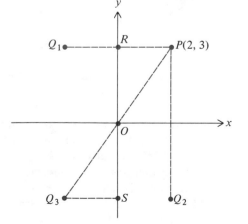

Using similar arguments, we can show that, if P and Q_2 are symmetric with respect to the x-axis, then Q_2 is the point $(2, -3)$.

If P and Q_3 are symmetric with respect to O, the origin of the coordinate system, then O is the point bisecting the line segment PQ_3. Furthermore, if S is the point of intersection of the y-axis and the line drawn through Q_3 perpendicular to the y-axis, then triangle Q_3OS is congruent to triangle POR. Hence, the lengths of OS and OR are equal and the lengths of Q_3S and PR are equal. Thus Q_3 is the point $(-2, -3)$.

Note that Q_1 and Q_3 are symmetric with respect to the x-axis, Q_2 and Q_3 are symmetric with respect to the y-axis, and Q_1 and Q_2 are symmetric with respect to the origin O.

If we are given an arbitrary point $P(a, b)$ and asked to find points Q_1, Q_2, and Q_3, such that P is symmetric to Q_1 with respect to the y-axis, P is symmetric to Q_2 with respect to the x-axis, and P is symmetric to Q_3 with respect to the origin, we could, using arguments similar to those used in Example 3, show that

$$Q_1 = (-a, b) \qquad Q_2 = (a, -b) \quad \text{and} \quad Q_3 = (-a, -b)$$

as illustrated in Figure 4.6.7.

Figure 4.6.7

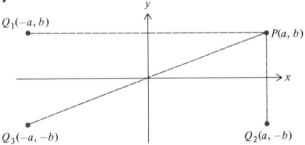

$Q_1(-a, b)$ $P(a, b)$ $Q_3(-a, -b)$ $Q_2(a, -b)$

Example 4

If P is the point $(-\pi, 2)$, the points symmetric to P with respect to the y-axis, x-axis, and origin are $Q_1(\pi, 2)$, $Q_2(-\pi, -2)$, and $Q_3(\pi, -2)$, respectively.

As suggested in the first paragraphs of this section, we would like to study symmetry in graphs. The next two definitions allow us to do this.

Definition

A plane graph C is **symmetric with respect to a line** l if whenever P is a point on C and P and Q are symmetric with respect to l then Q is also on C.

Definition

> A plane graph C is **symmetric with respect to a point** R if whenever P is a point on C and P and Q are symmetric with respect to R then Q is also on C.

The graph of a function f consists of the collection of points of the form $(x, f(x))$. Thus, the graph of f is symmetric with respect to the y-axis if and only if $(-x, f(x))$ lies on the graph whenever $(x, f(x))$ lies on the graph. (See Figure 4.6.8.) This is equivalent to saying that, if x is in D_f, then $-x$ is in D_f and (since f has one value at $-x$) $f(-x) = f(x)$.

Figure 4.6.8

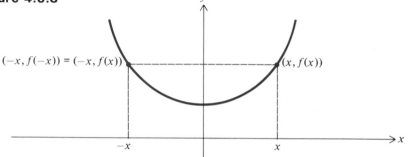

Using similar arguments, we can show that the graph of f is symmetric with respect to the origin if and only if, whenever x is in D_f, then $-x$ is in D_f and $f(-x)$ would equal $-f(x)$. (See Figure 4.6.9.)

Figure 4.6.9

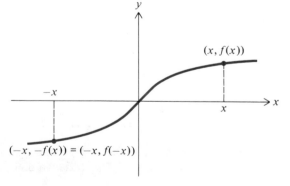

Note that if f is a function and $f(x) \neq 0$ at some point x in D_f, then f is not symmetric with respect to the x-axis. Indeed, if the graph of f were symmetric with respect to the x-axis, both $(x, f(x))$ and $(x, -f(x))$ would lie on the graph, and so $f(x) = -f(x)$, which is impossible if $f(x) \neq 0$.

Example 5

Consider the function defined in Example 2 of Section 4.6.1:

$$f(x) = \frac{x^2 - 4}{\sqrt{x^2 - 9}}$$

Since the domain of f consists of those numbers that are either greater than 3 or less than -3, we see $-x$ is in D_f whenever x is in D_f. Furthermore, if x is in D_f, then

$$f(-x) = \frac{(-x)^2 - 4}{\sqrt{(-x)^2 - 9}} = \frac{x^2 - 4}{\sqrt{x^2 - 9}} = f(x)$$

Hence, the graph of f is symmetric with respect to the y-axis.

Example 6

The graph of the function g defined by $g(x) = 1/x^2$ is symmetric with respect to the y-axis. D_g is the set of all real numbers except 0, so whenever x is in D_g so is $-x$. Furthermore,

$$g(-x) = \frac{1}{(-x)^2} = \frac{1}{x^2} = g(x)$$

The graph of g is sketched in Figure 4.6.10. Note that the graph to the right of the y-axis has been sketched by our "point plotting method." The portion of the graph to the left of the y-axis is sketched by drawing the mirror image of the right-hand portion in the y-axis.

Figure 4.6.10

x	$\frac{1}{4}$	$\frac{1}{3}$	$\frac{1}{2}$	1	$\frac{1}{4}$	$\frac{1}{9}$	$\frac{1}{16}$
$g(x)$	16	9	4	1	2	3	4

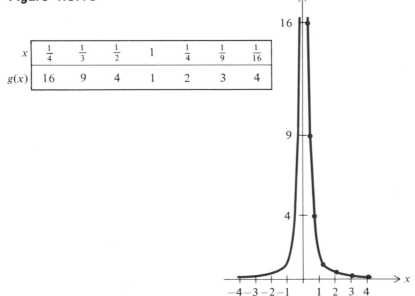

Example 7

The graph of f where $f(x) = 1/x$ is symmetric with respect to the origin. The graph of f was sketched in Example 3 of Section 4.4. D_f consists of all real numbers except 0, so if x is in D_f, then $-x$ is in D_f. Furthermore

$$f(-x) = \frac{1}{-x} = -\left(\frac{1}{x}\right) = -f(x)$$

Example 8

The graph of the function F given by

$$F(t) = t^3 + t$$

is symmetric with respect to the origin. The domain of F consists of all real numbers so $-t$ is in D_F whenever t is in D_F, and

$$F(-t) = (-t)^3 + (-t) = -t^3 - t = -(t^3 + t) = -F(t)$$

A sketch of the graph of F appears in Figure 4.6.11.

Figure 4.6.11

t	0	1	2	3
$F(t)$	0	2	10	30

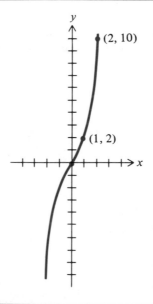

Suppose, as pictured in Figure 4.6.12, l is the line that passes through the origin and makes an angle of $45°$ with the positive x-axis. In the exercises at the end of this section, you will be asked to prove that the points (a, b) and (b, a) are symmetric with respect to the line l.

Figure 4.6.12

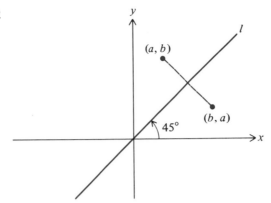

Knowing that (a, b) and (b, a) are symmetric with respect to l can help you sketch graphs of inverse functions. Recall that, if f is a function that has an inverse, then a point (a, b) is on the graph of f if and only if (b, a) is on the graph of f^{-1}. Hence, once we have drawn the graph of f, we can draw the graph of f^{-1} by sketching the mirror image of the graph of f in the line l. The sketch in Figure 4.6.13 indicates the symmetry in the graphs of a function f and its inverse.

Figure 4.6.13

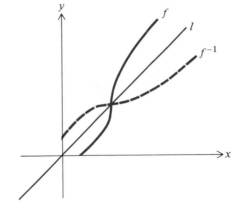

Example 9

The function f defined by

$$f(x) = x^3$$

has an inverse. (In fact, $f^{-1}(x) = \sqrt[3]{x}$.) The graphs of f and f^{-1} are given in Figure 4.6.14.

Figure 4.6.14

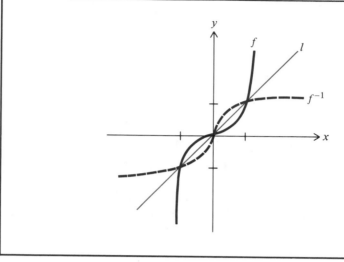

x	−3	−2	−1	0	1	2	3
f(x)	−27	−8	−1	0	1	8	27

Exercises 4.6.2

In Exercises 1–8, find the x and y intercepts of the graph of the given function.

1. $f(x) = 3x - 12$

2. $g(x) = 4x + 7$

3. $F(x) = x^2 + 2x - 15$

4. $G(x) = x^2 + 5x + 3$

5. $H(x) = \dfrac{x}{\sqrt{x^4 - 1}}$

6. $h(x) = \dfrac{x - 2}{\sqrt{4 - x^2}}$

7. $T(x) = x^3 + 3x^2 - 7x + 3$

8. $U(x) = x^2 + 3x + 1$

In Exercises 9–18, determine if the graph of the given function is symmetric with respect to the origin or y-axis.

9. $f(x) = 3x^2 + 4$

10. $g(x) = x^3 + 1$

11. $h(x) = \dfrac{x}{\sqrt{x^2 + 1}}$

12. $F(x) = \dfrac{x^2}{\sqrt{x^2 + 1}}$

13. $G(x) = x^5 + x^3$

14. $f(x) = x - 1$

15. $H(x) = x^5 + x^3 + 1$

16. $U(x) = x^6 + x^4 + x^2 + 1$

17. $T(x) = [[x]]$
(Recall that $[[x]]$ is the greatest integer less than or equal to x.)

18. $V(x) = x - [[x]]$

In Exercises 19–24, give the domain and the intercepts, and determine origin or y-axis symmetry for the given function. Sketch the graph of each function.

19. $f(x) = x^4 - x^2$ 　　　　　　　　　**20.** $g(x) = x/(x^2 + 1)$

21. $h(x) = \dfrac{x}{\sqrt{x^2 - 1}}$ 　　　　　　**22.** $F(x) = \dfrac{\sqrt{x^2 - 1}}{x}$

23. $G(x) = \dfrac{x^2}{x^2 + 1}$ 　　　　　　**24.** $H(x) = \dfrac{x^3}{x^2 + 1}$

 Each function, f, in Exercises 25–30 has an inverse. Sketch the graph of each function, f, and use symmetry to sketch the graph of f^{-1}.

25. $f(x) = 2x - 4$ 　　　　　　　　**26.** $f(x) = -3x + 9$
27. $f(x) = x^3 + 1$ 　　　　　　　　**28.** $f(x) = (x - 1)^3$
29. $f(x) = \dfrac{1}{x - 1}$ 　　　　　　　**30.** $f(x) = -x^3 - x$

31. The graph of an equation in two variables is the collection of points (x, y) that satisfy the equation. For example, the points $(0, 1)$ and $(-2, 3)$ are on the graph of $x^3 + y^2 = 1$ since $0^3 + 1^2 = 1$ and $(-2)^3 + (3)^2 = -8 + 9 = 1$. The graph of such an equation is symmetric with respect to the horizontal or x-axis if, whenever (x, y) lies on the graph, then $(x, -y)$ also lies on the graph. This is the same as saying that the equation obtained by replacing y by $-y$ in the original equation is equivalent to the original equation. In terms of the above example,

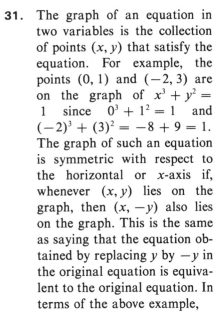

The graph of $x^3 + y^2 = 1$

$$x^3 + (-y)^2 = 1$$

is equivalent to

$$x^3 + y^2 = 1$$

So the graph of $x^3 + y^2 = 1$ is symmetric with respect to the x-axis.

We can make similar statements about symmetry with respect to the vertical or y-axis and the origin. In particular, the graph of an equation is symmetric with respect to the y-axis if an equivalent equation is obtained when x is replaced by $-x$ in the original equation, and the graph is symmetric with respect to the origin if an equivalent equation is obtained when x and y are replaced by $-x$ and $-y$, respectively, in the original. The graph of $x^3 + y^2 = 1$ is not symmetric with respect to the y-axis since

$$(-x)^3 + y^2 = 1$$

or equivalently

$$-x^3 + y^2 = 1$$

is not equivalent to $x^3 + y^2 = 1$. Since

$$(-x)^3 + (-y)^2 = 1$$

or equivalently

$$-x^3 + y^2 = 1$$

is not equivalent to $x^3 + y^2 = 1$, the graph of the original equation is not symmetric with respect to the origin, either.

Check to see if the graphs of the following equations are symmetric with respect to x-axis, y-axis, or origin:

a. $x^2 + y^2 = 1$
b. $x^3 + x^2 + y^3 = 0$
c. $x^4 + x^2 + (y + 1)^3 = 0$

32. Sketch the graphs of the following equations, find all intercepts, determine symmetry, and indicate whether an equation determines y as a function of x or x as a function of y.   (*Hint for item (d) only:* When you sketch this graph, be careful not to include any point (x, y) where $x + y = 0$.)

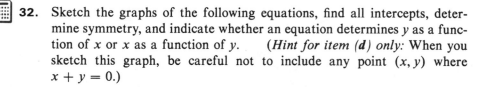

a. $x^4 - y = 0$ b. $x^5 - y^3 = 0$

c. $(x^2 - y)(x + y) = 0$ d. $\dfrac{2y - x}{y + x} = 1$

33. In this exercise we will show that the points with coordinates (a, b) and (b, a) are symmetric with respect to the line, l, passing through the origin and making an angle of $45°$ with the positive x-axis. Let O denote the origin, P the point (a, b), Q the point (b, a), and R the point of intersection of l and the line determined by P and Q.

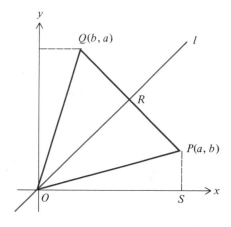

a. Prove triangle *POR* is congruent to triangle *QOR* by showing the length of *PO* is equal to the length of *QO* and that angle *POR* is congruent to angle *QOR*.

b. Use the congruence of triangles *POR* and *QOR* to show that angles *PRO* and *QRO* are right angles and that the length of *QR* equals the length of *PR*.

4.6.3

Families of Functions

A collection of functions that are related by some common characteristic is often referred to as a **family of functions.** For example, the family of polynomial functions consists of those functions that can be defined by polynomials; and the family of linear functions, which we will discuss in Chapter 5, consists of those polynomial functions that can be defined by polynomials of degree 1 (i.e., *f* is a linear function if there exist real numbers *a* and *b* such that $f(x) = ax + b$).

Let *f* be a function. We will investigate three families of functions that are closely related to *f*. The first family consists of those functions *F* defined by $F(x) = f(x) + r$ for some real number *r*; the second consists of those functions *G* defined by $G(x) = f(x - r)$ for some real number *r*; and the third consists of the functions *H* defined by $H(x) = rf(x)$ for some real number *r*. We will be particularly interested in the relationships that exist among the graphs in a given family.

Suppose *r* is a real number and *F* is the function defined by

$$F(x) = f(x) + r$$

We should specify the domain of *F*. Since $f(x) + r$ gives a unique number for each *x* in D_f, we will take $D_F = D_f$.

Note that, for each *x* in D_F, $F(x) - f(x) = r$, and so the graph of *F* may be obtained by shifting the graph of *f* vertically upward *r* units if *r* is positive and downward $-r$ units if *r* is negative. Figure 4.6.15 illustrates the relationship between the graphs of *f* and *F* when *r* is positive.

Figure 4.6.15   F(x) = f(x) + r

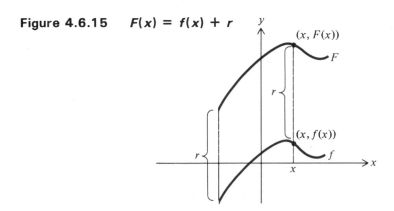

Example 10

The graph of the squaring function,

$$f(x) = x^2$$

was sketched in Example 2 of Section 4.4. If g and h are the functions defined by

$$g(x) = x^2 + 2 = f(x) + 2$$
$$h(x) = x^2 - 3 = f(x) - 3$$

then the graph of g can be found by raising the graph of f two units, and the graph of h can be found by lowering the graph of f three units. (Figure 4.6.16.)

Figure 4.6.16 $g(x) = f(x) + 2$
$h(x) = f(x) - 3$

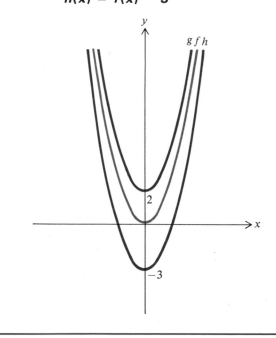

If G is the function defined by

$$G(x) = f(x - r)$$

we will take D_G to be those numbers x for which $x - r$ is in D_f. Furthermore, the value of G at x is the value of f at $x - r$. Hence, the graph of G can be sketched by shifting the graph of f horizontally. In particular, if r is positive, shift the graph of f r units to the right to get the graph of G. If r is negative, shift the graph of f $-r$ units to the left. Figure 4.6.17 illustrates the relationships between the domains and graphs of f and G when r is negative.

Figure 4.6.17 $G(x) = f(x - r)$

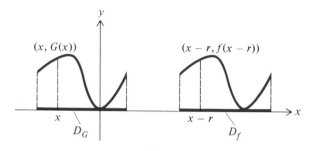

Example 11

The graphs of F, G_1, and G_2 where

$$F(x) = x^2$$
$$G_1(x) = (x - 2)^2 = F(x - 2)$$

and

$$G_2(x) = (x + 3)^2 = (x - (-3))^2 = F(x - (-3))$$

are sketched in Figure 4.6.18.

Figure 4.6.18

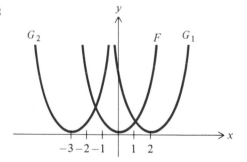

Example 12

To sketch the graph of the function f given by

$$f(x) = \frac{1}{(x - 2)}$$

note that $f(x) = g(x - 2)$ where g is the function given by

$$g(x) = \frac{1}{x}$$

The graph of g can be found in Example 3 of Section 4.4. The graph of f can be sketched by shifting the graph of g two units to the right as shown in Figure 4.6.19.

Figure 4.6.19

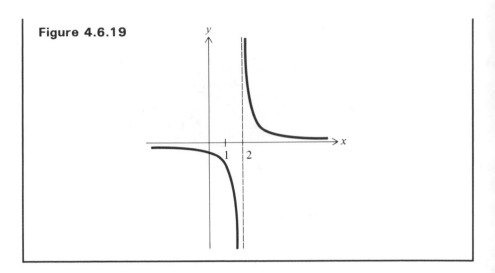

To analyze the graph of H in terms of the graph of f where $H(x) = rf(x)$, we will begin by considering the case where $r = -1$ so $H(x) = -f(x)$. The graphs of f and H in this case are mirror images of each other in the x-axis. To see this, note that if $(x, f(x))$ is on the graph of f, the reflection of this point in the x-axis is $(x, -f(x))$ or, equivalently, $(x, H(x))$, which is on the graph of H. Similarly, we can show that any point on the graph of H has its reflection in the x-axis on the graph of f. Figure 4.6.20 illustrates the relationship between f and H when $H(x) = -f(x)$.

Figure 4.6.20 $H(x) = -f(x)$

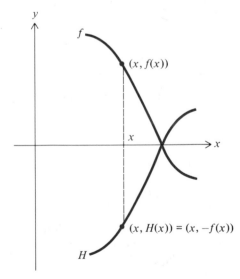

Example 13

The graphs of f and H where $f(x) = x^2$ and $H(x) = -x^2 = -f(x)$ are given in Figure 4.6.21. Note that the graphs of f and H are symmetric with respect to the x-axis.

Figure 4.6.21

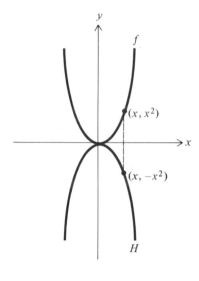

Now suppose r is positive and that

$$H(x) = rf(x)$$

Figure 4.6.22 illustrates how the graph of H can be obtained from the graph of f by appropriate stretchings or contractions. In particular it shows what occurs

Figure 4.6.22

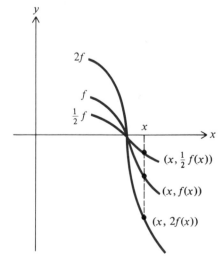

when r is $\frac{1}{2}$ and 2. If r is negative, view H as

$$H(x) = -[(-r)f(x)]$$

The graph of H may be found by stretching or contracting the graph of f to get the graph of H' where $H'(x) = (-r)f(x)$ as shown in Figure 4.6.22 ($-r$ is positive), and then reflecting the graph of H' in the x-axis.

Example 14

The graphs of the functions f, H_1, H_2, H_3, and H_4 where

$$f(x) = x^2$$
$$H_1(x) = 3x^2$$
$$H_2(x) = \tfrac{1}{3}x^2$$
$$H_3(x) = -3x^2$$

are sketched in Figure 4.6.23.

Figure 4.6.23

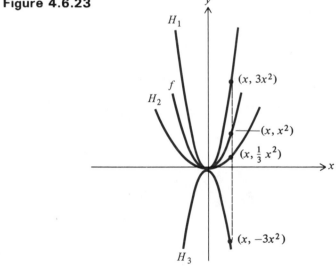

Exercises 4.6.3

1. The y-axis intercept of a function f is $(0, 3)$. What are the y-axis intercepts of the functions g and h defined by:

$$g(x) = f(x) - 7$$
$$h(x) = f(x) + 3$$

2. Suppose f is a function for which

    $f(-4) = 1$ $f(-3) = 3$ $f(-2) = 0$

    $f(-1) = 5$ $f(0) = 2$ $f(1) = 3$

    $f(2) = 4$ $f(3) = 6$ $f(4) = 4$

    a. What is the y-axis intercept of the graph of f?
    b. What are the y-axis intercepts of the graphs of the functions g and h
        if $g(x) = f(x - 2)$ and $h(x) = f(x + 3)$?

3. The x-axis intercepts of a function f are -2, 1, and 3. What are the x-axis
    intercepts of the functions g and h if $g(x) = f(x - 2)$ and $h(x) =$
    $f(x + 3)$?

4. The x-axis intercepts of the graph of a function g are 0, 4, and 5. What
    are the x-axis intercepts of the functions F and G if $F(x) = g(x + 3)$ and
    $G(x) = g(x - 3)$?

5. The following points are the intercepts of a function f: $(0, -5)$, $(2, 0)$,
    $(7, 0)$, $(-12, 0)$. What are the intercepts of the function g defined by

    $$g(x) = f(x + 7)$$

6. Sketch the graphs of the following functions using the methods discussed
    in this section.

    a. $f(x) = x^4$
    b. $g_1(x) = x^4 + 2 = f(x) + 2$
    c. $g_2(x) = x^4 - 3 = f(x) - 3$
    d. $g_3(x) = (x - 2)^4 = f(x - 2)$
    e. $g_4(x) = (x + 3)^4 = f(x + 3)$
    f. $g_5(x) = (x - 2)^4 + 3 = g_3(x) + 3 = f(x - 2) + 3$

7. Sketch the graphs of the following functions.

    a. $f(x) = x$
    b. $g_1(x) = 2x = 2f(x)$
    c. $g_2(x) = -2x = -2f(x)$
    d. $g_3(x) = 2x + 3 = g_1(x) + 3$
    e. $g_4(x) = 2x - 3 = g_1(x) - 3$

8. Sketch the graphs of the following functions.

    a. $f(x) = \dfrac{1}{x^2}$ (*Hint:* See Example 6, Section 4.6.2.)

    b. $g_1(x) = \dfrac{1}{x^2} + 4 = f(x) + 4$

    c. $g_2(x) = \dfrac{1}{x^2} - 3 = f(x) - 3$

d. $g_3(x) = \dfrac{1}{(x-3)^2} = f(x-3)$

e. $g_4(x) = \dfrac{1}{(x+4)^2} = f(x+4)$

f. $g_5(x) = \dfrac{-1}{x^2} = -f(x)$

9. Sketch the graphs of the following functions.

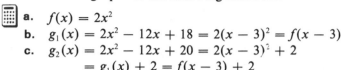

a. $f(x) = 2x^2$

b. $g_1(x) = 2x^2 - 12x + 18 = 2(x-3)^2 = f(x-3)$

c. $g_2(x) = 2x^2 - 12x + 20 = 2(x-3)^2 + 2$
$\qquad\qquad = g_1(x) + 2 = f(x-3) + 2$

10. Let f be the function defined by

$$f(x) = x^2 - 2x + 2$$

By completing the square of $x^2 - 2x$, we can define f as either

$$f(x) = x^2 - 2x + 1 - 1 + 2$$

or

$$f(x) = x^2 - 2x + 1 + 1$$

or

$$f(x) = (x-1)^2 + 1$$

Sketch the graph of f by using horizontal and vertical shifts of the graph of

$$g(x) = x^2$$

Note that $f(x) = g(x-1) + 1$.

SECTION 4.7   THE ARITHMETIC OF FUNCTIONS

Frequently, and especially in calculus, it is useful to recognize that a function is a sum or product of other (perhaps simpler) functions. We have seen, for example, that graphing $f(x) = x^2 + 2$ is equivalent to graphing $g(x) = x^2$ and raising the graph of g two units. In some sense, f can be thought of as the sum of the functions g and h where h is defined by $h(x) = 2$. ($f(x) = x^2 + 2 = g(x) + h(x)$.)

In this section, we examine five ways of combining functions: addition, subtraction, multiplication, division, and composition. The basic arithmetic properties of functional arithmetic are quite similar to those of rational expression arithmetic found in Section 3.3.

Example 1

Let us form the sum of the functions f and g where $f(x) = \dfrac{1}{x-1}$ and $g(x) = \sqrt{x}$. Intuitively, the value of the sum of f and g at a number x is to be the sum of the values of f and g at x. Symbolically,

$$(f+g)(x) = f(x) + g(x) = \frac{1}{x-1} + \sqrt{x}$$

We must be careful because we should specify or at least tacitly know the domain of the function when a function is defined. D_f consists of all real numbers $x \neq 1$ and $D_g = [0, +\infty)$. Since we cannot compute $f(x) + g(x)$ unless x is in D_f and D_g, it seems reasonable to take D_{f+g} to be those numbers that lie in *both* D_f and D_g. For this example, D_{f+g} is the set of real numbers x such that $x \geq 0$ and $x \neq 1$.

Suppose f and g are functions with domains D_f and D_g, respectively, and suppose D represents the set of points common to both D_f and D_g.

The function $f + g$ is that function with domain D, defined by

$$(f+g)(x) = f(x) + g(x) \tag{1}$$

The function $f - g$ is that function with domain D, defined by

$$(f-g)(x) = f(x) - g(x) \tag{2}$$

The function fg is that function with domain D, defined by

$$(fg)(x) = f(x)g(x) \tag{3}$$

The function f/g is that function whose domain consists of those real numbers x in D with $g(x) \neq 0$ and defined by

$$\frac{f}{g}(x) = \frac{f(x)}{g(x)} \tag{4}$$

Example 2

Let f and g be the functions defined by

$$f(x) = \frac{x-1}{x+1} \quad \text{and} \quad g(x) = \sqrt{x}$$

D_f consists of all real numbers $x \neq -1$ and $D_g = [0, +\infty)$, so $D = [0, +\infty)$.

$$(f+g)(x) = f(x) + g(x) = \frac{x-1}{x+1} + \sqrt{x}$$

$$(f-g)(x) = f(x) - g(x) = \frac{x-1}{x+1} - \sqrt{x}$$

$$(fg)(x) = f(x)g(x) = \left(\frac{x-1}{x+1}\right)\sqrt{x}$$

$$\left(\frac{f}{g}\right)(x) = \frac{f(x)}{g(x)} = \frac{\left(\dfrac{x-1}{x+1}\right)}{\sqrt{x}} = \frac{x-1}{\sqrt{x}(x+1)}$$

$D_{f+g} = D_{f-g} = D_{fg} = D = [0, \infty)$. Since $g(0) = 0$ and $g(x) \neq 0$ if $x > 0$, $D_{f/g} = (0, +\infty)$,

$$\left(\frac{g}{f}\right)(x) = \frac{g(x)}{f(x)} = \frac{\sqrt{x}}{\left(\dfrac{x-1}{x+1}\right)} = \sqrt{x} \cdot \frac{x+1}{x-1}$$

Note

$$0 = f(x) = \frac{x-1}{x+1}$$

if and only if $x - 1 = 0$ or $x = 1$. So $D_{g/f}$, which consists of those numbers in D for which $f(x) \neq 0$, is the set of non-negative real numbers different from 1.

Example 3

If f and g are functions given by

$$f(x) = \frac{1}{x} \quad \text{and} \quad g(x) = \frac{1}{x^2}$$

then $D_f = D_g = $ all real numbers $x \neq 0$.

The functions $f + g, f - g, fg$, and f/g are defined as follows:

$$(f + g)(x) = f(x) + g(x) = \frac{1}{x} + \frac{1}{x^2} = \frac{x+1}{x^2}$$

$$(f - g)(x) = f(x) - g(x) = \frac{1}{x} - \frac{1}{x^2} = \frac{x-1}{x^2}$$

$$fg(x) = f(x)g(x) = \frac{1}{x} \cdot \frac{1}{x^2} = \frac{1}{x^3}$$

$$(f/g)(x) = f(x)/g(x) = (1/x)/(1/x^2) = x$$

Furthermore

$$D_{f+g} = D_{f-g} = D_{fg} = D_{f/g} = D$$

Be careful! Even though $(f/g)(x) = x$, the domain of f/g is not $(-\infty, +\infty)$ since it must, by our definition, consist of those numbers in D for which $g(x) \neq 0$.

We now consider another way to combine functions. In Section 4.4, we graphed the function $h(x) = (x - 2)^2$ by sketching the graph of $f(x) = x^2$ and moving the graph of f two units to the right. In trying to understand how the graph of h compared with the graph of f, we noted that for each real number x,

$$h(x) = f(x - 2)$$

If we let g be the linear function defined by $g(x) = x - 2$, we can think of h as being composed of f and g:

$$h(x) = f(g(x))$$

Indeed $f(g(x)) = f(x - 2) = (x - 2)^2$.

Example 4

Let f and g be functions defined by

$$f(x) = \sqrt{x} \quad \text{and} \quad g(x) = \frac{1}{x - 1}$$

and suppose we wanted to consider the function h composed of f and g analogous to the composition discussed prior to this example.

$$h(x) = f(g(x)) = f\left(\frac{1}{x - 1}\right) = \sqrt{\frac{1}{x - 1}}$$

What set of numbers could we reasonably say is the domain of h? Note that in computing h, we must first compute $g(x)$ so x will have to lie in D_g. Second, we must compute $f(g(x))$ so $g(x)$ must lie in the domain of f. Thus, a reasonable domain of h is those real numbers x satisfying the following two conditions:

x is in D_g.

$g(x)$ is in D_f

For our example, $D_f = [0, +\infty)$ and D_g consists of all real numbers $x \neq 1$. Now

$$g(x) = \frac{1}{x - 1}$$

is positive if $x > 1$ and negative if $x < 1$, or $g(x)$ is in D_f if and only if $x > 1$. Hence, $D_h = (1, +\infty)$.

In general, if f and g are functions, then the function h whose domain consists of those numbers x satisfying

x is in D_g (5)

$g(x)$ is in D_f (6)

and is defined by

$$h(x) = f(g(x))$$

is called the **composite** or **composition** of f and g. We denote h by $f \circ g$.

Example 5

> If $f(x) = x^2 + 1$ and $g(x) = x - 1$, then $D_f = D_g = (-\infty, +\infty)$ so $D_{f \circ g} = (-\infty, +\infty)$, and
>
> $$(f \circ g)(x) = f(g(x)) = f(x - 1) = (x - 1)^2 + 1 = x^2 - 2x + 2$$
>
> Note that $(f \circ g)(x)$ may also be computed as
>
> $$(f \circ g)(x) = f(g(x)) = [g(x)]^2 + 1 = (x - 1)^2 + 1$$
>
> In general, $f \circ g \neq g \circ f$. For this example, $D_{g \circ f} = (-\infty, +\infty)$ and
>
> $$(g \circ f)(x) = g(f(x)) = g(x^2 + 1) = (x^2 + 1) - 1 = x^2$$
>
> Clearly, $g \circ f \neq f \circ g$.

Example 6

> Let f and g be defined by $f(x) = \sqrt{x}$ and $g(x) = x^2$. Then $D_f = [0, +\infty)$ and $D_g = (-\infty, +\infty)$. $D_{f \circ g}$ consists of those numbers x in $(-\infty, +\infty)$ for which $g(x) = x^2 \geq 0$. Hence, $D_{f \circ g} = (-\infty, +\infty)$ and
>
> $$f \circ g(x) = f(g(x)) = f(x^2) = \sqrt{x^2}$$
>
> Now $D_{g \circ f}$ consists of those numbers x in $[0, +\infty) = D_f$ for which $f(x) = \sqrt{x}$ is in $(-\infty, +\infty)$. Hence, $D_{g \circ f} = D_f = [0, +\infty)$ and
>
> $$(g \circ f)(x) = g(f(x)) = g(\sqrt{x}) = (\sqrt{x})^2 = x$$
>
> If $x \geq 0$, then $\sqrt{x^2} = x$ and $(f \circ g)(x) = (g \circ f)(x)$. Since $D_{f \circ g} \neq D_{g \circ f}$, we cannot say $f \circ g = g \circ f$.
>
> Note that, although $(g \circ f)(x) = x$, the domain of $g \circ f$ is not $(-\infty, +\infty)$. The domain of $g \circ f$ must be determined via rules (5) and (6) of our definition of composition.

Suppose the function f has an inverse. Then f is a one-to-one function; i.e., if x_1 and x_2 are two distinct numbers in D_f then $f(x_1) \neq f(x_2)$. An equivalent way to describe one-to-one functions is to say that the function f is one-to-one if whenever x_1 and x_2 are in D_f and $f(x_1) = f(x_2)$, then $x_1 = x_2$.

Consider the function $f^{-1} \circ f$. If x is in D_f and $y = f^{-1} \circ f(x) = f^{-1}(f(x))$ then $f(y) = f(x)$. (Recall that $b = f^{-1}(a)$ if and only if $f(b) = a$.) Since f is one-to-one, it follows that $x = y$. We have proven

$$f^{-1} \circ f(x) = x \text{ for each } x \text{ in } D_f \tag{7}$$

A similar argument may be used to prove

$$f \circ f^{-1}(x) = x \text{ for each } x \text{ in } D_{f^{-1}} \tag{8}$$

Although we will not prove it, properties (7) and (8) characterize inverse functions. To be precise, we mean that if f and g are functions and $f \circ g(x) = x$ for each x in D_g and $g \circ f(x) = x$ for each x in D_f then f has an inverse and $f^{-1} = g$.

Example 7

> If f is the function defined by
>
> $$f(x) = 3x - 7$$
>
> then by solving $y = 3x - 7$ for x in terms of y we see f has an inverse and $f^{-1}(x) = \dfrac{x + 7}{3}$. We will verify that properties (7) and (8) hold for f and f^{-1}.
>
> $$f \circ f^{-1}(x) = f(f^{-1}(x)) = f\left(\frac{x + 7}{3}\right)$$
>
> $$= 3\left(\frac{x + 7}{3}\right) - 7 = x + 7 - 7 = x$$
>
> $$f^{-1} \circ f(x) = f^{-1}(f(x)) = f^{-1}(3x - 7)$$
>
> $$= \frac{(3x - 7) + 7}{3} = x$$

Exercises 4.7

1. Let f and g be the functions defined by $f(x) = x^3$ and $g(x) = x^2$.

   a. Find D_{f+g} and $(f + g)(x)$. b. Find D_{f-g} and $(f - g)(x)$.

   c. Find D_{fg} and $fg(x)$. d. Find $D_{f/g}$ and $\dfrac{f}{g}(x)$.

2. Let g and h be the functions defined by

   $$f(x) = \frac{1}{x} \quad \text{and} \quad h(x) = \frac{1}{(x - 1)(x - 2)}$$

   a. Find D_{g+h} and $(g + h)(x)$. b. Find D_{gh} and $gh(x)$.

   c. Find $D_{g/h}$ and $\dfrac{g}{h}(x)$. d. Find $D_{h/g}$ and $\dfrac{h}{g}(x)$.

3. Let f and g be the functions defined by

   $$f(x) = x + 1 \quad \text{and} \quad g(x) = \sqrt{2 - x}$$

    a. Find D_{f-g} and $(f - g)(x)$.   **b.** Find D_{fg} and $fg(x)$.

    c. Find $D_{f/g}$ and $\dfrac{f}{g}(x)$.   **d.** Find $D_{g/f}$ and $\dfrac{g}{f}(x)$.

4. Let U and V be the functions defined by

$$U(x) = \frac{1}{x^2 - 1} \quad \text{and} \quad V(x) = (x - 1)^2$$

    a. Find D_{U+V} and $(U + V)(x)$.   **b.** Find D_{UV} and $UV(x)$.

    c. Find $D_{U/V}$ and $\dfrac{U}{V}(x)$.   **d.** Find $D_{V/U}$ and $\dfrac{V}{U}(x)$.

5. Let F and G be the functions defined by

$$F(x) = \sqrt{x - 3} \quad \text{and} \quad G(x) = \sqrt{x - 6}$$

    a. Find D_{F-G} and $(F - G)(x)$.

    b. Find D_{FG} and $FG(x)$.   **c.** Find $D_{G/F}$ and $\dfrac{G}{F}(x)$.

6. Let f and g be the functions defined by

$$f(x) = x^2 + x + 1 \quad \text{and} \quad g(x) = x^2 - 3x + 2$$

    a. Find D_{f+g} and $(f + g)(x)$.   **b.** Find D_{fg} and $fg(x)$.

    c. Find $D_{f/g}$ and $\dfrac{f}{g}(x)$.   **d.** Find $D_{g/f}$ and $\dfrac{g}{f}(x)$.

In Exercises 7–12 find:

    a. $D_{f \circ g}$ and $f \circ g(x)$ **b.** $D_{g \circ f}$ and $g \circ f(x)$

7. $f(x) = x - 2$ $g(x) = x^4 + x^2 + 1$

8. $f(x) = \dfrac{1}{x}$ $g(x) = \dfrac{1}{x}$

9. $f(x) = \dfrac{1}{x^3}$ $g(x) = \sqrt[4]{x}$

10. $f(x) = x^2 + 3x + 2$ $g(x) = \dfrac{1}{x^2}$

11. $f(x) = \sqrt{x - 2}$ $g(x) = x^2$

12. $f(x) = \dfrac{1}{\sqrt{3 - x}}$ $g(x) = \dfrac{1}{x}$

13. If $f(x) = x^3 + x^2 + x + 1$ and $g(x) = x - 1$, find $f \circ g(x)$ and $g \circ f(x)$.

14. If $f(x) = \dfrac{x}{\sqrt{x - 1}}$ and $g(x) = x^2 + 1$, find $f \circ g(x)$ and $g \circ f(x)$.

15. If $h(x) = \dfrac{1}{x - 3}$ and $g(x) = \dfrac{1}{x - 4}$, find $h \circ g(x)$ and $g \circ h(x)$.

16. If $U(x) = x^2 + 1$ and $V(x) = x^4 + x^2$, find $U \circ V(x)$ and $V \circ U(x)$.

17. If $f(x) = 10$ and $g(x) = -3$, find $f \circ g(x)$ and $g \circ f(x)$.

18. If $f(x) = x^2$ and $g(x) = x^2$, find $f \circ g(x)$.

19. Let f be the one-to-one function defined by $f(x) = 2x + 4$. Find f^{-1} and show $f^{-1} \circ f(x) = x$ and $f \circ f^{-1}(x) = x$.

20. Let g be the one-to-one function defined by $g(x) = x^3 + 1$. Find g^{-1} and show $g^{-1} \circ g(x) = x$ and $g \circ g^{-1}(x) = x$.

21. Let h be the one-to-one function defined by $h(x) = x^5 - 3$. Find h^{-1} and show $h^{-1} \circ h(x) = x$ and $h \circ h^{-1}(x) = x$.

22. Let F be the one-to-one function defined by $F(x) = \dfrac{x-1}{x-2}$. Find F^{-1} and show $F^{-1} \circ F(x) = x$ and $F \circ F^{-1}(x) = x$.

23. Let U be the one-to-one function defined by $U(x) = x^4$, $x \geq 0$. Find U^{-1} and show $U^{-1} \circ U(x) = x$ and $U \circ U^{-1}(x) = x$.

24. Let V be the one-to-one function defined by $V(x) = x^4$, $x \leq 0$. Find V^{-1} and show $V^{-1} \circ V(x) = x$ and $V \circ V^{-1}(x) = x$.

Review Exercises
Chapter 4

1. Let f be the function defined by

$$f(x) = x^4 + 3x^2$$

Find:

a. $f(1)$ b. $f(-2)$ c. $f(2)$

d. $f(-3)$ e. $f(t^2)$ f. $f(x^2)$

g. $f(x-2)$ h. $f(1/x)$ i. $f(f(1))$

j. $f(f(x))$

2. Let g be the function defined by

$$g(u) = \frac{u}{u+1}$$

Find:

a. $g(0)$ b. $g(2)$ c. $g(-1)$

d. $g(x+3)$ e. $g(3-x^2)$ f. $g(u^3)$

g. $g(u+1)$ h. $g(g(1))$ i. $g(g(x))$

j. $g((g(u))^2)$

3. Let h be the function defined by

$$h(x) = \sqrt{x^2 - 1}$$

Find:

a. $h(3)$ b. $h(1)$ c. $h(-1)$

d. $h(\sqrt{5})$ e. $h(x^2)$ f. $h(\sqrt{x})$

g. $h(2x+1)$ h. $h(h(2))$ i. $h(h(u))$

j. $h(\sqrt{x-1})$

4. Let F be the function defined by $F(x) = \dfrac{x}{\sqrt{x^2+9}}$.

   Find:

   a. $F(0)$ b. $F(4)$ c. $F(-4)$

   d. $F(3)$ e. $F(x-1)$ f. $F(x^2)$

   g. $[F(x)]^2$ h. $F(3x-3)$ i. $F(\sqrt{x})$

   j. $F(\sqrt{x-9})$

In Exercises 5–10, find the domain of the given function.

5. $f(x) = \sqrt{2x-32}$ 6. $g(x) = \sqrt{32-2x}$

7. $F(u) = -\dfrac{u}{\sqrt{u^2-9}}$ 8. $G(t) = t\sqrt{9-t^2}$

9. $H(x) = (\sqrt{x})^2$ 10. $U(x) = \dfrac{x}{x(x-1)}$

11. If f and g are the functions defined by $f(x) = x^2 + 7$ and $g(x) = x - 3$, find:

    a. D_{f+g} and $(f+g)(x)$ b. $D_{f/g}$ and $\dfrac{f}{g}(x)$

12. If f and g are the functions defined by $f(x) = \sqrt{x+5}$ and $g(x) = \sqrt{x-5}$, find:

    a. D_{fg} and $fg(x)$ b. $D_{f/g}$ and $\dfrac{f}{g}(x)$

    c. $D_{g/f}$ and $\dfrac{g}{f}(x)$

13. If g and h are the functions defined by

    $$g(t) = \frac{1}{t-3} \quad \text{and} \quad h(t) = t^2 - 5t + 4$$

    find:

    a. D_{g-h} and $(g-h)(t)$ b. $D_{g/h}$ and $\dfrac{g}{h}(t)$

    c. $D_{h/g}$ and $\dfrac{h}{g}(t)$

14. If F and G are the functions defined by

$$F(u) = 2u^2 + 3u + 2 \quad \text{and} \quad G(u) = u^2 - 10u - 11$$

find:

a. D_{F+G} and $(F + G)(u)$ b. $D_{F/G}$ and $\dfrac{F}{G}(u)$

c. $D_{G/F}$ and $\dfrac{G}{F}(u)$

15. If f and h are the functions defined by $f(x) = x^3 - 1$ and $h(x) = x - 1$, find:

a. D_{f-h} and $(f - h)(x)$ b. D_{fh} and $fh(x)$

c. $D_{f/h}$ and $\dfrac{f}{h}(x)$

16. Let f and g be the functions defined by

$$f(x) = \frac{2x}{x - 5} \quad \text{and} \quad g(x) = x^2 + 1$$

Find:

a. $f \circ g(x)$ b. $g \circ f(x)$ c. $f \circ g(3)$
d. $g \circ f(3)$ e. $f \circ g(x^2)$

17. Let F and G be the functions defined by $F(t) = 2t^2$ and $G(t) = t^3 + 3$. Find:

a. $F \circ G(t)$ b. $G \circ F(t)$ c. $F \circ G(1)$
d. $G \circ F(0)$ e. $G \circ F(x + 1)$

18. Let g and h be the functions defined by $g(u) = \sqrt{2u - 1}$ and $h(u) = u^2 + 5$. Find:

a. $g \circ h(u)$ b. $h \circ g(u)$ c. $g \circ h(0)$
d. $h \circ g(5)$ e. $g \circ h(\sqrt{x})$

19. Let f and g be the functions defined by $f(x) = 2x^2 + x + 5$ and $g(x) = 2x - 3$. Find:

a. $f \circ g(x)$ b. $g \circ f(x)$ c. $f \circ g(0)$
d. $f \circ g(2)$ e. $g \circ f(x^2)$

20. Let v and w be the functions defined by $v(t) = \sqrt[3]{t - 4}$ and $w(t) = 12 - t$. Find:

a. $v \circ w(t)$ b. $w \circ v(t)$ c. $v \circ w(0)$
d. $v \circ w(12)$ e. $w \circ v(12)$

21. If f and g are the functions defined by $f(x) = x^4$ and $g(x) = \sqrt[4]{x}$, find $D_{f \circ g}$ and $f \circ g(x)$.

22. If G and H are the functions defined by $G(t) = t^6$ and $H(t) = \sqrt[6]{t}$, find $D_{H \circ G}$ and $H \circ G(t)$.

23. If f and g are the functions defined by

$$f(x) = \frac{1}{x - 3} \quad \text{and} \quad g(x) = x^2 + x + 3$$

find $D_{f \circ g}$ and $f \circ g(x)$.

24. If g and h are the functions defined by $g(x) = \sqrt{x - 9}$ and $h(x) = x^2$, find $D_{g \circ h}$ and $g \circ h(x)$.

25. If U and V are the functions defined by $U(s) = s^3 + 1$ and $V(s) = \sqrt{s - 2}$, find $D_{U \circ V}$ and $U \circ V(s)$.

In Exercises 26–31, determine if the graph of the given function is symmetric with respect to the y-axis or origin, find all intercepts and sketch the graph of the function.

26. $f(x) = 2x - 3$

27. $g(x) = x^3 - 3x$

28. $h(x) = \dfrac{x^2}{x + 3}$

29. $F(x) = \begin{cases} \sqrt{x - 1} & \text{if } x \geq 1 \\ -\sqrt{-1 - x} & \text{if } x \leq -1 \end{cases}$

30. $G(x) = -2x^3 - 9$

31. $H(x) = -(x - 9)^2$

32. Sketch the graph of the function f defined by $f(x) = 2x^3$, and use your sketch of the graph of f to sketch the graphs of the functions defined by

  a. $g_1(x) = 2(x - 5)^3 = f(x - 5)$
  b. $g_2(x) = 2x^3 + 7 = f(x) + 7$
  c. $g_3(x) = 2(x - 5)^3 + 7 = g_1(x) + 7$

33. Sketch the graph of the function f defined by $f(x) = 2x$ and use your sketch of the graph of f to sketch the graphs of the functions defined by

  a. $g_1(x) = 2x - 4 = 2(x - 2) = f(x - 2)$
  b. $g_2(x) = -2x = -f(x)$
  c. $g_3(x) = -2x + 5 = g_2(x) + 5$

34. Sketch the graph of the function f defined by

$$f(x) = \frac{x}{x^2 + 1}$$

and use your sketch of the graph of f to sketch the graphs of the functions defined by

a. $g_1(x) = \dfrac{x-2}{x^2 - 4x + 5} = \dfrac{x-2}{(x-2)^2 + 1} = f(x-2)$

b. $g_2(x) = \dfrac{x^2 + x + 1}{x^2 + 1} = \dfrac{x}{x^2 + 1} + 1 = f(x) + 1$

c. $g_3(x) = \dfrac{x-2}{x^2 - 4x + 5} - 1 = g_1(x) - 1$

35. Sketch the graph of the function f defined by

$$f(x) = \frac{1}{x-2}$$

and use your sketch of the graph of f to sketch the graphs of the functions defined by

a. $g_1(x) = \dfrac{1}{2-x} = -f(x)$ b. $g_2(x) = \dfrac{2}{x-2} = 2f(x)$

c. $g_3(x) = \dfrac{1}{4-2x} = \tfrac{1}{2}g_1(x)$

36. Sketch the graph of the function f defined by $f(x) = x^2 - 4$, and use your sketch of the graph of f to sketch the graphs of the functions defined by

a. $g_1(x) = 4 - x^2 = -f(x)$ b. $g_2(x) = 2x^2 - 8 = 2f(x)$

c. $g_3(x) = 2 - \tfrac{1}{2}x^2 = -\tfrac{1}{2}f(x)$

In Exercises 37–41, determine if the given function f has an inverse. If a function does have an inverse, find f^{-1} and sketch the graphs of both f and f^{-1}.

37. $f(x) = 7x - 5$ **38.** $f(x) = x^2 + x + 3$

39. $f(x) = x^3 - 3x^2 + 3x - 1$ **40.** $f(x) = \dfrac{2}{(x-1)^3}$

41. $f(x) = \dfrac{2}{(x-1)^2}$

In Exercises 42–48, determine if the given equation determines y as a function of x and if the given equation determines x as a function of y.

42. $2x + 3y = 6$ **43.** $2y + 4x^2 - 6x + 2 = 0$

44. $x^2 - y^2 = 0$ **45.** $(2x^2 + y^2 + 5)(x - 3) = 0$

46. $(2x^2 + y^2 + 5)(y - 3) = 0$ **47.** $y^4 + 2x - 3 = 0$

48. $x^2 + y = 0$

49. **a.** Find the function f such that $f(x)$ gives the volume of a cube of edge x units.

  b. Find the function g such that $g(x)$ gives the surface area of a cube of edge x units.

  c. Show that f has an inverse and find $f^{-1}(V)$. Note that $f^{-1}(V)$ gives the length of an edge of a cube with volume V.

  d. Find $g \circ f^{-1}(V)$ and note that $g \circ f^{-1}(V)$ gives the surface area of a cube as a function of the cube's volume.

50. Water is being poured into a rectangular swimming pool with dimensions $25 \times 15 \times 10$ ft at the rate of 10 ft^3/min.

  a. Find a function f such that $f(t)$ gives the volume of water in the pool t minutes after water starts flowing into the empty pool.

  b. Find the volume of water in the pool after 10 minutes.

  c. Find the depth of the water in the pool after 100 minutes.

  d. How many minutes will it take to fill the pool?

51. The cost of setting up a plant to produce automobile engines is $500,000. The cost of labor and raw materials used to produce each engine is $1000.

  a. Give a function f such that $f(x)$ is the total cost of producing x engines.

  b. What is the cost of producing 100 engines?

  c. How many engines can be produced for $1,000,000?

52. Suppose that the number of 16-in. color television sets that will be purchased by consumers when the per-unit price of such a set is p dollars is given by the function of f where

$$f(p) = 10,000 - 5p$$

  a. How many sets would consumers purchase if the price of a set was $200?

  b. What would the price of a set be if the consumers were to purchase 5000 sets?

  c. Show f^{-1} exists and find $f^{-1}(x)$. Note that $f^{-1}(x)$ gives the price of a set when consumers are willing to buy x sets.

  d. Find the function g such that $g(p)$ represents the total revenue resulting from setting the unit price of a set at p dollars and selling $f(p)$ sets.

53. A box in the shape of a right circular cylinder is to be made so that the radius of its base is one-half its height.

  a. Find a function f so that $f(r)$ gives the lateral surface area if the radius of the base is r units.

  b. Find a function g so that $g(r)$ gives the volume of the box if the radius of the base is r units.

  c. Find a function h so that $h(V)$ gives the surface area of the box if its volume is V. Show that $h = f \circ g^{-1}$.

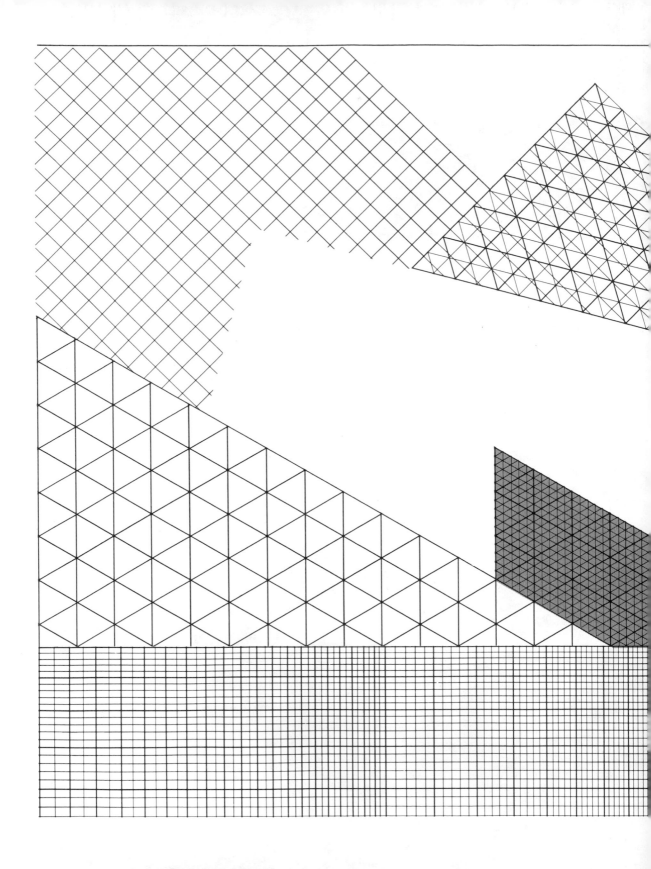

5 Linear and Quadratic Functions

INTRODUCTION

A function f defined by

$$f(x) = mx + b$$

is called a **linear function,** and a function g defined by

$$g(x) = ax^2 + bx + c \qquad a \neq 0$$

is called a **quadratic function.** Linear and quadratic functions are special types of polynomial functions. They are important for several reasons.

First, many phenomena can be described using these functions. For example, the cost function, $C(x) = 500{,}000 + 5000x$, defined in Example 2 of Section 4.2 and graphed in Example 4 of Section 4.3, is a linear function. Neglecting air resistance, the velocity in feet per second of an object falling near the surface of the earth can be given by the linear function $v(t) = -32t + v_0$ (v_0 represents its initial velocity and t is used to represent the time the object has been in free fall); the height of this object t seconds into its fall is given by the quadratic function

$$s(t) = -16t^2 + v_0 t + s_0$$

(s_0 represents the height at time zero or the object's initial height).

Second, linear and quadratic functions can serve as excellent approximations to more complicated functions. As an example, suppose an object moves on a ruled straight line in such a way that its position at time t is given by the function s where

$$s(t) = t^3$$

The graph of s appears in Figure 5.1.1. Using calculus we can show that the line tangent to the graph of s at the point $(1, 1)$ is the graph of the linear function

$$f(t) = 3t - 2$$

The graph of f, which is given in Figure 5.1.1, is a straight line. (We shall see this is true of all linear functions.) It has been shown through empirical investigations and mathematical reasoning that, for values of t near 1, the position and velocity of our object can be effectively analyzed using the linear function f rather than s itself. This should not be surprising since Figure 5.1.1 indicates that there is little difference between the graphs of s and f if we are close enough to the point $(1, 1)$.

Figure 5.1.1

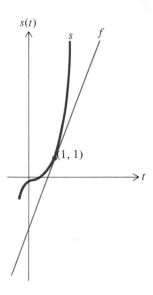

LINEAR FUNCTIONS

Consider a linear function

$$f(x) = mx + b$$

We will show that the graph of f is a straight line. By our work on families of functions, it will be sufficient to show that the graph of

$$g(x) = mx$$

is a straight line since the graph of f can be obtained by moving the graph of g upward b units if $b > 0$ or $-b$ units downward if $b < 0$.

Recall from plane geometry that three points are colinear if and only if there is some arrangement P_1, P_2, P_3 of these points such that

$$d(P_1, P_2) + d(P_2, P_3) = d(P_1, P_3)$$

(See Figure 5.2.1.) The points $P(0, 0)$ and $Q(1, m)$ are 'o distinct points on the graph of g. (Indeed $g(0) = m \cdot 0 = 0$ and $g(1) = m \cdot 1 = m$.)

We will first show that every point on the graph of g lies on the straight line determined by P and Q. Let $R(x, y)$ be a point on the graph of g (i.e., $y = mx$). (See Figure 5.2.2.) If $0 \leq x \leq 1$, then

$$d(P, R) + d(R, Q)$$
$$= \sqrt{(x - 0)^2 + (mx - 0)^2} + \sqrt{(1 - x)^2 + (m - mx)^2}$$
$$= x\sqrt{1 + m^2} + (1 - x)\sqrt{1 + m^2}$$
$$= \sqrt{1 + m^2} = \sqrt{(1 - 0)^2 + (m - 0)^2} = d(R, Q)$$

and so R is on the line determined by P and Q. Similar arguments when $x < 0$

Figure 5.2.1

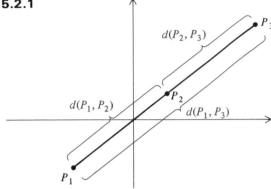

Figure 5.2.2

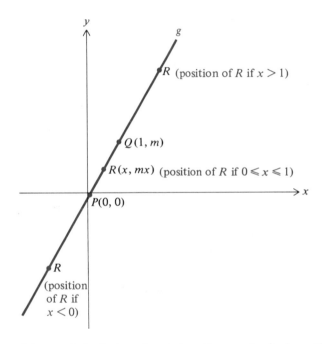

or $x > 1$ can be used to conclude that each point on the graph of g is on the line determined by P and Q. (See Exercise 30.)

To show that the graph of g is identical to the line passing through P and Q, we must also show that if $R(x, y)$ is a point on this line, then it lies on the graph of g (i.e., if (x, y) lies on the line then $y = g(x) = mx$).

If $R(x, y)$ is on the line through P and Q and $0 \le x \le 1$ then R lies between P and Q so

$$d(P, R) + d(R, Q) = d(P, Q)$$

that is,

$$\sqrt{x^2 + y^2} + \sqrt{(1 - x)^2 + (m - y)^2} = \sqrt{1 + m^2}$$

Hence,

$$\sqrt{x^2 + y^2} = \sqrt{1 + m^2} - \sqrt{(1 - x)^2 + (m - y)^2}$$

and squaring

$$x^2 + y^2 = 1 + m^2 - 2\sqrt{1 + m^2}\,\sqrt{(1 - x)^2 + (m - y)^2}$$
$$+ (1 - x)^2 + (m - y)^2$$

Simplifying we get

$$x + my - 1 - m^2 = \sqrt{1 + m^2}\,\sqrt{(1 - x)^2 + (m - y)^2}$$

Squaring again

$$(x + my - 1 - m^2)^2 = (1 + m^2)[(1 - x)^2 + (m - y)^2]$$

and simplifying

$$y^2 - 2mxy + (mx)^2 = 0$$

Hence,

$$(y - mx)^2 = 0$$

So $y = mx$, and $R(x, y)$ is a point on the graph of g.

Similar arguments can be used to show that if R is on the line determined by P and Q, and Q is between P and R (i.e., $x > 1$) or P is between Q and R (i.e., $x < 0$) then R lies on the graph of g. (See Exercise 31.)

Although this proof involves a fair amount of algebra, we have included it for two reasons.

The first is that the result is a fundamental one about linear functions. You have probably been using it, without proof, since you began studying elementary algebra.

The second is that it illustrates an important method of proof and one that you will surely see in calculus. If you want to prove that a curve C is the graph of a function f, you must prove two things: First, you must show that if (x, y) is on C then $y = f(x)$, and second you must show that if $y = f(x)$ then (x, y) lies on C.

Example 1

To sketch the graph of

$$f(x) = -3x + 1$$

it is only necessary to plot two distinct points on the graph. Once these points are plotted, use a straight edge to draw the line determined by these two points.

Note $f(0) = 1$ and $f(1) = -2$ so $(0, 1)$ and $(1, -2)$ are on the graph of f, as shown in Figure 5.2.3.

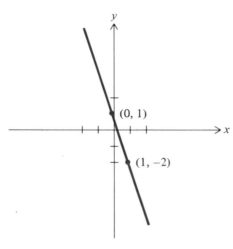

Figure 5.2.3 $f(x) = -3x + 1$

Example 2

To find the linear function

$$f(x) = mx + b$$

whose graph is the straight line through $(1, 4)$ and $(3, 6)$ note that

$$4 = f(1) = m(1) + b = m + b$$

and

$$6 = f(3) = m(3) + b = 3m + b$$

Hence, $b = 4 - m$ and $b = 6 - 3m$ so $4 - m = 6 - 3m$. Solving for m we see $m = 1$. Since $b = 4 - m$, $b = 4 - 1 = 3$.

The graph of $f(x) = x + 3$ is the straight line passing through $(1, 4)$ and $(3, 6)$.

Our discussions prior to Example 1 show that the graph of a linear function is a straight line. The wording of Example 2 suggests that each straight line is the graph of a linear function. Unfortunately, not every straight line is the graph of a linear function.

If l is a vertical line passing through the point $(c, 0)$ on the x-axis (see Figure 5.2.4), then P is a point on l if and only if P has coordinates of the form (c, y). There is no function whose graph is l, for if f were such a function, we would have, in particular, $f(c) = 0$ and $f(c) = 1$, which is impossible. However, since l is characterized by the fact that P is a point on l if and only if P has coordinates (c, y) for some real number y, l is the graph of the equation

$$x = c$$

Figure 5.2.4

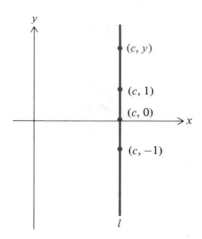

Example 3

An equation of the vertical line passing through $(2, 5)$ is $x = 2$, shown in Figure 5.2.5.

Figure 5.2.5

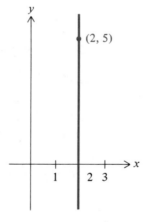

Although not every straight line is the graph of a linear function, we can show that, if l is not a vertical line, then there is a linear function whose graph is l.

Example 2 illustrates a method by which we can find a linear function $f(x) = mx + b$ with graph l. In particular, let (x_1, y_1) and (x_2, y_2) be two distinct points on l. The graph of f is l if and only if

$$y_1 = f(x_1) = mx_1 + b \tag{1}$$

and

$$y_2 = f(x_2) = mx_2 + b \tag{2}$$

Subtracting (1) from (2)

$$y_2 - y_1 = m(x_2 - x_1)$$

so

$$m = \frac{y_2 - y_1}{x_2 - x_1} \tag{3}$$

The number m is called the **slope** of the line. Notice that m is the ratio of the change in y-coordinates to the change in x-coordinates. (Since l is not vertical and (x_1, y_1) and (x_2, y_2) are distinct, $x_1 \neq x_2$ and we are not dividing by zero in (3).)

From (1) we can conclude $b = y_1 - mx_1$. Hence, an equation of the line is

$$y = mx + y_1 - mx_1$$

which is equivalent to

$$y - y_1 = m(x - x_1)$$

This last equation is known as the **point-slope form** for the equation of a line through (x_1, y_1) and with slope m. The point-slope form is often convenient to use when determining equations of lines and we suggest you memorize it.

Example 4

> We can use the point-slope form to find an equation of the line passing through $(2, 4)$ and $(5, 10)$. If we let $(x_1, y_1) = (2, 4)$ and $(x_2, y_2) = (5, 10)$, then
>
> $$m = \frac{10 - 4}{5 - 2} = \frac{6}{3} = 2$$
>
> and so an equation of the line passing through $(2, 4)$ and $(5, 10)$ is
>
> $$y - 4 = 2(x - 2)$$

It is not possible for two linear functions, $f(x) = m_1 x + b_1$ and $g(x) = m_2 x + b_2$, with $m_1 \neq m_2$ or $b_1 \neq b_2$, to have the same graph. If f and g did have the same graph, the y-intercept would be $(0, b_1) = (0, f(0)) = (0, g(0)) = (0, b_2)$, so b_1 would equal b_2. Furthermore, $(1, f(1))$ would equal $(1, g(1))$ so

$$m_1 + b_1 = m_1(1) + b_1 = f(1) = g(1) = m_2(1) + b_2 = m_2 + b_1$$

and hence, m_1 would equal m_2.

If l is a nonvertical line, our arguments above show that there is one and only one polynomial, $mx + b$, of degree one, which defines a linear function, $f(x) = mx + b$ whose graph is l. Since each line has a unique value of m corresponding to it, the same arguments show that m can be computed via formula (3) using any two arbitrarily chosen, distinct points, (x_1, y_1) and (x_2, y_2) on l.

Example 5

The line, l, which is the graph of $f(x) = -2x + 5$, has slope -2. Suppose (x_1, y_1) and (x_2, y_2) are two distinct points on l. Then

$$x_1 \neq x_2$$
$$y_1 = -2x_1 + 5$$
$$y_2 = -2x_2 + 5 \qquad \text{(See Figure 5.2.6.)}$$

Figure 5.2.6

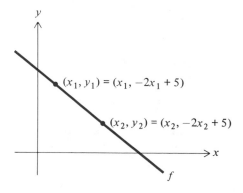

Using (x_1, y_1) and (x_2, y_2) in formula (3) to determine the slope of l, we get

$$m = \frac{y_2 - y_1}{x_2 - x_1} = \frac{(-2x_2 + 5) - (-2x_1 + 5)}{x_2 - x_1} = \frac{-2x_2 + 2x_1}{x_2 - x_1}$$
$$= \frac{-2(x_2 - x_1)}{x_2 - x_1} = -2$$

Thus, we see that the number m computed using formula (3) is independent of how the distinct points (x_1, y_1) and (x_2, y_2) are chosen on l.

Given a linear function $f(x) = mx + b$, we know that its graph is a line, l. We have called the number m the slope of l and seen that we can compute it by looking at the ratio

$$\frac{y_2 - y_1}{x_2 - x_1}$$

for any two distinct points (x_1, y_1) and (x_2, y_2) on l. (Sometimes the Greek letter delta (Δ) is used to represent the phrase "the change in" and a formula m is given by $m = \Delta y / \Delta x$ where $\Delta y = y_2 - y_1$ and $\Delta x = x_2 - x_1$.)

The slope gives us important information about a line: It tells us how steep it is. To be more precise:

> The slope of l represents the increase in the functional value of f for a unit increase in any value of x.

To see why this is so, recall that

$$m = \frac{y_2 - y_1}{x_2 - x_1}$$

for any two distinct points on the graph of f. Arbitrarily, choose a point (x_1, y_1) on l so that $y_1 = f(x_1)$. Now select a point (x_2, y_2) on l so the difference in x-coordinates, $x_2 - x_1$, is 1. (See Figure 5.2.7.) Then $(x_2, y_2) = (x_1 + 1, f(x_1 + 1))$ and, since we get the same value of m no matter which two points we choose,

$$m = \frac{f(x_1 + 1) - f(x_1)}{(x_1 + 1) - x_1} = f(x_1 + 1) - f(x_1)$$

But $f(x_1 + 1) - f(x_1)$ is precisely the increase in f corresponding to a unit increase in the value of x.

Figure 5.2.7

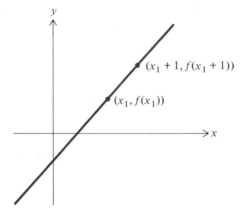

Example 6

For the linear function $f(x) = 5x + 3$, any increase of one unit in an x-value will cause a change of 5 units in the value of f. An increase of 12 units in any value of x will cause an increase of 60 units in the value of f since each unit of change in x causes 5 units of change in the value of f.

We mentioned earlier that slope gives us a measure of steepness. Figure 5.2.8 illustrates this. The graph of $f(x) = 6x$ is steeper than the graph of $g(x) = 2x$ since a unit change in x produces a larger change in the value of f than in the value of g.

Figure 5.2.8 $f(x) = 6x$
                      $g(x) = 2x$

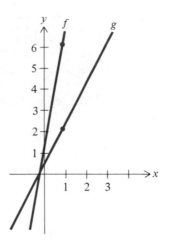

Note that, if a line has positive slope, then any point moving along the line from left to right must rise. If the line has negative slope, the point will fall as it moves along the line. Figure 5.2.9 illustrates the lines with positive and negative slopes.

Figure 5.2.9 (a) Lines with Positive Slopes
             (b) Lines with Negative Slopes

(a)

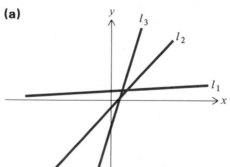

(b)

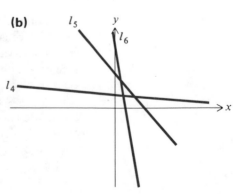

The following facts about straight lines and slopes are very useful. We will not prove either statement. Exercises 28 and 29 suggest how these facts can be proved.

> Let l_1 and l_2 be lines with slopes m_1 and m_2, respectively.
>
> A. l_1 is parallel or identical to l_2 if and only if $m_1 = m_2$.
> B. l_1 is perpendicular to l_2 if and only if $m_1 = -(1/m_2)$.

Example 7

To find the linear function g, whose graph is parallel to the line given by $f(x) = 3x + 4$ and passes through $(2, 5)$, note that the slope of the graph of f is 3, and hence, the slope of the graph of g is also 3. Therefore,

$$g(x) = 3x + b$$

Since $5 = g(2) = 3(2) + b = 6 + b$, $b = -1$ and $g(x) = 3x - 1$.

Example 8

To find the linear function $g(x) = mx + b$ whose graph is perpendicular to the line given by $f(x) = 3x + 4$ and passes through $(2, 5)$, note that $m = -\frac{1}{3}$ so

$$g(x) = -\tfrac{1}{3}x + b$$

Since $5 = g(2) = -\frac{1}{3}(2) + b = -\frac{2}{3} + b$, $b = \frac{17}{3}$ and $g(x) = -\frac{1}{3}x + \frac{17}{3}$.

Exercises 5.2

In Exercises 1–10, sketch the graph of the given linear function or equation. Find the slope and intercepts of each line, if possible.

1. $f(x) = 2x - 6$
2. $y = -x + 1$
3. $y = \frac{3}{2}x + 4$
4. $g(x) = -3x + 9$
5. $x = 5$
6. $h(x) = x$
7. $y = 3$
8. $2x + 3y - 7 = 0$
9. $-3x - 9y + 6 = 0$
10. $x = -3$

In Exercises 11–24, find the linear function whose graph is the straight line satisfying the given conditions. Sketch the graph.

11. The line passes through $(2, 1)$ and $(5, 3)$.
12. The line passes through $(-2, 4)$ and $(3, 0)$.
13. The line has slope -2 and y-intercept 3.
14. The line has slope 4 and y-intercept -2.
15. The line has slope 2 and x-intercept 1.
16. The line has slope -2 and x-intercept 5.

17. The line has slope -2 and passes through $(2, -4)$.

18. The line has slope π and passes through $(\sqrt{2}, \sqrt{3})$.

19. The line passes through $(5, 9)$ and is parallel to the line with equation $y = -2x + 4$.

20. The line passes through $(-2, 2)$ and is parallel to the line with equation $3x - y + 2 = 0$.

21. The line passes through $(-2, -7)$ and is perpendicular to the line with equation $y = 2x - 4$.

22. The line passes through $(4, -3)$ and is perpendicular to the line with equation $y = -5x + 2$.

23. The line is horizontal and passes through $(-2, -7)$.

24. The line is vertical and passes through $(6, 3)$.

25. We have discussed two forms of equations of lines in this section:

$$y = mx + b \qquad \text{(the slope-intercept form)}$$

and

$$y - y_1 = m(x - x_1) \qquad \text{(the point-slope form)}$$

In this exercise, you are asked to derive a third form. Show that if l is a line with y-intercept $(0, b)$, $b \neq 0$, and x-intercept $(a, 0)$, $a \neq 0$, then an equation of l is

$$\frac{x}{a} + \frac{y}{b} = 1$$

This form is called the **intercept form** of an equation of a line.

26. Use the intercept form (explained in Exercise 25) of an equation of a line to find an equation of the line with intercepts $(2, 0)$ and $(0, -4)$.

27. Use the intercept form (explained in Exercise 25) of an equation of a line to find an equation of the line with intercepts $(5, 0)$ and $(0, 7)$.

28. Let l_1 be a line given by $f_1(x) = m_1 x + b_1$ and l_2 a line given by $f_2(x) = m_2 x + b_2$. Show that l_1 is parallel to l_2 if and only if $m_1 = m_2$. (*Hint:* Choose two distinct numbers x_1 and x_2 and consider the quadrilateral with vertices $A(x_1, f_1(x_1))$, $B(x_1, f_2(x_1))$, $C(x_2, f_1(x_2))$, and $D(x_2, f_2(x_2))$ as illustrated in the figure.)

Note that line segment AB is parallel to line segment CD. Hence l_1 is parallel to l_2 if and only if the line segments AB, AC, CD, and BD are the edges of the parallelogram $ABCD$, or equivalently, $f_2(x_1) - f_1(x_1) = f_2(x_2) - f_1(x_2)$. Prove that $m_1 = m_2$ if and only if $f_2(x_1) - f_1(x_1) = f_2(x_2) - f_1(x_2)$.

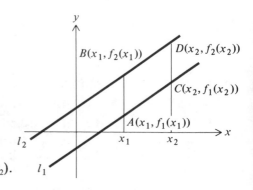

29. Let l_1 and l_2 be distinct lines given by $f_1(x) = m_1 x + b_1$ and $f_2(x) = m_2 x + b_2$. Show l_1 is perpendicular to l_2 if and only if $m_2 = -(1/m_1)$.   (*Hint*: Our work on families of functions suggests that l_1 and l_2 are perpendicular if and only if the graphs of $g_1(x) = m_1 x$ and $g_2(x) = m_2 x$ are perpendicular. The graphs of g_1 and g_2 are perpendicular if and only if the triangle OAB in the figure below is a right triangle with right angle at O. By the Pythagorean theorem OAB will be such a right triangle if and only if

$$d(A, B)^2 = d(O, A)^2 + d(O, B)^2 \tag{1}$$

Show (1) is true if and only if $m_2 = -(1/m_1)$.

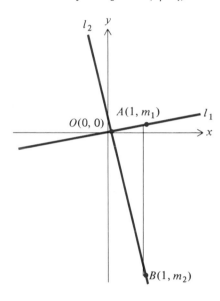

30. Complete the proof showing every point on the graph of $g(x) = mx$ is on the straight line determined by $P(0, 0)$ and $Q(1, m)$ by considering those points $R(x, y)$ on the graph of g with $x > 1$ or $x < 0$ and showing $d(PQ) + d(QR) = d(PR)$ or $d(RP) + d(PQ) = d(RQ)$, respectively.

31. Complete the proof showing every point on the straight line l determined by $P(0, 0)$ and $Q(1, m)$ lies on the graph of $g(x) = mx$ by considering points $R(x, y)$ on l with $x > 1$ or $x < 0$ and showing $y = mx$ in either case.

SECTION 5.3   **QUADRATIC FUNCTIONS**

We are familiar with the graph of

$$f(x) = mx^2$$

Figures 5.3.1 and 5.3.2 show the general shapes of the graphs of f when $m > 0$ and $m < 0$, respectively. Note that, if $m > 0$, $(0, 0)$ is the lowest point on the graph of f and if $m < 0$, $(0, 0)$ is the highest point.

Now let g be the quadratic function

$$g(x) = ax^2 + bx + c$$

Then,

$$g(x) = a\left(x^2 + \frac{b}{a}x\right) + c$$

Using the technique of completing the square

$$g(x) = a\left(x^2 + \frac{b}{a}x + \frac{b^2}{4a^2}\right) + c - \frac{b^2}{4a}$$

$$g(x) = a\left(x + \frac{b}{2a}\right)^2 + \left(c - \frac{b^2}{4a}\right)$$

$$g(x) = a\left[x - \left(\frac{-b}{2a}\right)\right]^2 + \left(c - \frac{b^2}{4a}\right)$$

Figure 5.3.1 $f(x) = mx^2$ $m > 0$

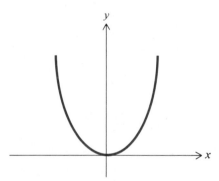

Figure 5.3.2 $f(x) = mx^2$ $m < 0$

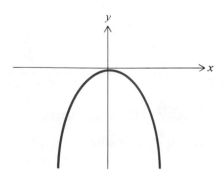

To find the graph of g, we may use the ideas suggested in Section 4.6.3 (Families of Functions). The graph of g is the graph of $f(x) = ax^2$ moved horizontally $-\dfrac{b}{2a}$ units and then vertically $c - \dfrac{b^2}{4a}$ units. Figure 5.3.3 indicates our reasoning when

$$a > 0 \qquad -\frac{b}{2a} > 0 \qquad \text{and} \qquad c - \frac{b^2}{4a} > 0$$

The above argument shows that the graph of a quadratic function, $f(x) = ax^2 + bx + c$, is identical to the graph of $g(x) = ax^2$ except for its position in the plane.

Figure 5.3.3   (a) $f(x) = ax^2$

(b) $f(x) = a\left(x + \dfrac{b}{2a}\right)^2$

(c) $f(x) = a\left(x + \dfrac{b}{2a}\right)^2 + \left(c - \dfrac{b^2}{4a}\right)$

(a)

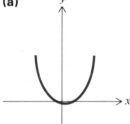

(b)

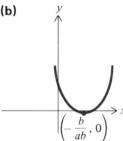

(c)

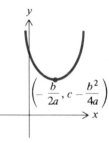

Example 1

To sketch the graph of

$$f(x) = x^2 + 2x + 3$$

we use the "completing the square" method described earlier.

$$f(x) = x^2 + 2x + 3$$
$$f(x) = x^2 + 2x + 1 - 1 + 3$$
$$f(x) = (x + 1)^2 + 2$$
$$f(x) = (x - (-1))^2 + 2$$

So the graph of f looks like the graph of $g(x) = x^2$ except that the low point is at $(-1, 2)$ instead of $(0, 0)$ (Figure 5.3.4).

Figure 5.3.4

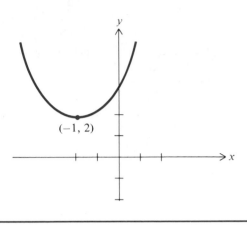

$(-1, 2)$

Example 2

To sketch the graph of

$$f(x) = -2x^2 + 8x - 5$$

we note

$$f(x) = -2(x^2 - 4x) - 5$$
$$f(x) = -2(x^2 - 4x + 4) - 5 + 8$$
$$f(x) = -2(x - 2)^2 + 3$$

The graph of a quadratic function is called a **parabola.** Exercises 16 and 17 outline a geometric approach to the definition of a parabola.

Figure 5.3.5 $f(x) = -2x^2 + 8x - 5$

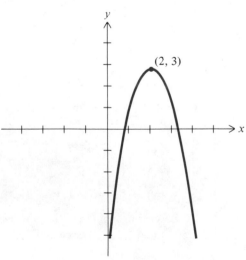

$(2, 3)$

Exercises 5.3

In Exercises 1–10, sketch the graph of the given quadratic function. Be sure to label the low or high point on the graph.

1. $f(x) = 2x^2 + 6x + 9$
2. $g(x) = -4x^2 + 8x - 12$
3. $h(x) = x^2 + 2x + 4$
4. $y = 3x^2 + 12x - 24$
5. $f(x) = -4x^2 - 8x - 8$
6. $h(x) = x^2 + x$
7. $y = -3x^2 + 6$
8. $y = 2(x + 5)^2$
9. $f(x) = -4x^2 - 12x + 5$
10. $g(x) = -x^2 + x + 2$

Each equation in Exercises 11–15 defines x as a quadratic function of y. Sketch the graph of each equation.

11. $x = y^2$
12. $x = -y^2$
13. $x = y^2 + 4y + 3$
14. $2x + 4y^2 + 8y - 7 = 0$
15. $y^2 + 8y - x = 0$

16. Show that the graph of $f(x) = ax^2$ $(a \neq 0)$ can be characterized as the set of points P for which the distance from P to $(0, \tfrac{1}{4}a)$ is equal to the distance from P to the line $y = -\tfrac{1}{4}a$.

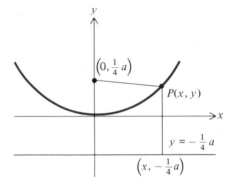

17. Graphs consisting of points equidistant from a fixed line and a fixed point not on the line are called **parabolas.** Use the ideas in Problem 16 to find the function $f(x) = ax^2$ whose graph consists of the points P that are equidistant from $(0, 1)$ and the line with equation $y = -1$.

SECTION 5.4

SYSTEMS OF EQUATIONS

When working with two functions f and g, it is often important to find all ordered pairs of real numbers (x_0, y_0) such that $y_0 = f(x_0)$ and $y_0 = g(x_0)$. Any such ordered pair is called a solution of the **system of (simultaneous) equations**

$$y = f(x)$$
$$y = g(x) \tag{1}$$

Now, (x_0, y_0) is a solution of the system (1) if and only if (x_0, y_0) gives the coordinates of a point that lies on the graphs of both f and g. Hence, we see that the problem of solving a system of equations (i.e., finding all solutions of the system) is equivalent to finding the points of intersection of the graphs of the two equations. (See Figure 5.4.1.)

Figure 5.4.1

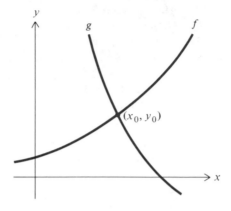

Examples 1 and 2 below illustrate how problems arise that require solving a system of equations.

Example 1

Suppose the cost, C, (in dollars) of producing x units of a product is given by

$$C = 20 + 5x$$

and the income, I, (in dollars) realized by selling x units is given by

$$I = 7x$$

How many units must be produced (and sold) in order for income to equal cost of production? To answer this question, we must find the value x_0 of x such that if $I_0 = 7x_0$ and $C_0 = 20 + 5x_0$ then $I_0 = C_0$. That is, we must solve the following system of equations

$$y = 20 + 5x$$
$$y = 7x$$

(2)

Now, (x, y) represents a solution of this system if and only if

$$7x = 20 + 5x$$

Solving for x we get

$$x = 10$$

Replacing x in either equation in system (2) by 10 will determine y. For example, using $y = 100 + 5x$, we get $y = 20 + 5(10) = 70$. Hence,

(10, 70) is the solution of (2); i.e., when 10 units are produced, the cost of production will be $70 and the income on selling 10 units will be $70.

The graph of the functions C and I are given in Figure 5.4.2. The point of intersection of the two graphs is (10, 70).

Figure 5.4.2

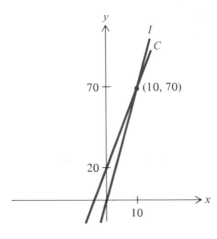

Example 2

Suppose the height, s, in feet of a projectile t seconds after it is fired vertically into the air is given by

$$s = -16t^2 + 100t$$

At what times will the projectile be at a height of 100 feet? To answer this question, we want to find the values of t when $s = 100$. This is equivalent to solving the system of equations

$$s = -16t^2 + 100t$$
$$s = 100$$

(3)

Now (t, s) is a solution of (3) if and only if

$$100 = -16t^2 + 100t$$

i.e.,

$$0 = -16t^2 + 100t - 100$$

Using the quadratic formula, we see

$$t = \frac{5}{4} \quad \text{or} \quad t = 5$$

(When $t = \frac{5}{4}$, the projectile is at a height of 100 feet and moving upward. When $t = 5$, the projectile is at a height of 100 feet and moving downward.)

The graphs of $s = -16t^2 + 100t$ and $s = 100$ are given in Figure 5.4.3. The points of intersection of the two graphs are $(\frac{5}{4}, 100)$ and $(5, 100)$.

Figure 5.4.3

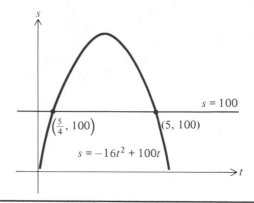

The problem of solving simultaneous equations can be generalized to the following problem: Given two equations in the variables x and y find all ordered pairs (x_0, y_0) whose coordinates satisfy both equations; i.e., find the points of intersection of the graphs of these equations. Note that, in this more general case, we are not required to work with functions.

Example 3

Consider the problem of solving the system of equations

$$x^2 + y^2 = 17$$
$$x + y = 3$$

(4)

If (x, y) is a solution of (4), then $x + y = 3$, and hence, $y = 3 - x$. Substituting $3 - x$ for y in the first equation of (4), we see that (x, y) is a solution of (4) if and only if

$$x^2 + (3 - x)^2 = 17 \quad \text{and} \quad y = 3 - x$$

Now

$$x^2 + (3 - x)^2 = 17$$
$$\Leftrightarrow x^2 + 9 - 6x + x^2 = 17$$
$$\Leftrightarrow 2x^2 - 6x - 8 = 0$$
$$\Leftrightarrow x^2 - 3x - 4 = (x + 1)(x - 4) = 0$$

Hence, $x = -1$ or $x = 4$. Substituting these values of x into $y = 3 - x$, we see that when $x = -1$, $y = 4$, and when $x = 4$, $y = -1$, and hence, the solutions of (4) are $(-1, 4)$ and $(4, -1)$. (See Figure 5.4.4.)

Figure 5.4.4

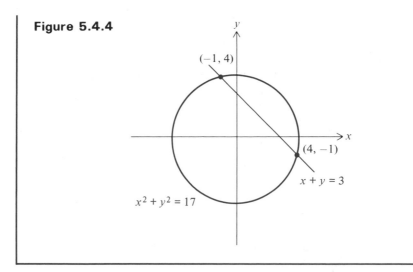

Examples 1–3 illustrate a method, called the **method of substitution,** which is frequently used to solve systems of equations. The method involves the following steps:

A.   Solve one equation for y in terms of x.
B.   Replace y in the second equation by the expression in x determined in (A). This will create an equation in x alone.
C.   Solve the equation obtained in (B) for x.
D.   Use the x-values found in (C) together with the first equation to determine the corresponding y-values of the solutions of the system.

Sometimes the substitution method is either impossible or inconvenient to use when solving systems of equations. An alternative method is the **method of elimination,** which is illustrated in the next examples.

Example 4

To solve the system

$$3x^2 + y^2 = 22$$
$$9x - y^2 = 8 \tag{5}$$

assume that (x, y) is a solution. Then (x, y) satisfies both equations of the system and must satisfy the equation obtained by "adding the two equations"; i.e.,

$$3x^2 + 9x = 30$$

This equation does not involve y (i.e., y has been eliminated) and may be solved for x either by using the quadratic formula or by factoring. The solutions of $3x^2 + 9x = 30$ are 2 and -5.

Note that our arguments to this point only show that, if (x, y) is a solution of the system (5), then $x = 2$ or $x = -5$. There is no guarantee that a solution (x, y) with $x = 2$ or $x = -5$ exists.

Replacing x by 2 in the second equation of (5), we have

$$9(2) - y^2 = 8$$
$$\Leftrightarrow y^2 = 10$$
$$\Leftrightarrow y = \pm \sqrt{10}$$

Hence $(2, \sqrt{10})$ and $(2, -\sqrt{10})$ satisfy the second equation of (5). Since $(2, \sqrt{10})$ and $(2, -\sqrt{10})$ satisfy the second equation of (5) and the equation derived by adding the two equations of (5), they also satisfy the first equation of (5).

When we replace x by -5 in the second equation of (5) we have

$$9(-5) - y^2 = 8$$
$$\Leftrightarrow y^2 = -53$$

Since the square of a number cannot be negative, it follows that the system has no solution (x, y) with $x = -5$.

Hence, the only solutions of system (5) are $(2, \sqrt{10})$ and $(2, -\sqrt{10})$.

Example 5

To solve the system

$$2x + 3y = 13$$
$$3x + 2y = 17$$

(6)

we create an equivalent system (i.e., one with the same set of solutions) by multiplying both sides of the first equation by 3 and both sides of the second equation by 2 to get

$$6x + 9y = 39$$
$$6x + 4y = 34$$

(7)

Subtracting the second equation from the first in (7), we eliminate the variable x and get $5y = 5$, and hence, $y = 1$. Replacing y by 1 in the first equation of (6), we get

$$2x + 3 = 13$$

and solving for x we have $x = 5$.

Hence $(5, 1)$ (i.e., $x = 5$ and $y = 1$) is the solution of (6).

The basic ideas behind the method of elimination illustrated in Examples 4 and 5 are:

A.   Add a nonzero multiple of one equation to a nonzero multiple of the other to obtain an equation in one of the variables, say x. (The second variable, y, has been eliminated.) The x-coordinates of the solutions of the system are solutions of this new equation.

B.   Solve the equation derived in (A).

C.   By substituting the x-values obtained in (B) into the equations of the system, determine the corresponding y-values of the solutions of the system.

The methods of substitution and elimination can be generalized to methods that can be used to solve systems of equations with three or more variables. However, when solving systems of linear equations, the method of elimination is generally more efficient than the method of substitution, and in the remainder of this section we will illustrate how to use the elimination method to find solutions of systems of linear equations.

Before considering the examples, consider the general problem of solving a system of linear equations in the variables x and y.

$$\begin{aligned} a_1 x + b_1 y &= c_1 \\ a_2 x + b_2 y &= c_2 \end{aligned} \tag{8}$$

The graphs of these two equations are straight lines. If the lines are parallel, they do not intersect; and hence, the system has no solutions. If the lines are distinct and not parallel, they have exactly one point of intersection; and hence, the system has one and only one solution. If the two equations have the same graph, then each point lying on this graph represents a solution of the system; and hence, the system has an infinite number of solutions.

A **linear equation in three variables** x, y, z is an equation of the form $ax + by + cz = d$ where a, b, c, and d are constants. The conclusions we derived about a system of linear equations in two variables can be applied to a system of linear equations in three or more variables. In particular, the system of equations

$$\begin{aligned} a_1 x + b_1 y + c_1 z &= d_1 \\ a_2 x + b_2 y + c_2 z &= d_2 \\ a_3 x + b_3 y + c_3 z &= d_3 \end{aligned}$$

has either no solutions, exactly one solution, or an infinite number of solutions.

In the following examples, we will show how to use the elimination method either to find the solutions of systems of linear equations or to show that a system of linear equations has no solutions.

Example 6

To solve the system

$$\begin{aligned} 2x - 5y &= 4 \\ -6x + 15y &= 0 \end{aligned} \tag{9}$$

eliminate the x-variable by multiplying both sides of the first equation by 3 and then adding as follows:

$$6x - 15y = 12$$
$$-6x + 15y = 0$$
$$0 = 12$$

If (x, y) satisfies system (9), it must also satisfy the equation $0 = 12$. This is impossible, since $0 \neq 12$. Hence, (9) has no solutions.

Example 7

Consider the following system of linear equations in three variables.

$$x + 2y + z = 8$$
$$-x + y - 2z = -5 \qquad (10)$$
$$-x + 3y + z = 8$$

Adding the first and second equations gives the equation $3y - z = 3$. Adding the first and third equations gives the equation $5y + 2z = 16$. If (x, y, z) is a solution of (10) then (y, z) must be a solution of

$$3y - z = 3$$
$$5y + 2z = 16 \qquad (11)$$

(Notice that the variable x has been eliminated.) Hence, to find the y- and z-coordinates of the solution of (10), we solve (11) as follows:

$$6y - 2z = 6 \qquad \text{(the first equation of (11) multiplied by 2)}$$
$$5y + 2z = 16$$

$$11y = 22$$
$$y = 2$$

Using the first equation of (11), we see $z = 3y - 3 = 3(2) - 3 = 3$; i.e., $y = 2$, $z = 3$ is the solution of (11). Then using the first equation of (10) we see $x = 8 - 2y - z = 8 - 2(2) - 3 = 1$. Hence, $x = 1$, $y = 2$, $z = 3$ is the solution of (10).

We know from our calculations that $(1, 2, 3)$ satisfies the first equation of (10). As a check against arithmetic errors we should see if $(1, 2, 3)$ satisfies the remaining equations as well.

Example 8

To solve the system

$$2x + y + 3z = 4$$
$$-x - y + z = 7 \qquad (12)$$
$$3x + y + 7z = 16$$

we eliminate the variable y by adding the first and second equations and by adding the second and third equations to get the system

$$x + 4z = 11$$
$$2x + 8z = 23 \tag{13}$$

We solve this system as follows:

$$2x + 8z = 22$$
$$2x + 8z = 23$$

Subtracting the first equation from the second we get

$$0 = 1$$

Since $0 \neq 1$, system (13), and hence system (12), has no solutions.

Example 9

To solve the system

$$2x + y - 3z = 4$$
$$x + 2y + 3z = 5 \tag{14}$$
$$3y + 9z = 6$$

eliminate the x-variable from (14) as follows:

$$\begin{cases} 2x + y - 3z = 4 \\ 2x + 4y + 6z = 10 \\ \qquad\quad 3y + 9z = 6 \end{cases}$$

$$3y + 9z = 6$$
$$3y + 9z = 6 \tag{15}$$

Since both equations of (15) are identical, it follows that (y, z) satisfies (15) if and only if it satisfies the equation $3y + 9z = 6$. Thus, if (x, y, z) is a solution of (14), then $3y + 9z = 6$, and hence, $y = 2 - 3z$. Furthermore, (x, y, z) must satisfy $x + 2y + 3z = 5$; i.e., the second equation of (14). Hence, $x = 5 - 2y - 3z$, and since $y = 2 - 3z$, we see $x = 5 - 2(2 - 3z) - 3z = 1 + 3z$. Notice that we can assign any value to z and then use the equations $x = 1 + 3z$ and $y = 2 - 3z$ to find corresponding values of x and y such that (x, y, z) is a solution of (14). It follows that (14) has an infinite number of solutions and that the set of solutions consists of all those triples (x, y, z) such that $x = 1 + 3z$ and $y = 2 - 3z$.

Exercises 5.4   In Exercises 1–10, find the solutions of the given system of equations using the substitution method.

1. $\begin{cases} x + 3y = 10 \\ 2x - 5y = -13 \end{cases}$
2. $\begin{cases} -4x + y = 6 \\ 2x + 3y = -10 \end{cases}$

3. $\begin{cases} 3x - 2y = 15 \\ 4x + y = 20 \end{cases}$
4. $\begin{cases} 5x + 4y = 2 \\ -2x + 6y = 3 \end{cases}$

5. $\begin{cases} -3x + y = 0 \\ x^2 + y^2 = 40 \end{cases}$
6. $\begin{cases} x + 2y = 9 \\ 2x^2 + x - y = -1 \end{cases}$

7. $\begin{cases} x^2 + 2x + y = 4 \\ x + y = 5 \end{cases}$
8. $\begin{cases} x^2 + y = 1 \\ 2x^2 + 3y = -1 \end{cases}$

9. $\begin{cases} 3x^3 + y = 8 \\ x^2 - y = -4 \end{cases}$
10. $\begin{cases} x^2 + y^2 + 3y = -3 \\ y - 3x = 0 \end{cases}$

In Exercises 11–27, find the solutions of the given system of equations using the method of elimination.

11. $\begin{cases} x + 2y = 4 \\ -x + 3y = 11 \end{cases}$
12. $\begin{cases} x + y = -2 \\ 3x + y = 8 \end{cases}$

13. $\begin{cases} 2x + y = 0 \\ 3x + 2y = -1 \end{cases}$
14. $\begin{cases} 3x - 2y = -2 \\ 6x - 2y = 4 \end{cases}$

15. $\begin{cases} 7x - 3y = 4 \\ -14x + 6y = 1 \end{cases}$
16. $\begin{cases} -3x + 5y = 6 \\ 10x = -20 \end{cases}$

17. $\begin{cases} 7x + 2y = 13 \\ 3x + 5y = 18 \end{cases}$
18. $\begin{cases} x^2 + 4y^2 = 17 \\ x^2 - 3y^2 = -11 \end{cases}$

19. $\begin{cases} 2x^3 + 3y^2 = -4 \\ x^3 + y^2 - y = -6 \end{cases}$
20. $\begin{cases} 3x^2 + y^2 - 2y = -13 \\ x^2 + y = 4 \end{cases}$

21. $\begin{cases} x + 3y - 2z = 7 \\ x + 2y + 3z = 5 \\ -x + 5y + z = 9 \end{cases}$
22. $\begin{cases} 3x + 4y + z = 5 \\ x - 2y + 3z = 1 \\ x + 8y - 5z = 0 \end{cases}$

23. $\begin{cases} 2x + y - 3z = 4 \\ x + y + z = 5 \\ 4x + 3y - z = 14 \end{cases}$
24. $\begin{cases} 3x - 2y + z = 11 \\ x + 3y = -1 \\ 2x + z = 7 \end{cases}$

25. $\begin{cases} 4x + y + z = -6 \\ -x + 2y + z = 3 \\ 3x + 2y = -8 \end{cases}$
26. $\begin{cases} 3x + y - z = 2 \\ -x - y + 3z = 4 \\ 5x + 3y - 7z = 6 \end{cases}$

27. $\begin{cases} y + 3z = 7 \\ x - 2z = 4 \\ 2x + y - z = 10 \end{cases}$

28. A company that produces car radios has found that the cost, C, in dollars of producing x car radios is given by $C = 5000 + 25x$. Each radio is sold for 50 dollars. How many radios should be produced so that the income from sales is equal to the cost of production?

29. Find the points on the line with equation $x + 3y = 15$ that are 5 units from the origin.

30. Find the points on the graph of $y = x^2$ that are 4 units from the origin.

31. A rectangular field has an area of 1000 ft^2 and a perimeter of 140 ft. What are the dimensions of the field?

32. Find two numbers whose sum is 19 and whose product is 60.

33. Bill has $2.00 in change consisting of nickels, dimes, and quarters. He has a total of 18 coins and twice as many nickels as dimes. How many of each type of coin does Bill have?

34. Find the quadratic function whose graph passes the points $(0, -3)$, $(-1, -4)$, and $(1, 0)$.

35. Find the quadratic function whose graph passes through the points $(-2, -2)$, $(0, 4)$, and $(1, 4)$.

Review Exercises
Chapter 5

In Exercises 1–12, find an equation of the line satisfying the given conditions and sketch its graph.

1. The line has slope -6 and y-intercept 4.
2. The line has slope 3 and y-intercept -5.
3. The line passes through $(2, 4)$ and $(-5, 6)$.
4. The line passes through $(-1, 0)$ and $(5, 8)$.
5. The line passes through $(2, 7)$ and is parallel to the line with equation $3x + 4y = 5$.
6. The line passes through $(-3, 4)$ and is parallel to the line with equation $x = 3y + 6$.
7. The line passes through $(2, 7)$ and is perpendicular to the line with equation $3x + 4y = 5$.
8. The line passes through $(-3, 4)$ and is perpendicular to the line with equation $x = 3y + 6$.
9. The line has intercepts $(2, 0)$ and $(0, -4)$.
10. The line has intercepts $(-6, 0)$ and $(0, -3)$.
11. The line is vertical and passes through $(2, 5)$.
12. The line is vertical and passes through $(-3, -9)$.

Each equation in Exercises 13–20 defines y as a quadratic function of x or x as a quadratic function of y. Sketch the graph of each equation using the methods of Section 5.3.

13. $y = x^2 - 2x + 2$ **14.** $y = x^2 + 6x + 5$
15. $y = -x^2 + 4x + 4$ **16.** $y = -2x^2 - 4x - 1$
17. $x = -3y^2 + 12y - 10$ **18.** $x = y^2 + 4y + 6$
19. $4x - 4y^2 - 4y = 14$ **20.** $4y - x^2 - 4x = 16$

In Exercises 21–30, find the points of intersection (if any) of the graphs of the given equations (i.e., solve the given system of equations) using the method of substitution.

21. $y = 3x - 4$ $y = 2x - 2$
22. $2x + 5y = 13$ $3x - 2y = -9$
23. $y = x^2 + 7x - 24$ $y = 4$
24. $y = -3x^2 + 4x + 24$ $y = 6$
25. $y - x^2 + 3x + 7 = 0$ $x = 3$
26. $y^2 + 3y + 2x = 0$ $x = 4$
27. $x^2 + 3y^2 = 7$ $x + y - 7 = 0$
28. $x^2 + y^3 + y^2 - 16 = 0$ $y = x - 4$
29. $x^4 + 2x^2 + y^2 = 3$ $y = x^2 - 1$
30. $y = \sqrt{25 - x^2}$ $2x + y = 11$

In Exercises 31–43, solve the given system of equations using the method of elimination.

31. $\begin{cases} 4x - 2y = 11 \\ x - y = 2 \end{cases}$

32. $\begin{cases} 6x + y = -14 \\ 2x + 5y = 14 \end{cases}$

33. $\begin{cases} 3x - 2y = 7 \\ -6x + 4y = 7 \end{cases}$

34. $\begin{cases} x + 2y = 6 \\ 2x + 4y = 12 \end{cases}$

35. $\begin{cases} 3x + 2y - 3z = 4 \\ x + 3y = 5 \\ y + 2z = 0 \end{cases}$

36. $\begin{cases} 2x - 4y + 6z = 10 \\ x + 2y + z = 3 \\ -x + 3y + z = -1 \end{cases}$

37. $\begin{cases} 2x - 3y + z = 0 \\ x + y - 2z = 0 \\ x - z = 0 \end{cases}$

38. $\begin{cases} x + 2y = 3 \\ 4x + z = 4 \\ 2x - 4y + z = 5 \end{cases}$

39. $\begin{cases} 3x - y - z = 2 \\ x + y + z = 4 \\ 4x + y + z = 8 \end{cases}$

40. $\begin{cases} x^2 + 3x + y^2 = 10 \\ x^2 + y^2 = 7 \end{cases}$

41. $\begin{cases} x^2 + 2x + y^2 = 3 \\ 2x^2 + y^2 = 2 \end{cases}$

42. $\begin{cases} x^3 + 4y^2 = 0 \\ x + y^2 = 0 \end{cases}$

43. $\begin{cases} x^2 + 2x + 2y^2 - 4y = 5 \\ x^2 + 3x + y^2 - 2y = 5 \end{cases}$

44. Find two numbers whose sum is 39 and whose difference is 15.

45. John is 25 years younger than his father and the sum of their ages is 59. How old is John?

46. Find the distance from the point $(2, 7)$ to the line $3x + 4y = 3$.

47. Find the points on the line $x + y = 3$ that are 5 units from the origin.

48. A company produces two types of televisions, A and B. It costs \$100 to produce one type A television and \$150 to produce one type B television. The total number of both types produced was 55 and the total cost of production was \$7000. How many televisions of each type were produced?

49. A collection of coins consisting of nickels, dimes, and quarters has a value of \$1.60. There are 13 coins in the collection, and the number of nickels is one less than the total number of dimes and quarters. How many of each type of coin is in the collection?

50. A 6000-gallon mixture of three types of gasoline, G_1, G_2, and G_3, has an octane rating of 0.9 and a value of \$8000. The octane ratings and cost per gallon of the three types of gasoline are given in the table.

	Cost/gallon	Octane rating
G_1	\$1.00	0.8
G_2	\$1.50	0.9
G_3	\$1.25	0.95

If x gallons of G_1, y gallons of G_2, and z gallons of G_3 are mixed, the octane rating of the mixture is computed using the formula

$$\frac{0.80x + 0.90y + 0.95z}{x + y + z}$$

How much of each type of gasoline is in the mixture?

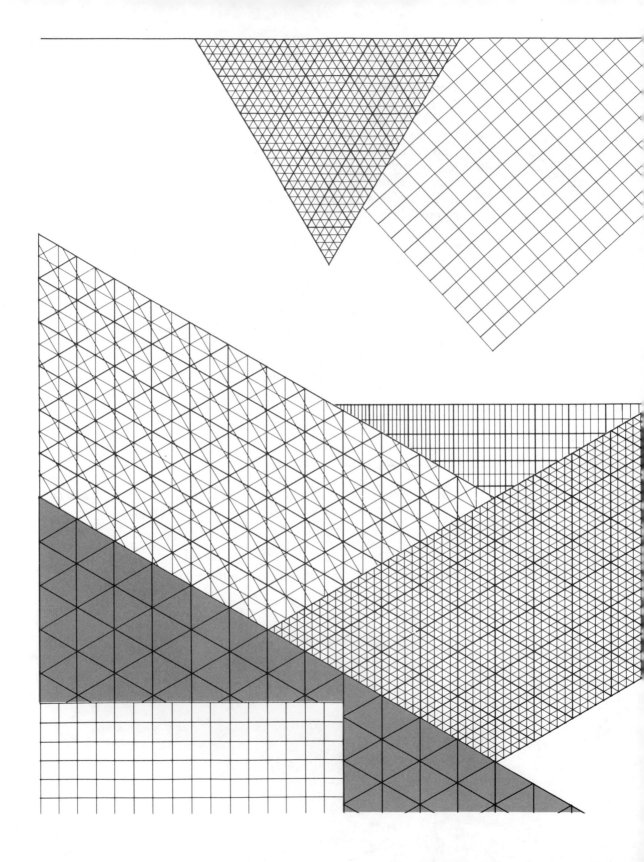

Trigonometric Functions

SECTION 6.1   INTRODUCTION

Trigonometry, literally the measurement of triangles, has a history going back at least as far as ancient Greece. Its earliest use, and still an important one, was indirect measurement—using certain ratios of the lengths of sides of right triangles to find distances that cannot be measured directly.

Our interest in trigonometry is very different. Many natural phenomena are periodic, recurring at regular intervals. One example of periodic behavior is the motion of planets around the sun. A less obvious example is the seasonal change in the price of certain consumer goods. Scientists use trigonometric functions to study this periodic behavior, and because of this widespread use, these functions are studied in calculus.

We begin our study by reviewing the trigonometry of the right triangle.

SECTION 6.2   THE TRIGONOMETRY OF RIGHT TRIANGLES

The mathematical study of right triangles had its beginnings in Ancient Egypt and Babylonia some 4,000 years ago and continues to be useful today.

Why is the study of right triangles so fundamental? We can suggest two possible reasons. First, right angles abound in nature and were used by humans long before any mathematical analysis was developed. Trees grow vertically upward and objects fall vertically down. Wells are dug and walls are constructed with the aid of plumblines, perpendicular to level ground. Distances from points to lines or from points to planes are measured along perpendiculars. These right angles are often viewed as parts of right triangles when measurements of related distances must be made indirectly.

Second, right triangles can be thought of as building blocks of arbitrary triangles. In Figure 6.2.1, triangle *ABC* is composed of two right triangles, *ABD* and *CDB*. By drawing an altitude—a perpendicular from a vertex to an opposite side—any triangle can be decomposed into two right triangles. If we develop a study of right triangles, we can apply it to arbitrary triangles.

Figure 6.2.1

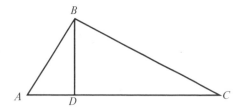

We begin this study by considering right triangles that contain a 30° angle. We will call such triangles "30° right triangles." We have drawn a few 30° right triangles in Figure 6.2.2.

Figure 6.2.2

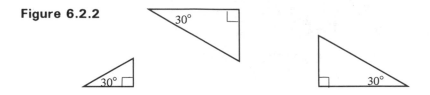

Although some 30° right triangles are larger than others, all seem to have the same shape. Look at a smaller one through a magnifying glass and it looks just like a larger one.

Triangles with the same shapes, although perhaps different sizes, are called **similar triangles.** You may recall from geometry that a sufficient condition for two triangles to be similar is for two angles of one triangle to be congruent to two angles of the other triangle. Since all 30° right triangles have two corresponding angles congruent (the 30° angles and the right, or 90°, angles), they are all similar.

Similar triangles have many interesting properties but by far the most important for us is this one:

> Corresponding sides of similar triangles are proportional.

To illustrate what this means, we will consider the following example.

Triangles *ABC* and *DEF* in Figure 6.2.3 are similar. (Notation: $\triangle ABC \sim \triangle DEF$.) Intuitively, all sides of *ABC* are twice as large as corresponding sides of *DEF*. Look at the ratio of $EF/DE = 3/5$. Now, look at the sides of *ABC* that correspond to *EF* and *DE*. *EF* corresponds to *BC* and *DE* corresponds to *AB*. $BC/AB = 6/10 = 3/5 = EF/DE$.

Figure 6.2.3

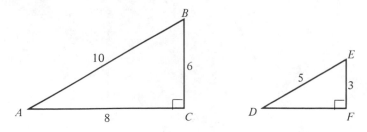

Example 1

Figure 6.2.4

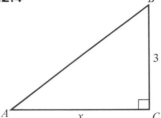

If we know that $\triangle ABC \sim \triangle DEF$, we can find AC. Since corresponding sides of similar triangles are proportional,

$$\frac{EF}{DF} = \frac{BC}{AC}$$

$$\frac{1}{3} = \frac{3}{x}$$

$$x = 9$$

It will be convenient to introduce some notation for right triangles. Suppose triangle ABC is a right triangle with right angle at C. (See Figure 6.2.5.) Then,

Figure 6.2.5

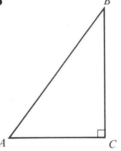

AB, the side opposite the right angle, is the **hypotenuse**

BC is **opposite** angle A

AC is **opposite** angle B

AC is **adjacent** to angle A

BC is **adjacent** to angle B

Although our use of the word opposite corresponds to ordinary usage, our use of the word adjacent does not. However we define adjacent, AB seems just as adjacent to angle A as AC. However, we agree never to consider the hypotenuse as an adjacent side to any angle. With this convention, there is no ambiguity.

In Figure 6.2.6, we have drawn two triangles, ABC and DEF. Triangle ABC is a 30°-60°-90° right triangle and the lengths of its sides are given.

Figure 6.2.6

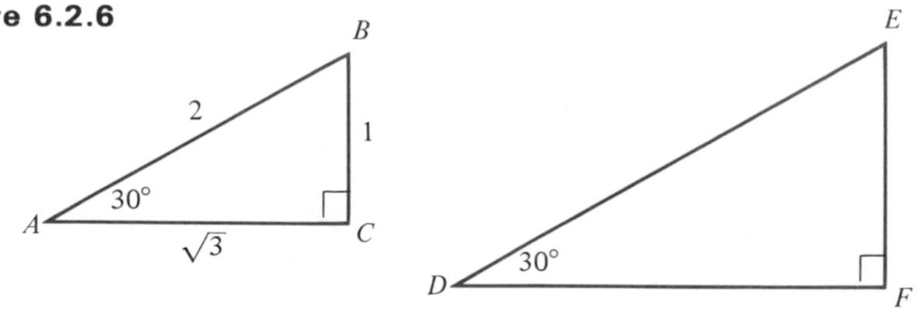

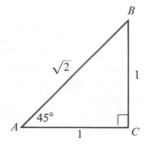

For each acute angle of right triangle ABC, consider the ratio of the length of the opposite side to the length of the hypotenuse, which we will write as opposite/hypotenuse. For angle A, this ratio is $\frac{1}{2}$ and for angle B this ratio is $\frac{\sqrt{3}}{2}$. What is this ratio for angle D?

Since $\triangle DEF$ is a 30° right triangle, it is similar to $\triangle ABC$. Since the ratios of corresponding sides in similar triangles are equal, the ratio of opposite/hypotenuse for angle D is also $\frac{1}{2}$.

There are two important ideas here. The first is that, although we do not know the lengths of DE or EF, we do know the ratio, EF/DE. The second is that this ratio is determined by acute angle D.

In our example, we know the ratio of opposite/hypotenuse for the 30° angle of one particular 30° right triangle. By similarity, we know the ratio of opposite/ hypotenuse for the 30° angle of any 30° right triangle.

This ratio, opposite/hypotenuse, is used often enough to have a name. It is called **sine,** which is traditionally abbreviated to sin. In our examples, we saw that sin 30° $= \frac{1}{2}$. If we look at angle B of triangle ABC, we see that sin 60° $= \frac{\sqrt{3}}{2}$.

Example 2

We can use $\triangle ABC$ (Figure 6.2.7) to find sin 45°. Looking at angle A, sin 45° $= \frac{1}{\sqrt{2}}$. Rationalizing the denominator,

$$\frac{1}{\sqrt{2}} \cdot \frac{\sqrt{2}}{\sqrt{2}} = \frac{\sqrt{2}}{2} \qquad \text{so sin 45°} = \frac{\sqrt{2}}{2}$$

Figure 6.2.7

Notice that although we used a particular triangle in Example 2 to compute sin 45°, we would get the same number if we found the sin 45° using *any* 45° right triangle.

We have used two triangles in our examples that will be referred to frequently in this chapter, the 30°-60°-90° and the 45°-45°-90° right triangles. We suggest that you remember these triangles, which are sketched in Figure 6.2.8. Using them you can read off the sines of 30°, 45°, and 60°.

Figure 6.2.8

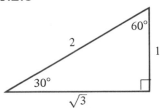
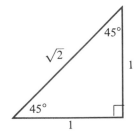

There are other ratios of sides in right triangles that are useful. We define them now, giving their usual abbreviations. In each definition, the words "opposite," "hypotenuse," and "adjacent" refer to the lengths of these sides of the triangle. If A is an acute angle of a right triangle,

sine A = sin A = opposite/hypotenuse

cosine A = cos A = adjacent/hypotenuse

tangent A = tan A = opposite/adjacent

cotangent A = cot A = adjacent/opposite

secant A = sec A = hypotenuse/adjacent

cosecant A = csc A = hypotenuse/opposite

The list of six trigonometric ratios will be easier to remember if you notice that the last three are reciprocals of the first three:

$$\cot A = \frac{1}{\tan A} \qquad \sec A = \frac{1}{\cos A} \qquad \csc A = \frac{1}{\sin A}$$

Example 3

Let us find all six trigonometric ratios of 30°. To do this we first sketch a 30°-60°-90° right triangle, as in Figure 6.2.9. From the diagram,

$$\sin 30° = \frac{1}{2} \qquad \cos 30° = \frac{\sqrt{3}}{2} \qquad \tan 30° = \frac{1}{\sqrt{3}} = \frac{\sqrt{3}}{3}$$

$$\cot 30° = \sqrt{3} \qquad \sec 30° = \frac{2}{\sqrt{3}} = \frac{2\sqrt{3}}{3} \qquad \csc 30° = 2$$

Figure 6.2.9

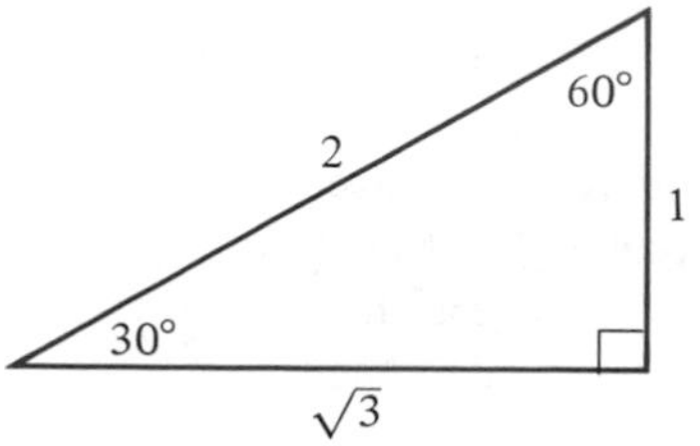

```
                              60°
                    2      /|
                       /   |
                    /      | 1
                 /         |
              30°          |
           /_____|
              √3
```

Examples 4 and 5 illustrate applications of the trigonometric ratios. Both use values of trigonometric ratios that appear in Appendix Table 1 found at the end of the text.

Example 4

From a point 75 feet from the base of a building, the angle of elevation to the top of the building is 50°. How tall is the building? To solve this problem, we let x represent the height of the building. Looking at the right triangle in Figure 6.2.10,

$$\tan 50° = \frac{x}{75}$$

But $\tan 50° = 1.192$. (See Appendix Table 1.)

$$1.192 = \frac{x}{75}$$

$$x = 89.4 \text{ feet}$$

Figure 6.2.10

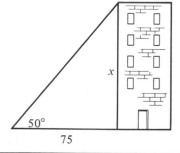

Example 5

What is the measure of the smallest acute angle in a right triangle whose sides measure 3, 4, and 5? Since the smallest angle is opposite the smallest side, we are trying to find the measure of angle A. Looking at $\triangle ABC$ (Figure 6.2.11),

$$\sin A = \frac{3}{5} = 0.6$$

From Appendix Table 1 at the back of this book we see that $\sin 36° = 0.588$ and $\sin 37° = 0.602$. To the nearest degree, the measure of angle A is $37°$.

Figure 6.2.11

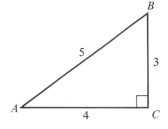

Exercises 6.2

1. **a.** Evaluate all 6 trigonometric ratios at angle A.
    b. Evaluate all 6 trigonometric ratios at angle B.

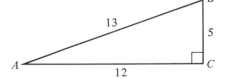

2. **a.** Evaluate all 6 trigonometric ratios at angle D.
    b. Evaluate all 6 trigonometric ratios at angle E.

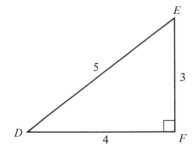

3. **a.** Evaluate all 6 trigonometric ratios at angle G.
    b. Evaluate all 6 trigonometric ratios at angle H.

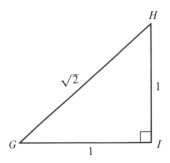

4. **a.** Evaluate all 6 trigonometric ratios at angle *J*.

  b. Evaluate all 6 trigonometric ratios at angle *K*.

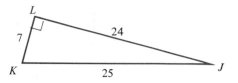

5. **a.** Evaluate all 6 trigonometric ratios at angle *M*.

  b. Evaluate all 6 trigonometric ratios at angle *N*.

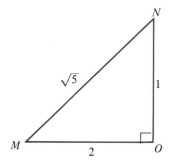

6. What is wrong with the following argument: "Looking at triangle *ABC*, shown on the right, it is clear that sin *A* = 5/7."

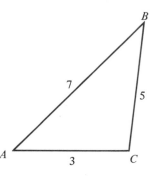

7. △ *EUC* ∼ △ *LID*. Find the length of *ID* if *EC* = 12, *LD* = 5, and *UC* = 8.

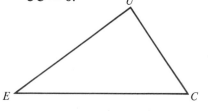
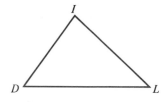

8. △ *ABC* ∼ △ *FED*. *AB* = 8, *AC* = 7, *EF* = 4. Find *DF*.

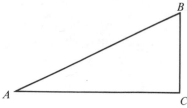
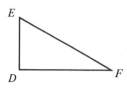

9. Angle *B* is congruent to angle *E*. *AB* is congruent to *BC* and *DE* is congruent to *EF*.
    a. Show that △ *ABC* ~ △ *DEF*. (This requires that you remember results from plane geometry.)
    b. *AC* = 6, *DF* = 4, *BC* = 7 Find *EF*.

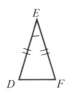

10. Use the table of trigonometric values to find the measure of the smallest angle in a right triangle with sides 5, 12, and 13, accurate to the nearest degree.

11. From a point 40 feet from the base of a tree, the angle of elevation to the top is 80°. How tall is the tree?

12. The hypotenuse of a 30°-60°-90° right triangle measures 8 feet. How long are the other sides?

13. The hypotenuse of a 45° right triangle measures 1 foot. How long are the other sides?

14. A 6-foot man looks at the top of a 70-foot building. The angle of elevation of his line of sight is 40°. How far from the building is he standing?

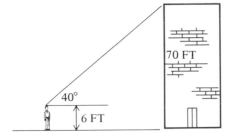

15. A ladder leans against a building, making an angle of 75° with the ground. The end of the ladder reaches a point 30 feet above ground on the building. How long is the ladder?

16. Two people stand on opposite banks of a river. A third person stands at point *C*, 200 feet from *B*. Angle *CBA* is a right angle, and *ACB* measures 75°. How wide is the river?

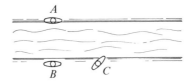

17. A lighthouse is opposite a straight shore 1 mile away. The beacon of the lighthouse rotates once every 20 seconds. Initially the beacon is shining on the point on shore nearest the lighthouse. Where is the beacon shining $2\frac{1}{2}$ seconds later?

APPENDIX 6.2   **SOME HISTORY AND AN APPLICATION**

Greek mathematicians, using mathematics no more complicated than the trigonometry presented in Section 6.2, were able to answer profound questions in astronomy. We will discuss one of them.

Figure 6.2.12   Four Views of Moon from Earth

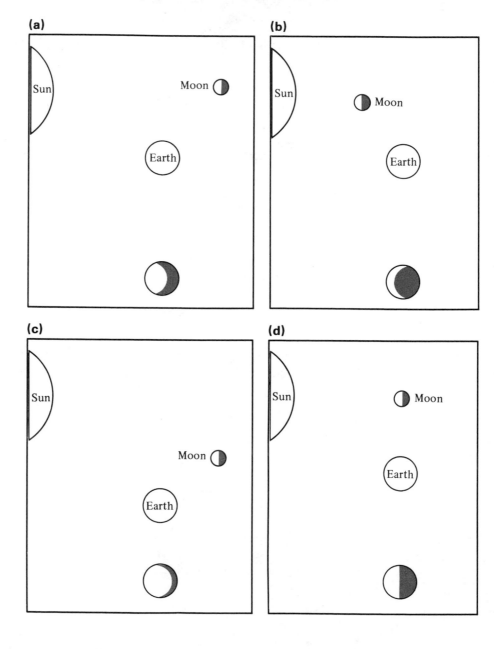

Aristarchus, a Greek mathematician and astronomer, was born in Samos about 310 B.C. In one of his works he considers the following question: Which is farther from earth, the sun or the moon? At the time this was a seemingly impossible question to answer. Although most people now know that the sun is farther away, it is unlikely that many people today have any idea how to verify the answer.

All of us are familiar with the phases of the moon. Figure 6.2.12 shows how the relative positions of earth, sun, and moon produce different lunar phases. Notice that when we see a half-moon the angle between sun, moon, and earth must be a right angle.

Figure 6.2.13

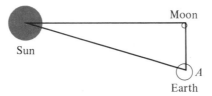

Sometimes, usually early in the morning or just before sunset, both the moon and the sun are visible, simultaneously, in the sky. Suppose this happens when the moon is a half-moon. We could then measure the angle between the moon and the sun (angle A in Figure 6.2.13). Then, using trigonometry,

$$\cos A = \frac{\text{Distance from moon to earth}}{\text{Distance from sun to earth}}$$

Aristarchus did just that. He found the angle to be 87°, from which he concluded that the sun is between 18 and 20 times farther away from earth than is the moon. (His trigonometry was similar to ours but slightly different—trigonometric tables would not be written for another 100 years until Hipparchus of Nicea.) Aristarchus' estimate is poor by today's standards. The angle between sun and moon when the moon is half full is about 89°50'. Since cos 89°50' is approximately 0.0029, the sun is approximately 345 times farther away from earth than is the moon (0.0029 = 1/345). The difficulty with his method was that he had to estimate, visually, when the moon was half full. But his logic was flawless and requires only elementary mathematics.

It is easy to think that the only value to what we are learning at the moment lies in its application to more advanced work to be done later. In fact, much elementary mathematics has great application. For the interested reader we suggest two books, Carl Boyer, *A History of Mathematics* (New York: Wiley & Sons, 1968); and George Polya, *Mathematical Methods in Science* (Washington: Mathematics Association of America, 1977), one of a series of books in the New Mathematical Library.

SECTION 6.3   RADIAN MEASUREMENT

In this section, we introduce a new unit of measurement for angles, the **radian.** We will postpone a discussion of the advantages of radian measurement until after we have had some practice using it.

Given an angle, θ, construct a circle with center at the vertex of θ, as in Figure 6.3.1. The angle θ is called a central angle of the circle. Suppose the radius of the circle is r and s is the length of arc that θ subtends along the circumference. We define the radian measure of θ as s/r.

Figure 6.3.1

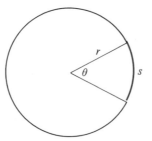

Our method for finding radian measure is independent of the size of the circle we use. Figure 6.3.2 shows θ as a central angle of three circles. Although s and r are different in each case, they are proportional, and the ratio s/r will be the same number each time.

Figure 6.3.2 $\dfrac{s_1}{r_1} = \dfrac{s_2}{r_2} = \dfrac{s_3}{r_3}$

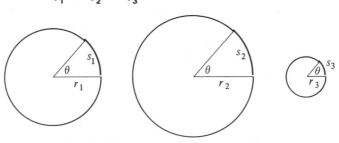

We will find the radian measure of some common angles using our definition. A 90° angle, if drawn as the central angle of a circle, will subtend $\frac{1}{4}$ of the circumference. (See Figure 6.3.3.) Since the entire circumference is $2\pi r$, a 90° angle subtends an arc whose length is $\frac{1}{4}(2\pi r) = \frac{1}{2}\pi r$. The radian measure of a 90° angle is $s/r = \frac{1}{2}\pi r/r = \pi/2$.

Figure 6.3.3   A right angle subtends $\frac{1}{4}$ of the circumference.

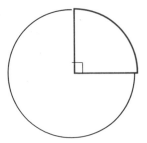

A 60° angle subtends $60/360 = 1/6$ of the circumference. Since $\frac{1}{6}(2\pi r) = \frac{1}{3}\pi r$, the radian measure of a 60° angle is $\frac{1}{3}\pi r/r = \pi/3$.

We can use these results, or the definition of radian measurement, to find the radian measure of the angles that will appear frequently in this chapter.

$$0° = 0 \text{ radians}$$
$$30° = \pi/6 \text{ radians}$$
$$45° = \pi/4 \text{ radians}$$
$$60° = \pi/3 \text{ radians}$$
$$90° = \pi/2 \text{ radians}$$
$$180° = \pi \text{ radians}$$
$$360° = 2\pi \text{ radians}$$

Occasionally, we want to convert a degree measurement to a radian measurement. Since $180° = \pi$ radians, $1° = \pi/180$ radians and

$$x° = \frac{\pi x}{180} \text{ radians} \tag{1}$$

Example 1

> To find the radian measure of a 40° angle, we write
>
> $$40° = \frac{\pi}{180} \cdot 40 \text{ radians}$$
>
> $$= \frac{2\pi}{9} \text{ radians}$$

Similarly, we can convert radian measurement to degree measurement. Since π radians $= 180°$, 1 radian $= 180/\pi$ degrees, and

$$x \text{ radians} = \frac{180x}{\pi} \text{ degrees} \tag{2}$$

Example 2

We will find the degree measurement of an angle whose radian measure is 4.

$$4 \text{ radians} = \left(\frac{180}{\pi} \cdot 4\right)^{\circ}$$

$$= \left(\frac{720}{\pi}\right)^{\circ}$$

remembering that $\pi \doteq 3.14$,

$$4 \text{ radians} \doteq \left(\frac{720}{3.14}\right)^{\circ} \doteq 229.3^{\circ}$$

Example 3

We will find the radian measurement of a 225° angle using two alternative methods.

A.   Using formula (1)

$$225^{\circ} = \left(\frac{\pi}{180} \cdot 225\right) \text{ radians}$$

$$= \frac{5\pi}{4} \text{ radians}$$

B. $225^{\circ} = 180^{\circ} + 45^{\circ} = \pi \text{ radians} + (\pi/4) \text{ radians}$
       $= (5\pi/4) \text{ radians.}$

Solution (B) illustrates a method we will frequently use. We will apply it to the problem of finding the radian measure of a 120° angle ($120^{\circ} = 2 \cdot 60^{\circ} = 2 \cdot (\pi/3)$ radians $= (2\pi/3)$ radians). We could also think of 120° as $4 \cdot 30^{\circ} = 4 \cdot (\pi/6)$ radians $= (2\pi/3)$ radians, or $120^{\circ} = 90^{\circ} + 30^{\circ} = (\pi/2)$ radians $+ (\pi/6)$ radians $= (2\pi/3)$ radians.

The same method can be used to convert radian measure to degree measure.

Example 4

We will find the degree measure of an angle whose radian measure is $7\pi/6$ using two alternative methods.

A.   Using the formula,

$$\frac{7\pi}{6} \text{ radians} = \left(\frac{180}{\pi} \cdot \frac{7\pi}{6}\right) = 210^{\circ}$$

B. $7\pi/6 \text{ radians} = 6\pi/6 \text{ radians} + \pi/6 \text{ radians}$
       $= \pi \text{ radians} + \pi/6 \text{ radians} = 180^{\circ} + 30^{\circ} = 210^{\circ}$

Most of the problems we will see that call for converting degrees to radians or radians to degrees will lend themselves to the type of solution shown in part (B) of Examples 3 and 4. We suggest you study these solutions carefully.

What are the advantages of radian measurement? There are formulas in calculus that are very simple if we use radian measurement but more complicated if we use angular measurement. Of course, this is not very satisfying to a student who has never seen the formulas in calculus! Fortunately, there are examples of this that do not require calculus, and we present two of them.

A. *The Area of a Circular Sector* Suppose we wish to find the area of circular sector *AOB* shown in Figure 6.3.4. Before deriving a formula for this area we will look at an example. Suppose the radius of the circle is 5, and angle *AOB* measures 40°. Since 40/360 = 1/9, the area we are looking for is 1/9 of the area of the entire circle. Since the area of the circle is 25π, the area of our sector is

$$\frac{1}{9} \cdot 25\pi = \frac{25\pi}{9}$$

Figure 6.3.4

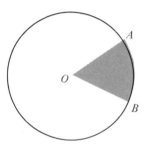

Now we will derive a formula for the area of circular sector *AOB*, first using degree measure for angle *AOB*, then radian measure. If the radius of the circle is *r* and angle *AOB* measures $t°$, the sector is $t/360$ of the entire circle. The formula for its area is $\pi r^2 t/360$.

If the measure of angle *AOB* is θ radians, then, since 360° = 2π radians, the sector is $\theta/2\pi$ of the entire circle, and the formula for its area is

$$\frac{\theta}{2\pi} \cdot \pi r^2 = \frac{1}{2}r^2\theta$$

The second formula for the area of a sector, Area = $\frac{1}{2}r^2\theta$, seems easier than the first, Area = $\pi r^2 t/360$. In later sections of this chapter, we will be working with a **unit circle,** a circle whose radius is 1. In trigonometry, we frequently work with unit circles and for these the area formulas are:

Area = $\pi t/360$ if t is the degree measure of angle *AOB*

Area = $\frac{1}{2}\theta$ if θ is the radian measure of angle *AOB*

B.   *Length of a Circular Arc*   Given a circle of radius r, and a central angle AOB, how can we find s, the length of arc AB? (Refer to Figure 6.3.5.)

Figure 6.3.5

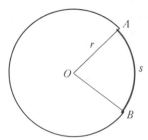

If angle AOB measures t degrees, s represents $t/360$ of the entire circumference. Since the circumference of a circle is $2\pi r$,

$$s = \frac{t}{360} \cdot 2\pi r = \frac{\pi r t}{180}$$

If angle AOB measures θ radians, the formula is particularly simple. By definition, $\theta = s/r$ so $s = r\theta$.

We will present one more example of how radian measurement simplifies a calculation, but it will appear in the last chapter of the book.

We now know what a radian is, what the radian measures of some common angles are, and some situations where radian measure simplifies calculations.

From now on all measures of angles will be in radians. The 30°-60°-90° and the 45° right triangles we spoke about in Section 1 are now $(\pi/6)$-$(\pi/3)$-$(\pi/2)$ and $\pi/4$ right triangles, respectively.

If x is any number in $(0, \pi/2)$, $\sin x$ is the sine of the angle whose radian measure is x. This rule defines a function, f, from $(0, \pi/2)$ to the real numbers given by

$$f(x) = \sin x$$

Similarly, we can use the five other trigonometric ratios to define five other functions from $(0, \pi/2)$ to the real numbers. In the remainder of this chapter we will enlarge the domains of these six trigonometric functions and study their properties.

Exercises 6.3

In Exercises 1–10, find the radian measure of the given angle. Whenever possible, use the method of Example 3B.

1. 135°	**2.** 315°	**3.** 15°	**4.** 250°
5. 225°	**6.** 330°	**7.** 23°	**8.** 75°
9. 210°	**10.** 120°		

In Exercises 11–20, find degree measurements for angles with the given radian measurements. Whenever possible, use the method of Example 4B.

11. 5   **12.** $\dfrac{3\pi}{4}$ **13.** 1   **14.** $\dfrac{7\pi}{4}$

15. $\dfrac{3\pi}{2}$ **16.** $\dfrac{\pi}{12}$ **17.** $\dfrac{5\pi}{4}$ **18.** $\dfrac{5\pi}{12}$

19. $\dfrac{2\pi}{3}$ **20.** $\dfrac{5\pi}{3}$

21. The central angle of a circular sector is $(\pi/3)$ radians. If the radius of the circle is 5, what is the area of the sector?

22. The area of a circular sector of a circle of radius 4 is 2π. What is the radian measure of the central angle of the sector?

23. Find the length of the circular arc subtended by an angle of $\pi/6$ in a circle of radius 4.

24. An angle of radian measure $\pi/3$ subtends an arc 6 units long. What is the radius of the circle?

 In Exercises 25–30, graph each equation for $0 < x < \pi/2$. Use the table of trigonometric values at the end of the book or a hand calculator.

25. $y = \sin x$ **26.** $y = \cos x$ **27.** $y = \tan x$

28. $y = \cot x$ **29.** $y = \sec x$ **30.** $y = \csc x$

SECTION 6.4

EXTENDING THE TRIGONOMETRIC FUNCTIONS

We have defined the trigonometric functions using ratios of lengths in right triangles. Using the terminology of Chapter 4, they are functions whose domains are the open interval $(0, \pi/2)$; i.e., the set of real numbers strictly between 0 and $\pi/2$. These are precisely the numbers that are radian measures of acute angles in right triangles.

We would like to enlarge the domains of the trigonometric functions. In the case of sine and cosine, for example, we would like the domain to include *all* real numbers, not just those between 0 and $\pi/2$.

To motivate our new definition, and to show why we would like to define $\sin t$ for all values of t, we will consider an example.

Suppose a particle is moving in a circle with center at the origin as in Figure 6.4.1. Circular motion occurs frequently in nature. For example, satellites revolving around the earth have approximately circular orbits.

Figure 6.4.1

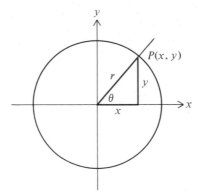

If the position of the particle is as shown in Figure 6.4.1, we can use the trigonometry of Section 6.2 to write $\cos \theta = x/r$; that is,

$$x = r \cos \theta \tag{1}$$

Similarly, $\sin \theta = y/r$; that is,

$$y = r \sin \theta \tag{2}$$

The importance of these equations is that if we know r and can measure θ, which is frequently the case, equations (1) and (2) tell us the position of the particle.

Suppose, however, that the position of the particle is as shown in Figure 6.4.2. Notice that θ is greater than $\pi/2$. It cannot appear in a right triangle, and $\sin \theta$ and $\cos \theta$ are not defined. We would still like equations such as (1) and (2) to give us the position of P as a function of θ.

Figure 6.4.2

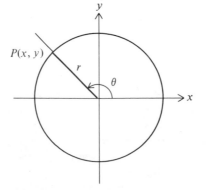

We will define $\sin \theta$ and $\cos \theta$ in such a way that $x = r \cos \theta$ and $y = r \sin \theta$ even when θ is not a number in $(0, \pi/2)$.

We first need to introduce some notation that we will use throughout the rest of the chapter.

An angle is determined by two rays sharing a common origin, called the **vertex.** Imagine that two rays initially coincide along the positive x-axis. Let one of the rays, called the **initial ray,** remain fixed and let the other ray, called the **terminal ray,** rotate in a counterclockwise direction through one-fourth a revolution. The final position of the rays is shown in Figure 6.4.3.

Figure 6.4.3

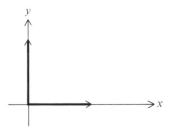

Now suppose the terminal ray had rotated in a *clockwise* direction through three-fourths a revolution or in a *counterclockwise* direction through $1\frac{1}{4}$ revolutions. The final positions of the two rays in these cases are also shown in Figure 6.4.3.

Observe that the same angle has been produced by different motions! It is often important to indicate the motion that produced an angle and we do this by assigning different measures to the same angle.

Returning to our example, we assign a measure of $\pi/2$ to the angle if it is produced by the first motion, $-3\pi/2$ if it is produced by the second motion, and $5\pi/2 \ [= 2\pi + (\pi/2)]$ if it is produced by the third motion.

We are using the following convention:

> If an angle is formed by a *counterclockwise* motion of the terminal ray, we assign it a **positive** measure. If the motion of the terminal ray is *clockwise,* we assign the angle a **negative** measure.

Example 1

The measures assigned to the angles determined by the motions shown in Figure 6.4.4 are

A. $\dfrac{3\pi}{2}$ B. $-\dfrac{\pi}{2}$ C. $2\pi + \dfrac{3\pi}{2} = \dfrac{7\pi}{2}$ D. $-2\pi + \dfrac{-\pi}{2} = -\dfrac{5\pi}{2}$

In (C) and (D), we used the fact that each revolution of the terminal ray changes the measure of the angle by 2π.

Figure 6.4.4

A.

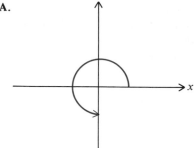

B.

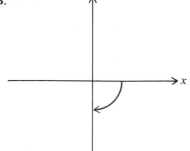

C.

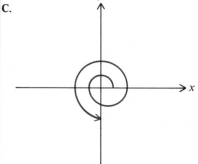

D.

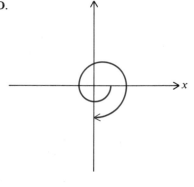

Example 2

Let us find all the measures we can assign to the angle in Figure 6.4.5.

Figure 6.4.5

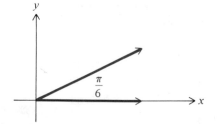

Certainly, one answer is $\pi/6$. But the same angle can be obtained by other counterclockwise rotations. In particular,

$$\frac{\pi}{6}, \frac{\pi}{6} + 2\pi, \frac{\pi}{6} + 4\pi, \ldots$$

Also, these angles could be produced by clockwise rotations.

$$-\frac{11\pi}{6}, \quad -\frac{11\pi}{6} - 2\pi, \quad -\frac{11\pi}{6} - 4\pi, \ldots$$

We can represent all these numbers with the notation

$$\frac{\pi}{6} + 2n\pi \qquad n = 0, \pm1, \pm2, \pm3, \ldots$$

We will be discussing angles frequently in this chapter, and it will be helpful to have a uniform way to draw them. An angle is said to be in **standard position** if its initial ray lies along the positive x-axis. All the angles we have drawn in our examples are in standard position. Figure 6.4.6 shows an angle that is not drawn in standard position.

Figure 6.4.6

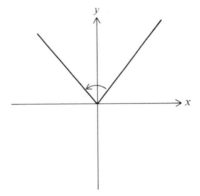

We now define $\sin \theta$ and $\cos \theta$ for all real numbers, θ. Let θ be any real number. Construct an angle with measure θ in standard position, as shown in Figure 6.4.7. Construct a circle of radius r, centered at O.

Figure 6.4.7

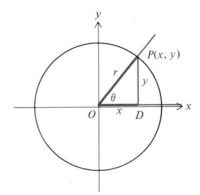

Suppose that the terminal ray intersects the circle at the point $P(x, y)$. Then we define

$$\sin \theta = \frac{y}{r}$$

$$\cos \theta = \frac{x}{r}$$

If we look at Figure 6.4.7, we notice that r is the length of the hypotenuse of $\triangle POD$ and y is the length of the side opposite angle θ. Our new formula, $\sin \theta = y/r$, is just an equivalent way of writing $\sin \theta = $ opposite/hypotenuse. For values of θ between 0 and $\pi/2$, the two formulas say exactly the same thing. The formulas differ only when θ is not between 0 and $\pi/2$. In this case, θ cannot be the measure of an acute angle in a right triangle and we have no opposite or hypotenuse to measure. We cannot apply the first definition. Our second definition, however, will make sense for any value of θ.

Example 3

We will use our definitions to find $\sin (\pi)$ and $\cos (\pi)$. In Figure 6.4.8, we have drawn a circle of radius 2. The terminal ray intersects the circle at $(-2, 0)$. Applying the formula,

$$\sin (\pi) = \frac{y}{r} = \frac{0}{2} = 0 \qquad \cos (\pi) = \frac{x}{r} = \frac{-2}{2} = -1$$

Figure 6.4.8

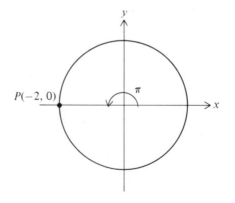

Our definitions allow us to use a circle of any radius to compute values of sine and cosine. Figure 6.4.9 shows a first quadrant angle of measure θ with two different circles centered at its vertex. If we use the first diagram, we find

$$\sin \theta = \frac{y_1}{r_1} \text{ and } \cos \theta = \frac{x_1}{r_1}$$

If we use the second diagram,

$$\sin \theta = \frac{y_2}{r_2} \text{ and } \cos \theta = \frac{x_2}{r_2}$$

However, $\triangle P_1OD_1 \sim \triangle P_2OD_2$, since each triangle has a right angle and an angle of measure θ. Since the triangles are similar, the lengths of corresponding sides are proportional, and

$$\frac{y_1}{r_1} = \frac{y_2}{r_2} \text{ and } \frac{x_1}{r_1} = \frac{x_2}{r_2}$$

This assures us that any choice of r will give us the same value for $\sin \theta$ and $\cos \theta$. Similar arguments can be used to verify this result when θ is not in $(0, \pi/2)$.

Figure 6.4.9

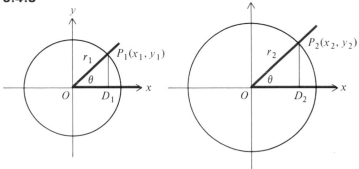

There is one particular value of r that will simplify our computations. It is $r = 1$, and a circle whose radius is 1 is called a **unit circle.**

If the terminal ray of angle θ crosses the unit circle at $P(x, y)$, then

$$\sin \theta = \frac{y}{r} = \frac{y}{1} = y$$

Similarly,

$$\cos \theta = \frac{x}{r} = \frac{x}{1} = x$$

$$\sin \theta = y \tag{3}$$

$$\cos \theta = x \quad \text{(See Figure 6.4.10 on next page.)} \tag{4}$$

The formulas tell us that to find sine or cosine of any number, θ, construct an angle in standard position whose measure is θ. Find the point, P, where the terminal ray of the angle crosses the unit circle. The x-coordinate of P is the cosine of θ and the y-coordinate of P is the sine of θ.

Since the choice of a unit circle makes the computation of sine and cosine easier, we will use unit circles and formulas (3) and (4) to compute sine and cosine throughout the rest of this section.

Figure 6.4.10 $\sin \theta = y$ $\cos \theta = x$

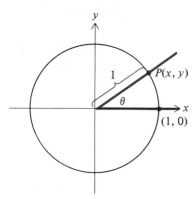

| Example 4 |

To find $\sin (3\pi/2)$ and $\cos (3\pi/2)$, we have drawn an angle with measure $(3\pi/2)$. (See Figure 6.4.11.) Its terminal ray intersects the unit circle at $(0, -1)$, so $\sin (3\pi/2) = -1$, $\cos (3\pi/2) = 0$.

Figure 6.4.11

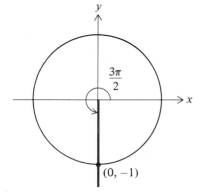

| Example 5 |

To find $\sin (-\pi)$, notice that the terminal ray of an angle of measure $-\pi$ intersects the unit circle at $(-1, 0)$. (See Figure 6.4.12.) Therefore, $\sin (-\pi) = 0$, $\cos (-\pi) = -1$.

Figure 6.4.12

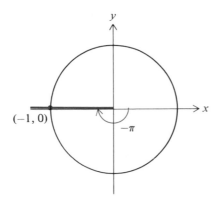

$(-1, 0)$

$-\pi$

Example 6

Suppose the measure of an angle is $\pi/6$. Where does the terminal ray of the angle intersect the unit circle? We recall that $\sin \pi/6 = 1/2$ and $\cos \pi/6 = \sqrt{3}/2$. Since the x- and y-coordinates of the intersection give us $\cos \theta$ and $\sin \theta$, respectively, the terminal ray intersects the unit circle (see Figure 6.4.13) at $\left(\dfrac{\sqrt{3}}{2}, \dfrac{1}{2}\right)$

Figure 6.4.13

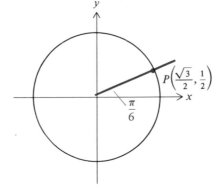

$P\left(\dfrac{\sqrt{3}}{2}, \dfrac{1}{2}\right)$

$\dfrac{\pi}{6}$

Having defined sine and cosine, we define the remaining trigonometric functions as follows: If the terminal ray of an angle of measure θ crosses the unit circle at $P(x, y)$, then

$$\tan \theta = \frac{\sin \theta}{\cos \theta} = \frac{y}{x} \qquad \sec \theta = \frac{1}{\cos \theta} = \frac{1}{x}$$

$$\cot \theta = \frac{\cos \theta}{\sin \theta} = \frac{x}{y} \qquad \csc \theta = \frac{1}{\sin \theta} = \frac{1}{y}$$

Notice that, although sine and cosine are defined for all values of θ, there are values of θ for which the other trigonometric functions are undefined.

Since

$$\tan \theta = \frac{\sin \theta}{\cos \theta} \quad \text{and} \quad \sec \theta = \frac{1}{\cos \theta}$$

$\tan \theta$ and $\sec \theta$ are not defined when $\cos \theta = 0$. Now, $\cos \theta = 0$ when

$$\theta = \frac{\pi}{2} + n\pi \qquad n = 0, \pm 1, \pm 2, \ldots$$

and for these values of θ, $\tan \theta$ and $\sec \theta$ are undefined. For all other values of θ, $\cos \theta \neq 0$, and $\tan \theta$ and $\sec \theta$ are defined. Therefore, the domain of tangent (and secant) is the set of all real numbers

$$\theta \neq \frac{\pi}{2} + n\pi \qquad n = 0, \pm 1, \pm 2, \ldots$$

Similarly, since $\cot \theta = \cos \theta / \sin \theta$ and $\csc \theta = 1/\sin \theta$, $\cot \theta$ and $\csc \theta$ are not defined when $\sin \theta = 0$. This happens only when $\theta = n\pi$, $n = 0, \pm 1, \pm 2, \ldots$. Thus, the domain of cotangent (and cosecant) is the set of all real numbers $\theta \neq n\pi$, $n = 0, \pm 1, \pm 2, \ldots$.

Example 7

Suppose we wish to find the values of all trigonometric functions at $3\pi/2$. From Example 4, we know that $\sin 3\pi/2 = -1$ and $\cos 3\pi/2 = 0$. Therefore,

$$\tan \frac{3\pi}{2} = \frac{\sin \dfrac{3\pi}{2}}{\cos \dfrac{3\pi}{2}} \quad \text{is undefined}$$

$$\cot \frac{3\pi}{2} = \frac{\cos \dfrac{3\pi}{2}}{\sin \dfrac{3\pi}{2}} = 0$$

$$\sec \frac{3\pi}{2} = \frac{1}{\cos \dfrac{3\pi}{2}} \quad \text{is undefined}$$

$$\csc \frac{3\pi}{2} = \frac{1}{\sin \dfrac{3\pi}{2}} = -1$$

We end this section with an important observation about sine and cosine. If θ is the measure of a first quadrant angle, the terminal ray of the angle intersects the unit circle at a point $P(x, y)$ where both x and y are positive. Therefore, sin θ and cos θ are both positive. If θ is the measure of a second quadrant angle, the terminal ray of the angle intersects the unit circle at $P(x, y)$ where x is negative and y is positive. Therefore, sin θ is positive but cos θ is negative. If θ measures a third quadrant angle, sin θ and cos θ are both negative since any point on the unit circle in the third quadrant will have negative x- and y-coordinates. Finally, if θ measures a fourth quadrant angle, sin θ is negative while cos θ is positive.

We summarize these results in the following table:

| | *Quadrant* | | | |
	I	*II*	*III*	*IV*
sin	+	+	−	−
cos	+	−	−	+

When working with trigonometric functions, it is important to know when they are positive and when they are negative. In particular, we will need this information in the next section.

Exercises 6.4

What measures should be assigned to the angles determined by the following motions?

1.

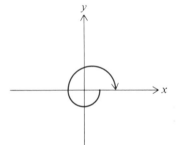

2.

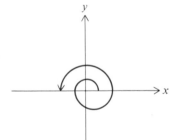

3.

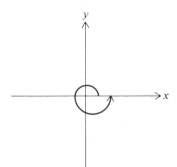

4.

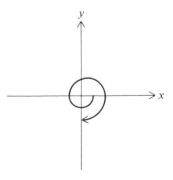

5.

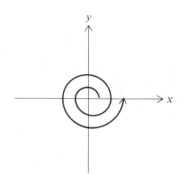

6.

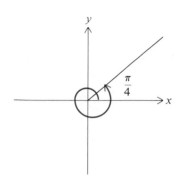

7. Find all possible measures we can assign to an angle with measure $\frac{\pi}{4}$.

8. Find all possible measures we can assign to an angle with measure $\frac{5\pi}{6}$.

In Exercises 9–18, find the coordinates of the point where the terminal ray of the angle with given measure intersects the unit circle.

9. 0

10. $\frac{\pi}{2}$

11. π

12. $\frac{3\pi}{2}$

13. 2π

14. $-\frac{\pi}{2}$

15. $-\pi$

16. $-\frac{3\pi}{2}$

17. -2π

18. $-\frac{5\pi}{2}$

In Exercises 19–28, use your answers to Exercises 9–18 to evaluate the following:

19. $\cos 0$

20. $\tan \frac{\pi}{2}$

21. $\sec \pi$

22. $\sin \frac{3\pi}{2}$

23. $\csc 2\pi$

24. $\tan \left(-\frac{\pi}{2}\right)$

25. $\cos (-\pi)$

26. $\sin \left(-\frac{3\pi}{2}\right)$

27. $\cos (-2\pi)$

28. $\cot \left(-\frac{5\pi}{2}\right)$

29. The table below is partially completed. It gives the signs of sine and cosine in the four quadrants. Use this information together with the definitions of the other trigonometric functions to complete the table:

	Quadrant			
	I	*II*	*III*	*IV*
sine	+	+	−	−
cosine	+	−	−	+
tangent				
cotangent				
secant				
cosecant				

In Exercises 30–38, decide if each of the given expressions represent a positive or negative number. (Do not try to evaluate the expression. Just decide if it is positive or negative.)

30. $\sin\left(\dfrac{4\pi}{3}\right)$ 　　　　 **31.** $\cos\left(\dfrac{4\pi}{3}\right)$ 　　　　 **32.** $\sec\left(-\dfrac{\pi}{4}\right)$

33. $\tan\left(\dfrac{5\pi}{6}\right)$ 　　　　 **34.** $\tan\left(-\dfrac{5\pi}{6}\right)$ 　　　　 **35.** $\sin\left(\dfrac{2\pi}{3}\right)$

36. $\cos\left(-\dfrac{\pi}{9}\right)$ 　　　　 **37.** $\sin(3)$ 　　　　 **38.** $\cos(4)$

39. A particle moves in a circular orbit in a counterclockwise direction with constant angular velocity. The radius of the circle is 100, and the angular velocity is $(3\pi/2)$ radians/minute. Initially, the particle is at $(100, 0)$. Find the coordinates of the particle:

　a. After 5 minutes 　　　　 **b.** After 1 hour
　c. After 20 seconds 　　　　 **d.** After 10 seconds

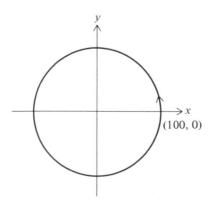

EVALUATING THE TRIGONOMETRIC FUNCTIONS

By now we have a small list of numbers whose trigonometric functional values we know without using a table. This list includes $\pi/6$, $\pi/4$, and $\pi/3$, numbers whose trigonometric functional values we have committed to memory (see Section 6.2), and numbers such as 0, $\pi/2$, π, and $3\pi/2$, whose trigonometric functional values we learned in the previous section.

The purpose of this section is to enlarge our list and learn to evaluate trigonometric functions of other numbers. Our discussion makes use of symmetry, studied in Chapter 4. Also, you may want to reread Section 4.6.2 before going on.

Before discussing the general case, we will look at an example. In Figure 6.5.1, an angle whose measure is $\pi/6$ is drawn in standard position. We know that it intersects the unit circle at the point $P_1\left(\dfrac{\sqrt{3}}{2}, \dfrac{1}{2}\right)$ since

$$\cos\frac{\pi}{6} = \frac{\sqrt{3}}{2} \quad \text{and} \quad \sin\frac{\pi}{6} = \frac{1}{2}$$

In Figure 6.5.2 we have drawn OP_2, the reflection of OP_1 in the y-axis. Since angle MOP_2 is congruent to angle LOP_1, it measures $\pi/6$. Therefore, a measure of angle LOP_2 is $\pi - (\pi/6) = 5\pi/6$.

Figure 6.5.1   **Figure 6.5.2**

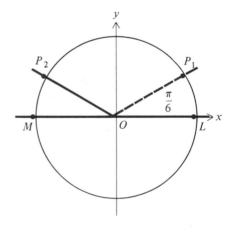

To evaluate the trigonometric functions at $5\pi/6$ we need to know the coordinates of P_2. Here again we make use of the fact that P_2 is the reflection of P_1 in the y-axis. Since the coordinates of P_1 are

$$\left(\frac{\sqrt{3}}{2}, \frac{1}{2}\right)$$

the coordinates of P_2 are

$$\left(-\frac{\sqrt{3}}{2}, \frac{1}{2}\right)$$

Since we know the coordinates of P_2, we know the values of the trigonometric functions at $5\pi/6$:

$$\sin \frac{5\pi}{6} = \frac{1}{2} \qquad \cot \frac{5\pi}{6} = \frac{-3}{\sqrt{3}} = -\sqrt{3}$$

$$\cos \frac{5\pi}{6} = \frac{-\sqrt{3}}{2} \qquad \sec \frac{5\pi}{6} = \frac{-2}{\sqrt{3}} = \frac{-2\sqrt{3}}{3}$$

$$\tan \frac{5\pi}{6} = \frac{-\sqrt{3}}{3} \qquad \csc \frac{5\pi}{5} = 2$$

Notice that the trigonometric functions evaluated at $5\pi/6$ and $\pi/6$ are closely related. They differ by, at most, a negative sign. These negative signs should not surprise us. An angle whose measure is $5\pi/6$ is a second quadrant angle, and cosine and tangent are both negative in the second quadrant.

It would be helpful to characterize angles with measures such as $\pi/6$ and $5\pi/6$, whose trigonometric functional values are related. The next definition helps us do that.

Definition

> The reference angle of an angle A, in standard position, is the smallest positive acute angle determined by the terminal ray of angle A and the x-axis.

Example 1

We will find measure of reference angles for angles of various measure as shown in Figure 6.5.3. The measure of the reference angle is shown in the cases where it differs from the measure of the angle.

Figure 6.5.3

(a) $\dfrac{2\pi}{3}$

(b) $\dfrac{\pi}{6}$

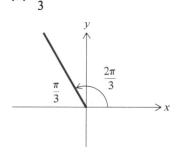

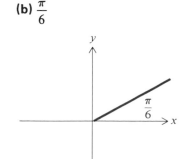

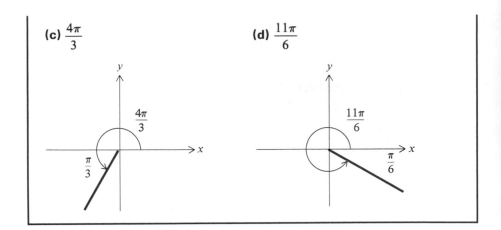

Prior to Example 1, we evaluated trigonometric functions at $\theta = 5\pi/6$. Notice that an angle with measure $5\pi/6$ has a reference angle with measure $\pi/6$.

Figure 6.5.4(a) shows OP_3, the reflection of OP_2 in the x-axis. The measure of angle LOP_3 is $\pi/6$ greater than π, or $7\pi/6$. Notice that the reference angle of angle LOP_3 is an angle with measure $\pi/6$.

Figure 6.5.4

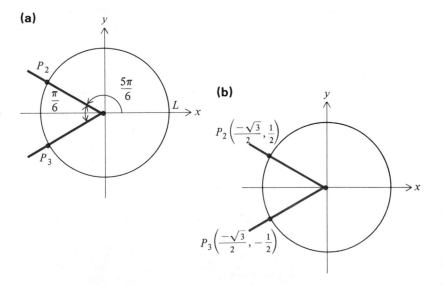

Using properties of reflection, the coordinates of P_3 are

$$\left(-\frac{\sqrt{3}}{2}, -\frac{1}{2}\right)$$

as shown in part (b) of Figure 6.5.4. Applying the definitions of sine and cosine,

$$\sin \frac{7\pi}{6} = -\frac{1}{2} \quad \text{and} \quad \cos \frac{7\pi}{6} = -\frac{\sqrt{3}}{2}$$

Let us examine the results we have obtained for $\pi/6$, $5\pi/6$, and $7\pi/6$.

	$\dfrac{\pi}{6}$	$\dfrac{5\pi}{6}$	$\dfrac{7\pi}{6}$
Sin	$\dfrac{1}{2}$	$\dfrac{1}{2}$	$-\dfrac{1}{2}$
Cos	$\dfrac{\sqrt{3}}{2}$	$\dfrac{-\sqrt{3}}{2}$	$\dfrac{-\sqrt{3}}{2}$
Tan	$\dfrac{\sqrt{3}}{3}$	$\dfrac{-\sqrt{3}}{3}$	$\dfrac{\sqrt{3}}{3}$

We chose a reference angle with measure $\pi/6$ for our examples only because we are familiar with the values of trigonometric functions at $\pi/6$. We would have obtained similar results for any reference angle, not just one with measure $\pi/6$.

Looking at the rows of our table we are led to the following conclusions:

A. If A is an angle with measure θ, and α is the measure of the reference angle of A, then either $\sin \theta = \sin \alpha$ or $\sin \theta = -\sin \alpha$.
B. We can determine if $\sin \theta = \sin \alpha$ or $\sin \theta = -\sin \alpha$ by observing in which quadrant the terminal ray of angle A lies.

Statements analogous to (A) and (B) can be made for the other trigonometric functions.

Our discussion shows that if we know the value of a trigonometric function on $[0, \pi/2]$, we can find the value of the trigonometric function at any number where the trigonometric function is defined.

We will illustrate these ideas in the following examples.

Example 2

To find $\cos (7\pi/4)$ we sketch in Figure 6.5.5 an angle with measure $7\pi/4$. It is a fourth quadrant angle and its reference angle measures $\pi/4$. Since the cosine is positive in the fourth quadrant,

$$\cos \frac{7\pi}{4} = \cos \frac{\pi}{4} = \frac{\sqrt{2}}{2}$$

Figure 6.5.5

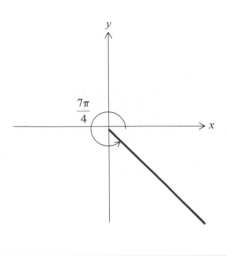

Example 3

To find $\sin 7\pi/6$ we sketch in Figure 6.5.6 an angle with measure $7\pi/6$. It is a third quadrant angle, and the sine is negative in the third quadrant. The reference angle measures $\pi/6$, so

$$\sin \frac{7\pi}{6} = -\sin \frac{\pi}{6} = -\frac{1}{2}$$

Figure 6.5.6

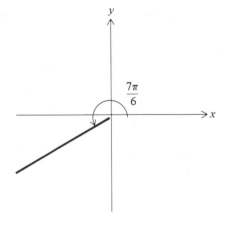

Example 4

To find $\sin 5$, we notice in Figure 6.5.7 that $3\pi/2 \doteq 4.711$ and $2\pi \doteq 6.281$, so an angle with measure 5 is a fourth quadrant angle with reference angle measuring approximately 1.281. Using the table, $\sin 1.281 \doteq 0.958$. Since sine is positive in the fourth quadrant, $\sin 5 \doteq 0.958$.

Figure 6.5.7

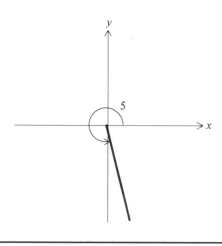

Example 5	To find $\sin 13\pi/6$, we notice that $13\pi/6 = 2\pi + \pi/6$. The terminal ray of an angle with measure $13\pi/6$ is in the same position as the terminal ray of an angle with measure $\pi/6$. Therefore,

$$\sin \frac{13\pi}{6} = \sin \frac{\pi}{6} = \frac{1}{2}$$

(See Figure 6.5.8.)

Figure 6.5.8

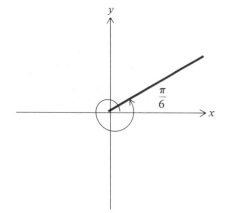

Example 5 illustrates an important fact about the sine function: For any number, t, $\sin(t + 2\pi) = \sin t$. Since $t + 4\pi = (t + 2\pi) + 2\pi$, it follows that $\sin(t + 4\pi) = \sin(t + 2\pi) = \sin t$. By repeating this process, we see that the numbers t, $t + 2\pi$, $t + 4\pi$, $t + 6\pi$, etc., all have the same sines.

The same argument applies if we subtract multiples of 2π. (Draw an angle with measure

$$\frac{\pi}{6} - 2\pi = -\frac{11\pi}{6}$$

and it will have the same terminal ray as an angle with measure $\pi/6$.) In general, t, $t - 2\pi$, $t - 4\pi$, $t - 6\pi$, etc., all have the same sines. A convenient notation for numbers of the form t, $t + 2\pi$, $t - 2\pi$, $t + 4\pi$, $t - 4\pi$, etc., is $t + 2n\pi$, $n = 0, \pm 1, \pm 2, \ldots$.

Although our illustration used the sine function, analogous results hold for all the trigonometric functions. We will use these results in the next examples.

Example 6

We will find all numbers, t, such that $\sin t = \frac{1}{2}$.

We first notice that sine is positive in the first and second quadrants. Therefore, we are looking for measures of first and second quadrant angles whose sines are $\frac{1}{2}$. In the first quadrant, $\sin\frac{\pi}{6} = \frac{1}{2}$. If the measure of any other angle has sine equal to $\frac{1}{2}$, that angle must have a reference angle with measure $\frac{\pi}{6}$. In the second quadrant, this angle has measure $\frac{5\pi}{6}$, so $\sin\frac{5\pi}{6} = \frac{1}{2}$. Finally, we use the discussion prior to this example to say that

$$\left.\begin{array}{c} \dfrac{\pi}{6} + 2n\pi \\[2ex] \dfrac{5\pi}{6} + 2n\pi \end{array}\right\} n = 0, \pm 1, \pm 2, \ldots$$

are the numbers that satisfy $\sin t = \frac{1}{2}$.

Example 7

We will solve the equation

$$\cos t = -\frac{\sqrt{2}}{2}$$

We know that cosine is negative in the second and third quadrants. Since

$$\cos\frac{\pi}{4} = \frac{\sqrt{2}}{2}$$

we are looking for measures of second- and third-quadrant angles whose reference angles measure $\frac{\pi}{4}$. These measures are $\frac{3\pi}{4}$ and $\frac{5\pi}{4}$, so

$$\cos\frac{3\pi}{4} = -\frac{\sqrt{2}}{2} \quad \text{and} \quad \cos\frac{5\pi}{4} = -\frac{\sqrt{2}}{2}$$

The numbers, t, satisfying

$$\cos t = -\frac{\sqrt{2}}{2}$$

are numbers of the form

$$\left.\begin{array}{l} \dfrac{3\pi}{4} + 2n\pi \\[2mm] \dfrac{5\pi}{4} + 2n\pi \end{array}\right\} n = 0, \pm 1, \pm 2, \ldots$$

Exercises 6.5

In Exercises 1–15, find the measures of reference angles for the angles with given measure:

1. $\dfrac{5\pi}{3}$ **2.** $\dfrac{7\pi}{3}$ **3.** $\dfrac{3\pi}{4}$ **4.** $\dfrac{\pi}{4}$

5. $-\dfrac{\pi}{6}$ **6.** $\dfrac{7\pi}{4}$ **7.** $\dfrac{5\pi}{6}$ **8.** $\dfrac{5\pi}{4}$

9. $\dfrac{7\pi}{6}$ **10.** 3   **11.** $\dfrac{13\pi}{12}$ **12.** $-\dfrac{4\pi}{3}$

13. $-\dfrac{7\pi}{3}$ **14.** $-\dfrac{2\pi}{3}$ **15.** 4

In Exercises 16–30, use your answers to questions 1–15 to evaluate each expression.

16. $\cos\left(\dfrac{5\pi}{3}\right)$ **17.** $\sin\left(\dfrac{7\pi}{3}\right)$ **18.** $\tan\left(\dfrac{3\pi}{4}\right)$

19. $\sec\left(\dfrac{\pi}{4}\right)$ **20.** $\csc\left(-\dfrac{\pi}{6}\right)$ **21.** $\tan\left(\dfrac{7\pi}{4}\right)$

22. $\sin\left(\dfrac{5\pi}{6}\right)$ **23.** $\cos\left(\dfrac{5\pi}{4}\right)$ **24.** $\cos\left(\dfrac{7\pi}{6}\right)$

25. $\cot(3)$ **26.** $\cos\left(\dfrac{13\pi}{12}\right)$ **27.** $\tan\left(-\dfrac{4\pi}{3}\right)$

28. $\tan\left(-\dfrac{7\pi}{3}\right)$ **29.** $\sec\left(-\dfrac{2\pi}{3}\right)$ **30.** $\sin(4)$

In Exercises 31–39, find all solutions of the given equations:

31. $\sin x = \dfrac{1}{2}$ **32.** $\sin x = -\dfrac{1}{2}$

33. $\sin x + \cos x = 0$ **34.** $\tan^2 x = 1$
                                                        (*Note:* $\tan^2 x$ means $(\tan x)^2$.)

35. $\cos x = -\dfrac{1}{2}$ **36.** $\sec x = -1$

37. $\sin(x + 2) = -\dfrac{\sqrt{3}}{2}$

38. $\csc x = \dfrac{2}{\sqrt{3}}$

39. $\tan x^2 = 1$

 In Exercises 40–45, sketch the graphs of the given equations for $0 \le x \le 2\pi$.

40. $y = \sin x$ **41.** $y = \cos x$ **42.** $y = \tan x$

43. $y = \cot x$ **44.** $y = \sec x$ **45.** $y = \csc x$

46. A particle moves along a number line in such a way that its position, p, at any time, t, is given by $p = 3 \cos 2t$. (*Note:* This type of motion is an example of **simple harmonic motion**.)

     a. What is its position when $t = 0$?

     b. What is its position when $t = \dfrac{\pi}{12}$?

     c. What is its position when $t = 10\pi$?

     d. What is its position when $t = 5$?

SECTION 6.6

GRAPHS OF THE TRIGONOMETRIC FUNCTIONS

We have defined six trigonometric functions and have learned to evaluate them at different numbers. In this section, we consider their graphs.

We will begin with the sine function. One way to graph $y = \sin \theta$ is to plot a large number of points, $(\theta, \sin \theta)$, and see what smooth curve is suggested. Before plotting points, however, we will consider another approach to graphing the sine function. This approach uses the definition of the sine function and gives us information about the graph that might not be apparent if we had only plotted points.

Imagine a ray, OA, with initial point at the origin, rotating in a counter-clockwise direction, as in Figure 6.6.1. At each moment, OA determines an angle in standard position with measure θ, having OA as its terminal ray. (You can think of OA as something like a sweep-second hand of a clock, only moving counterclockwise.)

In Figure 6.6.2, we have superimposed a unit circle on our diagram. The ray OA intersects the unit circle at $P(x, y)$. Recall that $\sin \theta$ is by definition the y-coordinate of P.

Figure 6.6.3 shows different positions of ray OA in the first quadrant. In its initial position, when $\theta = 0$, the y-coordinate of P is 0. As θ increases from 0 to $\pi/2$, the y-coordinate of P increases from 0 to 1. Since the y-coordinate is $\sin \theta$, we see that as θ increases from 0 to $\pi/2$, $\sin \theta$ increases from 0 to 1.

We will continue looking at the y-coordinate of P (and therefore at $\sin \theta$) as OA moves through the quadrants, as illustrated in Figure 6.6.4.

Figure 6.6.1

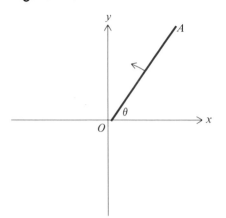

Figure 6.6.2

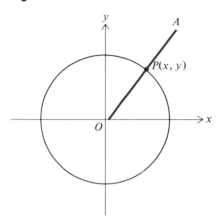

Figure 6.6.3

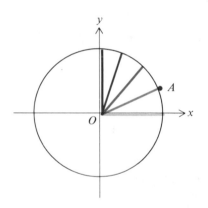

Figure 6.6.4

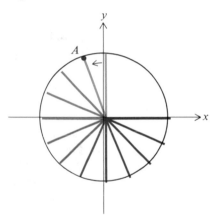

As OA moves through the second quadrant, θ increases from $\pi/2$ to π. The y-coordinate of P (and, therefore, $\sin \theta$) decreases from 1 to 0.

As OA moves through the third quadrant, θ increases from π to $3\pi/2$ and $\sin \theta$ decreases from 0 to -1.

As OA moves through the fourth quadrant, θ increases from $3\pi/2$ to 2π and $\sin \theta$ increases from -1 to 0.

In Figure 6.6.5 (next page), we have graphed the equation $y = \sin \theta$ for values of θ between 0 and 2π.

In the previous section, we noted that $\sin (t + 2n\pi) = \sin t$ if n is any integer. For example,

$$\sin \frac{13\pi}{6} = \sin \frac{\pi}{6} \qquad \text{since} \qquad \frac{13\pi}{6} = 2\pi + \frac{\pi}{6}$$

Figure 6.6.5 $y = \sin \theta$ $0 \leq \theta \leq 2\pi$

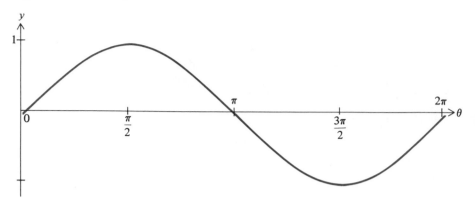

We will use this property to help us sketch the graph of $y = \sin \theta$ when $\theta < 0$ or $\theta > 2\pi$. It will be convenient to have the following definition:

Definition

> A function f is **periodic with period** p, if p is the smallest positive number such that $f(x + p) = f(x)$ for each x in D_f.

With this definition, we may say that the sine function is periodic with period 2π. (Although it may be clear that $\sin(x + 2\pi) = \sin x$, it is not clear that 2π is the smallest positive number with this property. We outline a proof of this fact in the exercises.)

Since sine is periodic with period 2π, the graph of the sine function between 2π and 4π or between -2π and 0 should look just like the graph of the sine function between 0 and 2π. Figure 6.6.6 shows the graph of $y = \sin \theta$, including the values of greater than 2π and less than 0. Notice that the graph shows the periodicity of the function.

Figure 6.6.6 $y = \sin \theta$

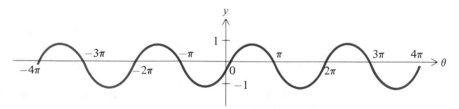

To sketch $y = \cos \theta$, we recall that $\cos \theta$ is the x-coordinate of the intersection of OA with the unit circle (see Figure 6.6.7). As θ varies from 0 to $\pi/2$, the

Figure 6.6.7

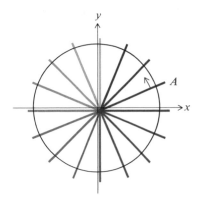

x-coordinate (which equals cos θ), decreases from 1 to 0. Still referring to Figure 6.6.7, if we follow *OA* as it moves through the quadrants, then we see the following:

As θ Varies From	x = cos θ *Varies From*
0 to $\pi/2$	1 to 0
$\pi/2$ to π	0 to -1
π to $3\pi/2$	-1 to 0
$3\pi/2$ to 2π	0 to 1

Cosine is also a periodic function with period 2π. As a result, if we know what the graph of the cosine function looks like between 0 and 2π, we know what it looks like everywhere. (See Figure 6.6.8.)

Figure 6.6.8   **y = cos θ**

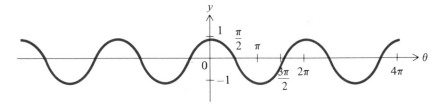

To help sketch the graph of $y = \tan \theta$, we use a geometric argument. Figure 6.6.9 shows a unit circle with a vertical tangent line drawn at $(1, 0)$. Corresponding to a number, θ, between 0 and $\pi/2$, we construct *OP* such that θ is the measure of angle *AOP*. Applying the trigonometry of right triangles that we studied in Section 6.2, $\tan \theta = AP/OA$. However, *OA* is 1 since *A* is a point on the unit circle. Therefore, $\tan \theta = AP$. We notice that $\tan 0 = 0$, and that, as θ increases from 0 to $\pi/2$, *AP*, which is $\tan \theta$, increases without bound. (See

Figure 6.6.9

Figure 6.6.10

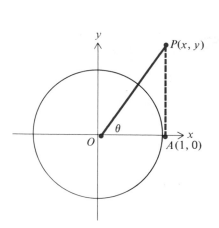

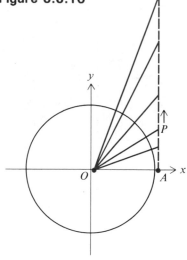

Figure 6.6.11 $y = \tan \theta$ $0 \le \theta < \pi/2$

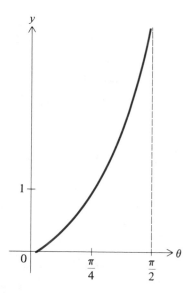

(Figure 6.6.10.) That is, if we let θ approach $\pi/2$, then the corresponding values of AP, and therefore, $\tan \theta$, increase without bound. In Figure 6.6.11, we have sketched the graph of $y = \tan \theta$ for $0 \le \theta < \pi/2$. Notice that $\tan \theta$ is not defined when $\theta = \pi/2$.

We use a similar argument to help sketch $y = \tan \theta$ for values of θ between $\pi/2$ and π. In Figure 6.6.12, we have constructed a vertical tangent line to the

Figure 6.6.12

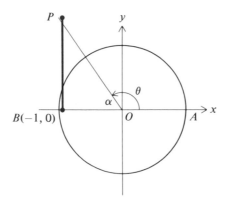

unit circle at $(-1, 0)$. The length of PB is the tangent of α, the measure of the reference of angle AOP.

Since tangent is negative in the second quadrant, $\tan \theta = -\tan \alpha = -PB$. As θ approaches π, PB decreases to 0. Therefore $\tan \theta (= -PB)$ increases to zero. When θ is close to $\pi/2$, PB is very large. Since $\tan \theta = -PB$, $\tan \theta$ is the negative of a very large number. As θ approaches $\pi/2$, PB increases without bound and $\tan \theta = -PB$ decreases without bound. Figure 6.6.13 shows the graph of $\tan \theta$ for values of θ between 0 and π.

Figure 6.6.13 $y = \tan \theta$ $0 \leq \theta \leq \pi$

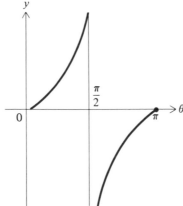

In the third quadrant, $\tan \theta$ is positive and the graph of $y = \tan \theta$ for $\theta \leq \theta < 3\pi/2$ is identical to the graph for $0 \leq \theta < \pi/2$. In the fourth quadrant, $\tan \theta$ is negative and the graph of $y = \tan \theta$ for $3\pi/2 < \theta \leq 2\pi$ is identical to the graph for $\pi/2 < \theta \leq \pi$. Figure 6.6.14 shows the graph of $y = \tan \theta$ for $0 \leq \theta \leq 2\pi$.

Figure 6.6.14 $y = \tan \theta$ $0 \le \theta \le 2\pi$

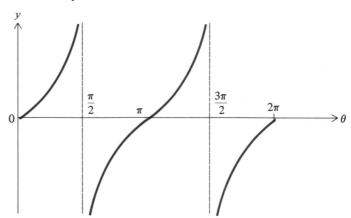

Comparing this graph with the sine or cosine graph, we see two differences.

A. The tangent graph is not a "smooth" curve. It has "breaks," called **discontinuities,** when $\theta = \pi/2$ and $3\pi/2$.

B. Sine and cosine are never greater than 1 or less than -1. By contrast, tangent becomes larger than any given positive number and smaller than any given negative number.

We see another difference if we continue the graph of tangent for values of θ greater than 2π or less than 0. Looking at Figure 6.6.15, we notice that tangent is periodic with period π, unlike sine and cosine, which are periodic with period 2π.

Figure 6.6.15 $y = \tan \theta$

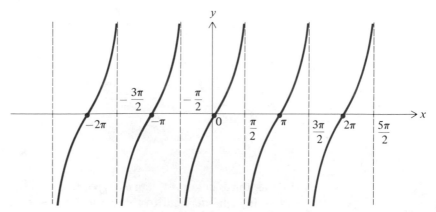

The graphs of the three remaining trigonometric functions (shown in Figure 6.6.16) are used less frequently. We will make two observations about our graphs.

Figure 6.6.16

(a) $y = \cot \theta$

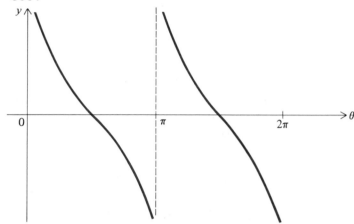

(b) $y = \sec \theta$

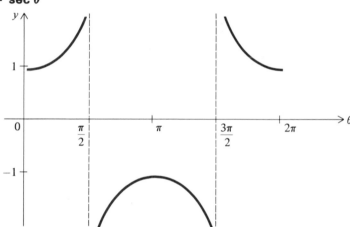

(c) $y = \csc \theta$

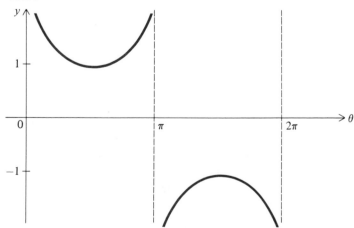

The first is that, if we draw the graph of $y = \sin \theta$ and shift the graph to the left by $\pi/2$ units, we have the graph of $y = \cos \theta$. A similar comment applies to the graphs of cosecant and secant. If we draw the graph of $y = \csc \theta$ and shift the graph to the left by $\pi/2$ units, we have the graph of $y = \sec \theta$. Since $\csc \theta = 1/\sin \theta$ and $\sec \theta = 1/\cos \theta$, it is not surprising that, if the graphs of sine and cosine are related by a translation, so are the graphs of cosecant and secant.

The second observation is that the graphs of cotangent, secant, and cosecant are determined by the reciprocal relationships they have with sine, cosine, and tangent. Figure 6.6.17 shows the graphs of cosine and secant in the first quadrant. Recall that $\sec \theta = 1/\cos \theta$. Since $\cos 0 = 1$, $\sec 0 = 1/1$. As θ approaches $\pi/2$, $\cos \theta$ gets small, approaching 0. What happens to a fraction, $1/x$, as the denominator approaches zero? When we let $x = 1/2$, $1/5$, $1/10$, or $1/100$, we see that $1/x = 2$, 5, 10, or 100, respectively. The fraction increases without bound. As θ increases to $\pi/2$, $\cos \theta$ decreases to 0 and $\sec \theta = 1/\cos \theta$ increases without bound. If we wanted to, we could sketch the entire secant graph by observing the cosine graph and seeing what happens to $1/\cos \theta$.

Figure 6.6.17

(a) y = cos θ

(b) y = sec θ

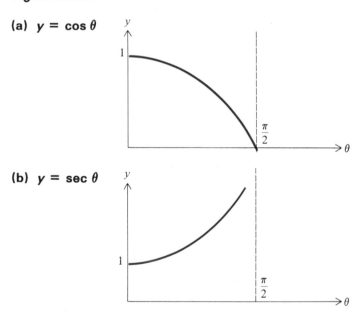

This section is a lengthy one and contains a large number of graphs. By far the most important for you to remember are the sine, cosine, and tangent graphs between 0 and 2π, which are shown in Figure 6.6.18 (a), (b), and (c), respectively. Although you should be familiar with the graphs of all the trigonometric functions, these graphs are so important that we suggest you memorize them.

Figure 6.6.18

(a) y = sin θ

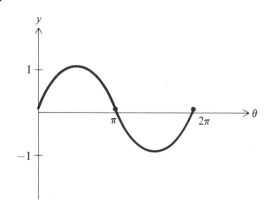

(b) y = cos θ

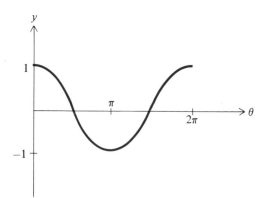

(c) y = tan θ

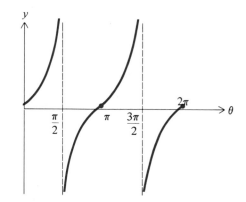

   To understand why they are important, try to answer the following questions without looking at the sketches:

A.   Where are sine, cosine, and tangent = 0?
B.   Where are they positive? Negative?

Table 6.6.1   **Properties of the Trigonometric Functions**

Trigonometric Function	Domain	Range	Period	Sketch Between 0 and 2π
Sine	All real numbers	$[-1, 1]$	2π	
Cosine	All real numbers	$[-1, 1]$	2π	
Tangent	Real numbers $\neq \pi/2 + n\pi$ $n = 0, \pm 1, \pm 2, \ldots$	All real numbers	π	
Cotangent	Real numbers $\neq n\pi$ $n = 0, \pm 1, \pm 2, \ldots$	All real numbers	π	
Secant	Real numbers $\neq \pi/2 + n\pi$ $n = 0, \pm 1, \pm 2, \ldots$	Real numbers ≥ 1 or ≤ -1	2π	
Cosecant	Real numbers $\neq n\pi$ $n = 0, \pm 1, \pm 2, \ldots$	Real numbers ≥ 1 or ≤ -1	2π	

C.   Where is tangent undefined?
D.   In what quadrants do sine and cosine have the same sign?

Remembering the sketches makes answering such questions as these much easier. Table 6.6.1 summarizes the work we have done in this section.

Exercises 6.6

In Exercises 1–13, sketch the graphs of each equation. (*Hint:* You should use the methods of Section 4.6.3.)

1. $y = 3 \sin x$

2. $y = 2 \cos x$

3. $y = \sin \left(x - \dfrac{\pi}{2} \right)$

4. $y = \tan \left(x + \dfrac{\pi}{2} \right)$

5. $y = -\tan x$

6. $y = 2 \sec x$

7. $y = -\cot x$

8. $y = \cos \left(x + \dfrac{\pi}{2} \right)$

9. $y = \sin (x + \pi)$

10. $y = -2 \sin (x - \pi)$

11. $y = -\sin \left(x - \dfrac{\pi}{2} \right)$

12. $y = 3 \csc x$

13. $y = \cot \left(x - \dfrac{3\pi}{2} \right)$

14.   Draw the graph of $y = \sin^2 x + \cos^2 x$. Plot points. (*Note:* $\sin^2 x$ means $(\sin x)^2$.)

15.   Draw the graph of $y = \sin 2x$.

16.   Draw the graph of $y = \sin \dfrac{1}{2}x$.

17.   Draw the graph of $y = \cos 3x$.

18.   Draw the graph of $y = \cos \dfrac{1}{3}x$.

19.   We saw that the graphs of sine and cosine are related by a translation, as are the graphs of cosecant and secant. How are the graphs of tangent and cotangent related?

20.   In this exercise, we show that the period of the sine function is 2π. (We know that $\sin (x + 2\pi) = \sin x$ for all numbers, x. We must show that 2π is the smallest number with this property.) Suppose p is a positive number less than or equal to 2π such that

$$\sin (x + p) = \sin x \text{ for all } x \tag{1}$$

a.   Let $x = 0$ in (1) to show that p must be either π or 2π.
b.   Let $x = \pi/6$ in (1) to show that $p = 2\pi$.

21. A particle moves on a number line in such a way that its position, p, at any time, t, is given by $p = 2 \cos t$.

   a. Where is the particle at $t = 0$, $t = \pi/2$, and $t = 11$?
   b. When is the particle moving from right to left?
   c. When is the particle moving from left to right?

SECTION 6.7   **IDENTITIES**

Exercise 14 in the previous section asked you to graph the equation $y = \sin^2 x + \cos^2 x$. You may have begun this problem by constructing a table of values such as Table 6.7.1.

Table 6.7.1

x	$\sin x$	$\cos x$	$\sin^2 x + \cos^2 x$
0	0	1	$0 + 1 = 1$
$\dfrac{\pi}{6}$	$\dfrac{1}{2}$	$\dfrac{\sqrt{3}}{2}$	$\dfrac{1}{4} + \dfrac{3}{4} = 1$
$\dfrac{\pi}{4}$	$\dfrac{\sqrt{2}}{2}$	$\dfrac{\sqrt{2}}{2}$	$\dfrac{1}{2} + \dfrac{1}{2} = 1$
$\dfrac{\pi}{3}$	$\dfrac{\sqrt{3}}{2}$	$\dfrac{1}{2}$	$\dfrac{3}{4} + \dfrac{1}{4} = 1$
$\dfrac{\pi}{2}$	1	0	$1 + 0 = 1$

Based on these observations, we might think that $\sin^2 x + \cos^2 x = 1$ for all values of x. In fact, we can prove this result easily. Figure 6.7.1 shows a unit circle. Construct an angle, in standard position, with measure x. Let its terminal ray intersect the circle at $P(a, b)$. Since the circle has radius 1, the distance

Figure 6.7.1

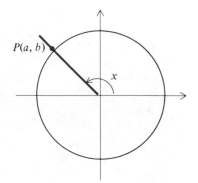

from $P(a, b)$ to the origin is 1. Using the distance formula, $\sqrt{a^2 + b^2} = 1$. Squaring both sides, $a^2 + b^2 = 1$. Since $a = \cos x$ and $b = \sin x$, rearranging terms we have

$$\sin^2 x + \cos^2 x = 1 \tag{1}$$

Formula (1) is an example of an identity, the subject of this section. An **identity** is a mathematical equality that is true for all values of independent variable(s) for which both sides are defined. In this section, we will discuss a number of trigonometric identities.

Starting with (1), we can easily obtain two more identities. Divide each term of (1) by $\cos^2 x$ to get

$$\tan^2 x + 1 = \sec^2 x \tag{2}$$

Divide each term of (1) by $\sin^2 x$ to get

$$1 + \cot^2 x = \csc^2 x \tag{3}$$

In identities (2) and (3), there are values of x for which the right and left sides are undefined. However, when both sides are defined, they are equal.

Example 1

> We will use identity (1) to find $\sin t$ if $\cos t = 1/3$ and $0 < t < \pi/2$. We know that $\sin t = \pm \sqrt{1 - \cos^2 t}$ from (1). Since $0 < t < \pi/2$, $\sin t > 0$ and we choose the positive square root.
>
> $$\sin t = \sqrt{1 - \frac{1}{9}} = \frac{\sqrt{8}}{9} = \frac{\sqrt{8}}{3} = \frac{2\sqrt{2}}{3}$$

We are going to develop a list of identities—(a long list)—but most of them are related to one particular identity, much as (2) and (3) are related to (1). To understand this new identity, consider the problem of finding $\sin 7\pi/12$.

Since $7\pi/12 = (\pi/4) + (\pi/3)$, we might think (incorrectly, it turns out) that $\sin 7\pi/12 = \sin \pi/4 + \sin \pi/3$. It is easy to see that this is false, since $\sin \pi/4 = \sqrt{2}/2$, $\sin \pi/3 = \sqrt{3}/2$, and if $\sin 7\pi/12 = \sin \pi/3 + \sin \pi/4$, it would be greater than 1, which is impossible.

Nevertheless, it is reasonable to think that $\sin 7\pi/12$ is related to the values of trigonometric functions at $\pi/4$ and $\pi/3$. What is the relationship? To phrase the question more generally, can we find a formula for $\sin (A + B)$ in terms of the values of trigonometric functions at A and B? We can, but the formula, which follows, is more complicated than you might have guessed.

$$\sin (A + B) = \sin A \cos B + \cos A \sin B \tag{4}$$

This result is one of our most important identities, and we will soon use it to derive other identities.

Our proof of this identity is not a traditional one but we chose it for two reasons.* First, although it is long, it is fairly easy to follow. Second, our proof uses a number of important results from trigonometry that you may not be familiar with, and this is a good time to introduce them. The results we need are these:

R1.   In triangle ABC, $c = a \cos B + b \cos A$. (See Figure 6.7.2.)
R2.   In triangle ABC, $[\sin C]/c = [\sin B]/b = [\sin A]/a$. This result is
      The Law of Sines.
R3.   In triangle ABC, $A + B + C = \pi$.
R4.   For any number, θ, $\sin(\pi - \theta) = \sin \theta$.

You may remember R3 as a theorem in plane geometry: The sum of the angles in a triangle is $180°$. Here are the proofs of the three other results.

R1.   *In triangle* ABC, c = a *cos* B + b *cos* A.   In Figure 6.7.2, $c = x + y$. From triangle BCD, $\cos B = x/a$ and, therefore, $x = a \cos B$. From triangle ACD, $\cos A = y/b$ and, therefore, $y = b \cos A$. Since $x = a \cos B$ and $y = b \cos A$, $x + y = c = a \cos B + b \cos A$. This result is sometimes called the **projection law for triangles.**

R2.   *The Law of Sines. In triangle* ABC, [*sin* C]/c = [*sin* B]/b = [*sin* A]/a. The area of a triangle is $\frac{1}{2}$ (base) (height). Since $\sin A = h/b$, $h = b \sin A$, and the area of triangle $ABC = \frac{1}{2}bc \sin A$. Looking at triangle BCD, $\sin B = (h/a)$, $h = a \sin B$, and the area of triangle ABC is $\frac{1}{2}ac \sin B$. If we draw an altitude from A to BC and consider BC to be the base of triangle ABC, we find the area of the triangle to be $\frac{1}{2}ab \sin C$. (See Figure 6.7.3.) However we compute the area, we should get the same number as our answer. So, $\frac{1}{2}ab \sin C = \frac{1}{2}ac \sin B = \frac{1}{2}bc \sin A$. Dividing all three terms by $\frac{1}{2}abc$ we get

$$\frac{\sin C}{c} = \frac{\sin B}{b} = \frac{\sin A}{a}$$

Figure 6.7.2   **Figure 6.7.3**

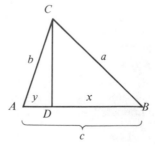   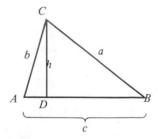

---

*An outline of this proof appeared in *The Mathematics Magazine,* vol. 35 (1962), p. 229.

R4. $sin\ (\pi - \theta) = sin\ \theta$.   In Figure 6.7.4, angle *MON* and its mirror image in the *y*-axis, angle *POQ*, are shown intersecting the unit circle.

Figure 6.7.4

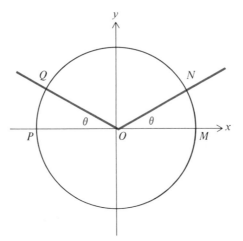

Angle *MOQ* has measure $\pi - \theta$. Since Q is the reflection of N in the *y*-axis, the *y*-coordinate of Q is the same as the *y*-coordinate of N. But the y coordinate of Q is $\sin (\pi - \theta)$ and the *y*-coordinate of N is $\sin \theta$. Therefore, $\sin (\pi - \theta) = \sin \theta$. Although our diagram shows θ as the measure of a first quadrant angle, our proof is valid for any value of θ.

Now that we have our preliminary results, we are ready to prove identity (4): $\sin (A + B) = \sin A \cos B + \cos A \sin B$.

We start with the projection theorem (R1)

$$C = a \cos B + b \cos A$$

Multiply the left side by $[\sin C]/c$, the first term of the right side by $[\sin A]/a$, and the second term of the right side by $[\sin B]/b$. By the law of sines, we are actually multiplying both sides of the equation by equal quantities. We get

$$\frac{\sin C}{c} \cdot c = \frac{\sin A}{a} \cdot a \cos B + \frac{\sin B}{b} \cdot b \cos A$$

Simplifying,

$$\sin C = \sin A \cos B + \cos A \sin B$$

Now $C = \pi - (A + B)$ by R3, since it is the third angle in a triangle whose other angles are A and B. Applying (R4),

$$\sin C = \sin (\pi - (A + B)) = \sin (A + B)$$

Therefore

$$\sin (A + B) = \sin A \cos B + \cos A \sin B$$

In our proof, A and B are measures of angles in triangles and, therefore, satisfy the conditions $A > 0$, $B > 0$, and $A + B < \pi$. Identity (4) is true for all values of A and B, and we indicate how to extend our proof in the exercises.

Let us return to the problem of finding $\sin (7\pi/12)$. Using identity (4),

$$\sin \frac{7\pi}{12} = \sin \left(\frac{\pi}{3} + \frac{\pi}{4}\right) = \sin \frac{\pi}{3} \cos \frac{\pi}{4} + \cos \frac{\pi}{3} \sin \frac{\pi}{4}$$

$$= \frac{\sqrt{3}}{2} \cdot \frac{\sqrt{2}}{2} + \frac{1}{2} \cdot \frac{\sqrt{2}}{2}$$

$$= \frac{\sqrt{6}}{4} + \frac{\sqrt{2}}{4}$$

$$= \frac{\sqrt{6} + \sqrt{2}}{4}$$

Many other identities follow from (4), but we will wait until the next section to derive them. In the exercises, you are asked to apply identities (1)–(4), but you are also asked to derive some new identities yourself. Try to resist the temptation to look ahead.

Exercises 6.7

In Exercises 1–6, use identities (1), (2), or (3) (page 241).

1. Find all possible values of $\cos t$ if $\sin t = 1/3$.
2. Find all possible values of $\tan x$ if $\sin x = 1/4$.
3. Find $\tan t$ if $\sec t = 4$ and t is the measure of a fourth quadrant angle.
4. Find $\cos t$ if $\sin t = 1/4$ and t is the measure of a second quadrant angle.
5. Find $\sin t$ if $\tan t = -3$ and t is the measure of a fourth quadrant angle.
6. Find $\csc t$ if $\cos t = 1/4$ and t is the measure of a first quadrant angle.
7. Find $\sin (5\pi/12)$ using identity (4).
8. Find $\cos (5\pi/12)$. (*Hint:* Use your answer to 7 and identity (1).)
9. Show that $\sin (\pi + t) = -\sin t$.
10. Show that $\sin \left(\dfrac{3\pi}{2} + t\right) = -\cos t$.

11. Express $\tan t$ in terms of $\sin t$.
12. Express $\tan t$ in terms of $\cos t$.
13. Express $\cos t$ in terms of $\cot t$.
14. Express $\sec t$ in terms of $\sin t$.

 15. The measures of the angles in a triangle are $\pi/5$, $\pi/5$, and $3\pi/5$. If the longest side is 10 inches long, find the lengths of the other sides accurate to the nearest 1/10 of an inch.

 16. The largest angle in a parallelogram measures $120°$. The longer diagonal of the parallelogram is 10 inches long and makes an angle of $20°$ with the base of the parallelogram. Find the dimensions of the parallelogram.

17. a. Draw angles of measure A and $-A$. (A can be any number.) What symmetry do the terminal rays possess?

b. If the terminal ray of A crosses a unit circle at $P(x, y)$, where does the terminal ray of $-A$ cross the unit circle?

c. What is the relationship between $\sin A$ and $\sin (-A)$?

d. What is the relationship between $\cos A$ and $\cos (-A)$?

18. Think of $A - B$ as $A + (-B)$. Use identity (4) and your answers to 17(c) and (d) to derive an identity for $\sin (A - B)$.

19. Think of $2A$ as $A + A$. Use identity (4) to derive an identity for $\sin (2A)$.

20. In this exercise, we indicate how to extend identity (4) to all values of A and B.

a. Show that identity (4) is true when $A = 0$ or $B = 0$.

b. Use the following result: $\cos (\pi - A) = -\cos A$, to show that the identity is true when $A + B = \pi$, $A, B > 0$.

c. Prove that identity (4) is true if $0 < A < \pi$, $0 < B < \pi$, and $A + B > \pi$, by applying (4) to the numbers $\pi - A$, $\pi - B$.

d. Prove that identity (4) is true if $0 \leq A < 2\pi$, $0 \leq B < 2\pi$. Use the identities: $\sin (A - \pi) = -\sin A$, $\cos (A - \pi) = -\cos A$.
If $A, B \geq \pi$, apply (4) to the numbers $A - \pi$, $B - \pi$.
If $A \geq \pi$, $B < \pi$, apply (4) to the numbers $A - \pi$, B.
If $A < \pi$, $B \geq \pi$, apply (4) to the numbers A, $B - \pi$.

e. Parts (a)–(d) of this exercise show that identity (4) holds if $0 \leq A < 2\pi$, $0 \leq B < 2\pi$. How do we know that the identity holds for all values of A and B?

SECTION 6.8 **OTHER IDENTITIES**

If we draw angle LOP_1 with measure θ and angle LOP_2 with measure $-\theta$ in standard position, their terminal rays are mirror images of each other with respect to the x-axis. (See Figure 6.8.1.)

If $P_1(x, y)$ is the intersection of OP_1 with the unit circle, then $P_2(x, -y)$ is the intersection of OP_2 with the unit circle. So,

A. $x = \cos \theta$
B. $y = \sin \theta$
C. $x = \cos (-\theta)$
D. $-y = \sin (-\theta)$

Comparing (B) and (D), and replacing θ with A, we have our next identity:

$$\sin (-A) = -\sin A \qquad (1)$$

Figure 6.8.1

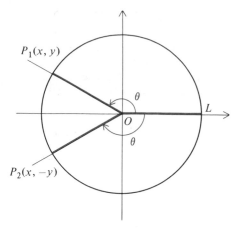

Comparing (A) and (C), we have another identity:

$$\cos(-A) = \cos A \tag{2}$$

We will use identities (1) and (2) together with $\sin(A + B) = \sin A \cos B + \cos A \sin B$, to derive other identities.

Since $\sin(A - B) = \sin[A + (-B)]$,

$$\sin(A - B) = \sin A \cos(-B) + \cos A \sin(-B)$$

Using (1) and (2) we have

$$\sin(A - B) = \sin A \cos B - \cos A \sin B \tag{3}$$

If we let $A = \pi/2$ in identity (3), we obtain $\sin(\pi/2 - B) = \sin(\pi/2)\cos B - \cos(\pi/2)\sin B$. Since $\sin(\pi/2) = 1$ and $\cos(\pi/2) = 0$, we have our next identity.

$$\sin\left(\frac{\pi}{2} - B\right) = \cos B \tag{4}$$

If, in identity (4), we let $B = (\pi/2) - A$, we have

$$\sin\left[\frac{\pi}{2} - \left(\frac{\pi}{2} - A\right)\right] = \cos\left(\frac{\pi}{2} - A\right)$$

Since

$$\frac{\pi}{2} - \left(\frac{\pi}{2} - A\right) = A$$

we have the following identity:

$$\cos(\pi/2 - A) = \sin A \tag{5}$$

From identity (4) we notice that

$$\cos(A + B) = \sin\left[\frac{\pi}{2} - (A + B)\right] = \sin\left[\left(\frac{\pi}{2} - A\right) - B\right]$$

Applying identity (3), we have

$$\cos (A + B) = \sin \left(\frac{\pi}{2} - A\right) \cos B - \cos \left(\frac{\pi}{2} - A\right) \sin B$$

Using identities (4) and (5) we obtain our next identity:

$$\cos (A + B) = \cos A \cos B - \sin A \sin B \tag{6}$$

An identity that follows immediately from (6) is

$$\cos (A - B) = \cos A \cos B + \sin A \sin B \tag{7}$$

Since $2A = A + A$, we can use the identities for $\sin (A + B)$ and $\cos (A + B)$ to derive identities for $\sin (2A)$ and $\cos (2A)$.

$$\sin 2A = 2 \sin A \cos A \tag{8}$$

$$\cos 2A = \cos^2 A - \sin^2 A \tag{9}$$

Identity (9) together with our first identity, $\sin^2 A + \cos^2 A = 1$, can be used to derive two other important identities. Since $\sin^2 A + \cos^2 A = 1$, $\sin^2 A = 1 - \cos^2 A$. Substituting this result into identity (9), we have $\cos 2A = \cos^2 A - (1 - \cos^2 A)$ or $\cos 2A = 2 \cos^2 A - 1$. Solving for $\cos^2 A$, we have this identity:

$$\cos^2 A = \frac{1 + \cos 2A}{2} \tag{10}$$

We obtained this result by substituting for $\sin^2 A$ in identity (9). Had we substituted for $\cos^2 A$ instead, we would have obtained:

$$\cos 2A = (1 - \sin^2 A) - \sin^2 A = 1 - 2 \sin^2 A$$

Solving for $\sin^2 A$, we have our next identity:

$$\sin^2 A = \frac{1 - \cos 2A}{2} \tag{11}$$

We can use our identities for sine and cosine to derive identities for the other trigonometric functions. For example,

$$\tan (A - B) = \frac{\sin (A - B)}{\cos (A - B)} = \frac{\sin A \cos B - \cos A \sin B}{\cos A \cos B + \sin A \sin B}$$

If we divide numerator and denominator by $\cos A \cos B$, we obtain

$$\tan (A - B) = \frac{\tan A - \tan B}{1 + \tan A \tan B} \tag{12}$$

Letting $A = 0$ in (12), and remembering that $\tan 0 = 0$, we have

$$\tan (-B) = -\tan B \tag{13}$$

Since $A + B = A - (-B)$, $\tan (A + B) = \tan [A - (-B)]$. Using identities

(12) and (13) we have the identity

$$\tan (A + B) = \frac{\tan A + \tan B}{1 - \tan A \tan B} \qquad (14)$$

We are going to end our list here, but it should be clear that the list can be made much longer. For convenience, we list all the identities we have derived. Although all the identities are useful in calculus, the identities preceded by dots are used so often that we suggest you memorize them. Actually, if you work through the exercises at the end of this section, you will have used the identities often enough to remember most of them.

$\bullet \sin^2 A + \cos^2 = 1$

$\bullet \tan^2 A + 1 = \sec^2 A$

$\bullet 1 + \cot^2 A = \csc^2 A$

$\bullet \sin (-A) = -\sin A$

$\bullet \cos (-A) = \cos A$

$\bullet \tan (-A) = -\tan A$

$\bullet \sin (A + B) = \sin A \cos B + \cos A \sin B$

$\quad \sin (A - B) = \sin A \cos B - \cos A \sin B$

$\bullet \sin (2A) = 2 \sin A \cos B$

$\bullet \cos (A + B) = \cos A \cos B - \sin A \sin B$

$\quad \cos (A - B) = \cos A \cos B + \sin A \sin B$

$\bullet \cos (2A) = \cos^2 A - \sin^2 A$

$\quad \tan (A + B) = \dfrac{\tan A + \tan B}{1 - \tan A \tan B}$

$\quad \tan (A - B) = \dfrac{\tan A - \tan B}{1 + \tan A \tan B}$

$\quad \sin \left(\dfrac{\pi}{2} - A \right) = \cos A$

$\quad \cos \left(\dfrac{\pi}{2} - A \right) = \sin A$

$\bullet \sin^2 A = \dfrac{1 - \cos 2A}{2}$

$\bullet \cos^2 A = \dfrac{1 + \cos 2A}{2}$

It is difficult to explain how trigonometric identities are used in calculus before you have seen any calculus. Instead, we will look at an example in algebra to indicate one application.

Suppose we wanted to find the values of θ that satisfy the equation

$$\cos 2\theta - \sin \theta = 0$$

Identity (11) is

$$\sin^2 \theta = \frac{1 - \cos 2\theta}{2}$$

Multiplying out,

$$2 \sin^2 \theta = 1 - \cos 2\theta$$

and so

$$\cos 2\theta = 1 - 2 \sin^2 \theta$$

Substituting this into the original equation,

$$1 - 2 \sin^2 \theta - \sin \theta = 0 \quad \text{or} \quad 2 \sin^2 \theta + \sin \theta - 1 = 0$$

This is a quadratic equation with $\sin \theta$ as the variable. (Think of the equation $2x^2 + x - 1 = 0$.) We can solve it by factoring.

$$2 \sin^2 \theta + \sin \theta - 1 = 0$$
$$\Leftrightarrow (2 \sin \theta - 1)(\sin \theta + 1) = 0$$
$$\Leftrightarrow 2 \sin \theta - 1 = 0 \quad \text{or} \quad \sin \theta + 1 = 0$$
$$\Leftrightarrow 2 \sin \theta = 1 \quad \text{or} \quad \sin \theta = -1$$
$$\Leftrightarrow \theta = \left\{ \begin{array}{l} \dfrac{\pi}{6} + 2n\pi \\[2mm] \dfrac{5\pi}{6} + 2n\pi \end{array} \right\} \; n = 0, \pm 1, \pm 2, \ldots \quad \text{or} \quad \theta = \dfrac{3\pi}{2} + 2n\pi$$

$$n = 0, \pm 1, \pm 2, \ldots$$

Exercises 6.8

In Exercises 1–14, $A = \pi/4$, $B = \pi/3$, and $C = \pi/6$. Use the identities in this section to evaluate each expression.

1. $\sin (A + C)$	**2.** $\cos (A + B)$	**3.** $\tan (B - A)$
4. $\sin (2A)$	**5.** $\cos (2B)$	**6.** $\cos (A - C)$
7. $\sin (C - B)$	**8.** $\tan (A + C)$	**9.** $\tan (-A)$
10. $\cos (2C)$	**11.** $\sin (2B)$	**12.** $\sin^2 B$
13. $\cos^2 A$	**14.** $\tan (2A)$	

In Exercises 15–20, use the identities to evaluate the given expression.

15. $\sin \left(\dfrac{5\pi}{12} \right)$ **16.** $\cos \left(\dfrac{7\pi}{12} \right)$ **17.** $\tan \left(\dfrac{7\pi}{12} \right)$

18. $\sin\left(\dfrac{5\pi}{6}\right)$ **19.** $\cos\left(\dfrac{7\pi}{6}\right)$ **20.** $\tan\left(\dfrac{\pi}{12}\right)$

21. Evaluate $\sin(\pi/12)$ two different ways.

    a. By using the identity for $\sin(A - B)$.
    b. By using the identity $\sin^2 A = (1 - \cos 2A)/2$.
    c. Show algebraically that your answers to (a) and (b) are equal.

22. Evaluate $\cos(\pi/12)$ two different ways.

    a. By using the identity for $\cos(A - B)$.
    b. By using the identity $\cos^2 A = (1 + \cos 2A)/2$.

23. Prove identity (7).
24. Prove identity (8).
25. Prove identity (9).
26. Derive a formula for $\tan 2A$.
27. Derive an identity for $\sin A \cos B$. (*Hint:* Use the identities for $\sin(A + B)$ and $\sin(A - B)$.)

In Exercises 28–31, evaluate each expression without using a table of trigonometric values.

28. $\sin\dfrac{\pi}{5}\cos\dfrac{4\pi}{5} + \cos\dfrac{\pi}{5}\sin\dfrac{4\pi}{5}$ **29.** $2\sin\dfrac{\pi}{12}\cos\dfrac{\pi}{12}$

30. $\cos^2\dfrac{\pi}{12} - \sin^2\dfrac{\pi}{12}$ **31.** $\dfrac{\tan\dfrac{\pi}{5} + \tan\dfrac{3\pi}{10}}{1 - \tan\dfrac{\pi}{5}\tan\dfrac{3\pi}{10}}$

32. Let $f(x) = \cos x$.

    a. What is $f\left(x - \dfrac{\pi}{2}\right)$?

    b. How are the graphs of $y = f(x)$ and $y = f\left(x - \dfrac{\pi}{2}\right)$ related?

    c. Use the identity for $\cos(A - B)$ to show that $f\left(x - \dfrac{\pi}{2}\right) = \sin x$.

33. Let $g(x) = \sin x$.

    a. What is $g\left(x + \dfrac{\pi}{2}\right)$?

    b. How are the graphs of $y = g(x)$ and $y = g\left(x + \dfrac{\pi}{2}\right)$ related?

    c. Use the identity for $\sin(A + B)$ to show that $g\left(x + \dfrac{\pi}{2}\right) = \cos x$.

In Exercises 34–39, show that each equation is an identity.

34. $(\sin A + \cos A)^2 = \sin 2A + 1$
35. $\sin 3A = 3 \sin A - 4 \sin^3 A$
36. $\cos 2A = (\cos^2 A \csc^2 A - \sin^2 A \sec^2 A)/\sec^2 A \csc^2 A$
37. $\sin^3 A = \sin A - \sin A \cos^2 A$
38. $\sin^2 A = \tan^2 A/(\tan^2 A + 1)$
39. $(\tan A + \cot A)^2 = \sec^2 A + \csc^2 A$

SECTION 6.9   INVERSE TRIGONOMETRIC FUNCTIONS

In Section 4.5, we learned that if a function f is one-to-one or equivalently, if a horizontal line crosses the graph of f in at most 1 point, f has an inverse, f^{-1}, defined by

$$y = f^{-1}(x) \Longleftrightarrow x = f(y)$$

If we look at the graph of $y = \sin x$ (Figure 6.9.1), we see that the sine function has no inverse, since there are horizontal lines that cross the graph infinitely many times. Consider, however, the graph of the function f, defined by

$$f(x) = \sin x, \; -\frac{\pi}{2} \le x \le \frac{\pi}{2}$$

This is the sine function whose domain is restricted to values of x between $-\pi/2$ and $\pi/2$. Its graph, shown in Figure 6.9.2, has the property that horizontal lines cross it at most once; and, therefore, the function f defined by

$$f(x) = \sin x, \; -\frac{\pi}{2} \le x \le \frac{\pi}{2}$$

has an inverse.

In Section 4.5, we learned that we can sometimes get a formula for f^{-1} by solving $y = f(x)$ for x in terms of y. In the case of $y = \sin x$, we cannot solve this equation for x using any methods we have previously learned. Nevertheless, the inverse exists and the name we choose for f^{-1} is $\sin^{-1}$. (Read: inverse sine.

Figure 6.9.1 $y = \sin x$

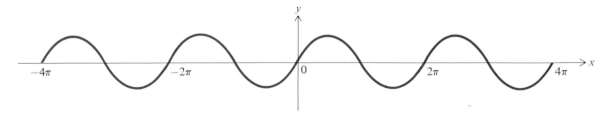

Figure 6.9.2 $y = \sin x$ $-\dfrac{\pi}{2} \le x \le \dfrac{\pi}{2}$

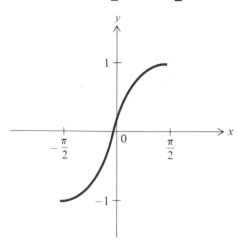

Some texts call this function arcsine.) By definition,

$$y = \sin^{-1} x \Longleftrightarrow -\frac{\pi}{2} \le y \le \frac{\pi}{2} \text{ and } x = \sin y$$

Example 1

> **A.** $\sin^{-1}\left(\dfrac{1}{2}\right) = \dfrac{\pi}{6}$ since $\sin\left(\dfrac{\pi}{6}\right) = \dfrac{1}{2}$
>
> **B.** $\sin^{-1}(-1) = -\dfrac{\pi}{2}$ since $\sin\left(-\dfrac{\pi}{2}\right) = -1$
>
> **C.** $\sin^{-1}\left(\dfrac{\sqrt{2}}{2}\right) = \dfrac{\pi}{4}$ since $\sin\left(\dfrac{\pi}{4}\right) = \dfrac{\sqrt{2}}{2}$

A convenient way to think about $\sin^{-1} x$ is to regard $\sin^{-1} x$ as the number between $-\pi/2$ and $\pi/2$ whose sine is x. The phrase "between $-\pi/2$ and $\pi/2$" is important. If $\sin^{-1} x$ just meant the number whose sine is x, $\sin^{-1}(1/2)$ could equal $\pi/6$ or $5\pi/6$ or $17\pi/6$, since all of these numbers have the property that their sines are $1/2$. By restricting our attention to values between $-\pi/2$ and $\pi/2$ we are guaranteeing that $\sin^{-1}(x)$ will be one unique number, and $\sin^{-1}$ will be a function. Another reason we chose the interval $[-\pi/2, \pi/2]$ is that between $-\pi/2$ and $\pi/2$, the sine function takes on all values between -1 and 1; i.e., all the numbers in the range of the sine function.

We can use some of our results about inverse functions to get results about the inverse sine function. We know that for any one-to-one function, f, and its inverse, f^{-1}, $D_f = R_{f^{-1}}$ and $R_f = D_{f^{-1}}$. We have restricted the domain of the

sine function to $[-\pi/2, \pi/2]$. The range of the sine function is $[-1, 1]$. Therefore the domain of the inverse sine function is $[-1, 1]$ and its range is
$[-\pi/2, \pi/2]$. The graph of $y = \sin^{-1} x$ can be obtained from the graph of
$y = \sin x$ by reflection in the line $y = x$ as in Figure 6.9.3.

Figure 6.9.3

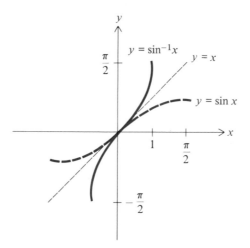

The other inverse trigonometric functions are defined using similar methods.
Figure 6.9.4 shows the graph of $y = \cos x$. If we restrict our attention to values
of x in $[0, \pi]$ we have a function that has an inverse. The inverse is called $\cos^{-1}$
(read: inverse cosine) and is defined by

$$y = \cos^{-1} x \quad \Leftrightarrow \quad x = \cos y \quad \text{and} \quad 0 \le y \le \pi$$

In words, $\cos^{-1} (x)$ is the number between 0 and π whose cosine is x.

Figure 6.9.4   **y = cos x**

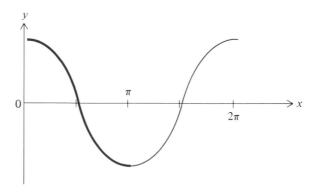

Example 2

$$\cos^{-1}(1) = 0 \qquad \text{since } \cos(0) = 1$$

$$\cos^{-1}\left(\frac{-\sqrt{3}}{2}\right) = \frac{5\pi}{6} \qquad \text{since } \cos\left(\frac{5\pi}{6}\right) = \frac{-\sqrt{3}}{2}$$

$$\cos^{-1}(0) = \frac{\pi}{2} \qquad \text{since } \cos\left(\frac{\pi}{2}\right) = 0$$

Notice that, when calculating $\cos^{-1} x$, we will always get a number between 0 and π.

Figure 6.9.5 shows the graph of $y = \cos^{-1} x$. Notice that the domain of the inverse cosine function is $[-1, 1]$ and the range is $[0, \pi]$.

For the tangent function, we restrict our attention to values of x in $(-\pi/2, \pi/2)$. (See Figure 6.9.6.)

Figure 6.9.5 $y = \cos^{-1} x$

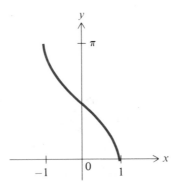

Figure 6.9.6 $y = \tan x$

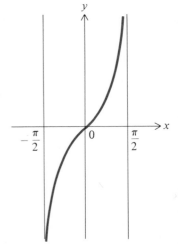

On this interval tan x takes on all values in the range of the tangent function, and the function has an inverse, $\tan^{-1}$, defined by

$$y = \tan^{-1} x \Leftrightarrow x = \tan y \quad \text{and} \quad -\frac{\pi}{2} < y < \frac{\pi}{2}$$

As with the previous inverse trigonometric function, we can think of $y = \tan^{-1} x$ as "y is the number between $-\pi/2$ and $\pi/2$ whose tangent is x."

Example 3

A. $\tan^{-1}(1) = \dfrac{\pi}{4}$ since $\tan\left(\dfrac{\pi}{4}\right) = 1$

B. $\tan^{-1}(-1) = -\dfrac{\pi}{4}$ since $\tan\left(-\dfrac{\pi}{4}\right) = -1$

C. $\tan^{-1}\left(\dfrac{\sqrt{3}}{3}\right) = \dfrac{\pi}{6}$ since $\tan\left(\dfrac{\pi}{6}\right) = \dfrac{\sqrt{3}}{3}$

Figure 6.9.7 shows the graph of $y = \tan^{-1} x$ indicating its range, $(-\pi/2, \pi/2)$ and its domain $(-\infty, \infty)$.

Figure 6.9.7 $y = \tan^{-1} x$

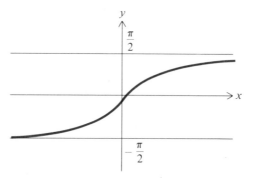

We summarize the results of this section in Table 6.9.1, including the remaining three inverse trigonometric functions.

Table 6.9.1   **Properties of Inverse Trigonometric Functions**

Function	Domain	Range	Graph
$\sin^{-1}$	$[-1, 1]$	$\left[-\dfrac{\pi}{2}, \dfrac{\pi}{2}\right]$	
$\cos^{-1}$	$[-1, 1]$	$[0, \pi]$	
$\tan^{-1}$	$(-\infty, +\infty)$	$\left(-\dfrac{\pi}{2}, \dfrac{\pi}{2}\right)$	
$\cot^{-1}$	$(-\infty, +\infty)$	$(0, \pi)$	
$\sec^{-1}$	$x \geq 1$ or $x \leq -1$	All real numbers x such that $0 \leq x < \dfrac{\pi}{2}$ or $\dfrac{\pi}{2} < x \leq \pi$	
$\csc^{-1}$	$x \geq 1$ or $x \geq -1$	All real numbers x such that $-\dfrac{\pi}{2} \leq x < 0$ or $0 < x \leq \dfrac{\pi}{2}$	

Exercises 6.9   In Exercises 1–20, evaluate each expression.

1. $\sin^{-1}\left(-\dfrac{1}{2}\right)$ 2. $\cos^{-1}\left(\dfrac{\sqrt{2}}{2}\right)$ 3. $\tan^{-1}\left(\dfrac{\sqrt{3}}{3}\right)$

4. $\tan^{-1}\left(-\dfrac{\sqrt{3}}{3}\right)$ 5. $\sin^{-1}\left(\dfrac{\sqrt{3}}{2}\right)$ 6. $\sin^{-1}\left(-\dfrac{\sqrt{2}}{2}\right)$

7. $\cos^{-1}(-1)$ 8. $\cos^{-1}(1)$ 9. $\sin^{-1}(0)$

10. $\csc^{-1}(1)$ 11. $\sec^{-1}(2)$ 12. $\csc^{-1}(-2)$

13. $\tan^{-1}(-1)$ 14. $\cot^{-1}(\sqrt{3})$ 15. $\cot^{-1}(-\sqrt{3})$

16. $\sec^{-1}(-\sqrt{2})$ 17. $\sin^{-1}\left(\dfrac{1}{2}\right)$ 18. $\tan^{-1}(0)$

19. $\cos^{-1}(0)$ 20. $\cos^{-1}\left(-\dfrac{\sqrt{3}}{2}\right)$

21. Evaluate $\sin\left[\cos^{-1}\left(\dfrac{\sqrt{3}}{2}\right)\right]$

22. Evaluate $\tan[\tan^{-1}(1)]$

23. Evaluate $\sec[\tan^{-1}(1)]$

24. Evaluate $\cos\left[\sin^{-1}\left(\dfrac{1}{2}\right)\right]$

25. Evaluate $\cot\left[\sin^{-1}\left(\dfrac{\sqrt{2}}{2}\right)\right]$

26. Evaluate $\sin^{-1}\left(\sin\dfrac{\pi}{2}\right)$

27. Evaluate $\sin^{-1}\left(\sin\dfrac{2\pi}{3}\right)$

28. Evaluate $\cos\left[\cos^{-1}\left(\dfrac{8}{9}\right)\right]$

29. Evaluate $\cos^{-1}\left(\cos\dfrac{11\pi}{4}\right)$

30. We know that $\csc x = \dfrac{1}{\sin x}$. Show by example that $\csc^{-1}x \neq \dfrac{1}{\sin^{-1}x}$.

31. Prove that $\csc^{-1}x = \sin^{-1}(1/x)$. (*Hint:* Let $\csc^{-1}x = a$ and find $\csc a$.)

32. Prove that $\sec^{-1}x = \cos^{-1}(1/x)$.

33. Prove that $\tan^{-1}x = (\pi/2) - \cot^{-1}x$.

34. Show that $\sin^{-1}(\sin x)$ does not always equal x. (*Hint:* See Exercise 27.) What conditions on x would guarantee that $\sin^{-1}(\sin x) = x$?

258 6 Trigonometric Functions

Review Exercises
Chapter 6

 1. The sides of a right triangle measure 8, 10, and $\sqrt{164}$. Find the degree measure of the smallest angle of the triangle.

 2. What is the degree measure of the largest acute angle in a right triangle whose sides measure 7, 24, and 25?

 3. From a point 60 feet from the base of a building, the angle of elevation to the top measures 75°. Find the height of the building to the nearest $\frac{1}{10}$ of a foot.

4. How long are the sides of a 45° right triangle whose hypotenuse is 8 inches?

In Exercises 5–16, a degree measure of an angle is given. Convert it to a radian measure.

5. 315°		**6.** 45°		**7.** 120°	
8. 300°		**9.** 1125°		**10.** −225°	
11. 7°		**12.** 135°		**13.** 400°	
14. 270°		**15.** 180°		**16.** 500°	

In Exercises 17–26, a radian measure of an angle is given. Convert it to a degree measure.

17. 7π **18.** $\dfrac{5\pi}{4}$ **19.** $\dfrac{-3\pi}{2}$ **20.** $\dfrac{4\pi}{3}$

21. $\dfrac{5\pi}{6}$ **22.** 7 **23.** 20π **24.** $\dfrac{\pi}{12}$

25. 1 **26.** $\dfrac{13\pi}{4}$

27. A circle has radius $r = 5$. What is the area of a sector whose central angle measures $\pi/4$?

28. A sector with central angle measuring $\pi/12$ has area $A = \pi$ square inches. What is the radius of the circle?

29. In what quadrant(s) do sine and cosine have the same sign?

30. In what quadrant(s) is tangent negative?

31. In what quadrant(s) do sine and tangent have the same sign?

In Exercises 32–38, evaluate the given expression.

32. $\sin \dfrac{3\pi}{4}$ **33.** $\cos\left(-\dfrac{\pi}{6}\right)$ **34.** $\cos 12\pi$

35. $\tan\left(-\dfrac{5\pi}{2}\right)$ **36.** $\sec\dfrac{5\pi}{4}$ **37.** $\cot\left(\dfrac{-4\pi}{3}\right)$

38. $\sin 10$

39. Express $\sin x$ in terms of $\cos x$.
40. Express $\tan x$ in terms of $\cos x$.
41. Evaluate $\sin(\pi/12)$ without using a table of values of the trigonometric functions.
42. Evaluate $\sin(\pi/8)$ without using a table of values of the trigonometric functions.
43. Evaluate $\sin^{-1}\left(\dfrac{\sqrt{2}}{2}\right)$.

44. Evaluate $\tan\left(\cos^{-1}\dfrac{1}{2}\right)$.

45. $\sin\dfrac{5\pi}{6} = \dfrac{1}{2}$. Why doesn't $\sin^{-1}\dfrac{1}{2} = \dfrac{5\pi}{6}$?

46. Find all values of x that satisfy $2\sin x - \sqrt{3} = 0$.
47. Find all values of x that satisfy $2\sin^2 x + 3\sin x - 2 = 0$.
48. Express $\sin 4\theta$ in terms of $\sin\theta$ and $\cos\theta$.

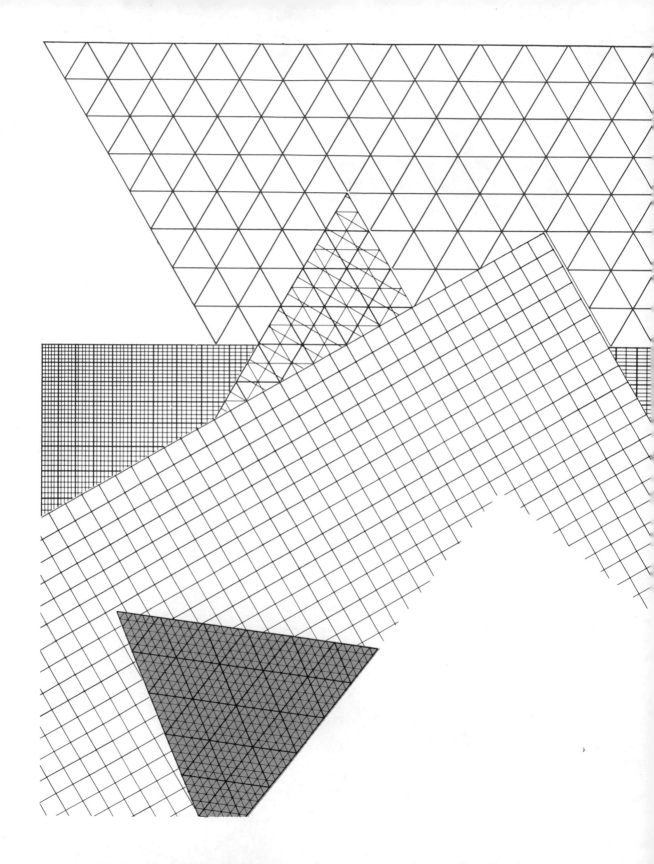

7

Exponential and Logarithmic Functions

INTRODUCTION

Imagine a sheet of paper so thin that 1000 sheets form a stack only one inch tall. Take this sheet of paper, tear it in half, and place the two sheets on top of each other. Now tear the two sheets in half and form a stack of four sheets. Repeat this process a total of 50 times. Is the height of the stack nearest

A. One inch tall?
B. One foot tall?
C. As tall as a six-story building?
D. More than a million miles tall?

Answer and then read on.

The correct answer is (D), although even 1 million miles is an underestimate. The stack of paper would actually be over 17 million miles tall! While this answer may seem unbelievable, you can convince yourself with some relatively simple calculations that it is true.

As a first step, we need to know how many sheets of paper we have. After the first tear, we have 2 sheets. After the second tear, we have $2 \times 2 = 4$ sheets. After the third tear, we have $2 \times 2 \times 2 = 8$ sheets.

A convenient mathematical notation allows us to avoid writing expressions with long chains of common factors, such as $2 \times 2 \times 2 \times 2 \times 2 \times 2 \times 2 \times 2 \times 2 \times 2$. If b is a natural number, a^b represents the product of a with itself b times. The number a^b is called an **exponential.** We call a the **base** and b the **exponent** of a. With this notation $2 \times 2 = 2^2$, $2 \times 2 \times 2 = 2^3$, and $2 \times 2 \times 2 \times 2 \times 2 \times 2 \times 2 \times 2 \times 2 \times 2 = 2^{10}$.

Returning to our problem: After the third tear we had 2^3 sheets; after the fourth tear, 2^4 sheets; and after the fiftieth tear, we will have 2^{50} sheets. How large a number is 2^{50}?

There is a multiplicative property of exponentials that can help us compute 2^{50}. We know that $2 \times 2 \times 2 \times 2 \times 2 \times 2 \times 2 \times 2 \times 2 \times 2 = 2^{10}$. We can break up the product on the left side of the equality in a number of ways:

$$(2 \times 2 \times 2 \times 2) \times (2 \times 2 \times 2 \times 2 \times 2 \times 2) = 2^{10}$$

or

$$(2 \times 2 \times 2 \times 2 \times 2) \times (2 \times 2 \times 2 \times 2 \times 2) = 2^{10}$$

or

$$(2 \times 2) \times (2 \times 2 \times 2 \times 2 \times 2 \times 2 \times 2 \times 2) = 2^{10}$$

Using our exponential notation, we can write these equalities as follows:

$$2^4 \times 2^6 = 2^{10}$$
$$2^5 \times 2^5 = 2^{10}$$
$$2^2 \times 2^8 = 2^{10}$$

Note that the sum of the exponents on the left-hand side of each of these equalities is 10. This illustrates the following general property of exponentials. To multiply exponentials having the same base, add the exponents. In exponential notation,

$$a^b \times a^c = a^{b+c}$$

To compute 2^{50}, we could start writing a table

$2^1 = 2$	$2^6 = 64$
$2^2 = 4$	$2^7 = 128$
$2^3 = 8$	$2^8 = 256$
$2^4 = 16$	$2^9 = 512$
$2^5 = 32$	$2^{10} = 1024$

and continue until we get to 2^{50}, but we would prefer to use

$$2^{50} = 2^{10} \times 2^{10} \times 2^{10} \times 2^{10} \times 2^{10}$$
$$= (1024) \times (1024) \times (1024) \times (1024) \times (1024)$$

If you multiply these numbers together, you will find that $2^{50} = $ 1,125,899,906,842,642. This number is read one quadrillion, one hundred twenty-five trillion, eight hundred ninety-nine billion, nine hundred six million, eight hundred forty-two thousand, six hundred forty-two, and represents the number of sheets of paper. How tall is the stack?

1,125,899,906,842,642 sheets

$\doteq$ 1,125,899,906,843 inches   (Divide by 1000 since there are 1000 sheets/inch.)

$\doteq$ 93,824,992,236 feet   (Divide by 12.)

$\doteq$ 17,769,885 miles   (Divide by 5280.)

Our stack of paper would be over 17 million miles tall, approximately 75 times the distance between the earth and the moon. The computation contains no tricks. Seventeen million miles is the correct, if unbelievable, answer.

For our purposes, the notation of exponents and the rule for multiplication are just as important as the example, and we will use them in the next section. So far, a^b makes sense only when b is a natural number. We have not defined symbols such as $a^{1/2}$, a^0, or a^{-3}. Our next task is to extend the definition of exponentials so that a^b is defined for any real number, b.

Exercises 7.1   One of the reasons the problem in the text has such a surprising answer is that we are asked to do something that seems possible—cutting paper 50 times—yet we get an impossible result. This exercise shows that it's not really possible to make the 50 cuts.

Suppose we start carrying out the experiment with a sheet of paper 100 meters × 100 meters, about the area of two football fields. To simplify the problem, keep one dimension fixed. After the first cut, the paper is 100 m × 50 m. After the second cut, it is 100 m × 25 m, etc.

How wide is the paper after the fiftieth cut? (For comparison, the radius of the nucleus of an atom is on the order of $1/2^{46}$ meter.)

SECTION 7.2   **EXTENDING EXPONENTS**

In the previous section, we defined a^b when b was a positive integer. In this section, we extend the definition of a^b to include all real values of b. We will first define a^b when b is zero, then when b is rational, and finally when b is irrational.

To motivate our new definition, recall that, when b and c are positive integers, we have a useful multiplication rule:

$$a^b \cdot a^c = a^{b+c}$$

Suppose we would like $a^b \cdot a^c$ to equal a^{b+c} for all real numbers b and c. How should we define a^b for $b = 0$?

If a^0 is to satisfy the multiplication rule, then,

$$a^1 \cdot a^0 = a^1$$

If we assume $a \neq 0$, we can divide both sides of this equation by a^1 to find $a^0 = 1$! *If we want the multiplication rule to hold, we have no choice for the definition of a^0—it has to be 1.*

Definition

> If $a \neq 0$, $a^0 = 1$.

To motivate our definition for rational exponents, we will first consider a particular example, $a^{1/3}$.

If $a^{1/3}$ is to obey our multiplication rule,

$$a^{1/3} \cdot a^{1/3} \cdot a^{1/3} = a^1 = a$$

The number $a^{1/3}$ should have the property that, when multiplied by itself three times, the result is a. But this means that $a^{1/3} = \sqrt[3]{a}$.

A similar argument tells us that $a^{1/c} = \sqrt[c]{a}$. Finally, we want

$$a^{b/c} = \underbrace{(a^{1/c})(a^{1/c})(a^{1/c}) \cdots (a^{1/c})}_{b \text{ times}} = (a^{1/c})^b$$

and this leads to our next definition.

Definition

> If b and c are positive integers, $a^{b/c} = (\sqrt[c]{a})^b$ whenever this expression is defined.

Example 1

To illustrate this definition,

A. $8^{2/3} = (\sqrt[3]{8})^2 = 2^2 = 4$
B. $16^{3/4} = (\sqrt[4]{16})^3 = 2^3 = 8$
C. $(25)^{3/2} = (\sqrt[2]{25})^3 = 5^3 = 125$

Notice that an expression such as $(-25)^{3/2}$ is not defined. If we want to apply our definition to $(-25)^{3/2}$, we have to find $\sqrt{-25}$ and then raise it to the third power. But there is no real number that is $\sqrt{-25}$! In general, we must be careful when we raise negative numbers to rational exponents. If we have to take even roots (square roots, 4th roots, 6th roots, etc.) of negative numbers, the expressions are not defined. There is no problem taking odd roots, however, as illustrated in parts (A) and (C) of the next example.

Example 2

A. $(-8)^{2/3} = (\sqrt[3]{-8})^2 = (-2)^2 = 4$.
B. $(-16)^{3/4} = (\sqrt[4]{-16})^3$ is not defined, since $\sqrt[4]{-16}$ is not a real number.
C. $(-32)^{3/5} = (\sqrt[5]{-32})^3 = (-2)^3 = -8$.

We come next to the meaning of negative exponents. Suppose b is a positive rational number. If a^{-b} is to satisfy the multiplication rule, then

$$a^{-b} \cdot a^b = a^0$$

Since $a^0 = 1$, we have

$$a^{-b} \cdot a^b = 1$$

If we assume that $a \neq 0$, we can divide by a^b to find

$$a^{-b} = 1/a^b$$

Definition

> If $b > 0$, $a^{-b} = 1/a^b$.

Notice that negative exponents are not used to represent negative numbers. (Do not confuse a^{-b} with $- (a^b)$!) Rather, they are a convenient notation for representing reciprocals.

Example 3

To illustrate the definition of a^{-b}

A. $8^{-2} = \dfrac{1}{8^2} = \dfrac{1}{64}$

B. $5^{-1} = \dfrac{1}{5^1} = \dfrac{1}{5}$

C. $16^{-(3/2)} = \dfrac{1}{16^{3/2}} = \dfrac{1}{(\sqrt{16})^3} = \dfrac{1}{4^3} = \dfrac{1}{64}$

In (C) we had to use two of the definitions to evaluate $16^{-(3/2)}$. (We had to use the definition of exponentials with negative exponents and the definition of exponentials with rational exponents.)

We still have to define a^b when b is irrational. For example, what can $4^{\sqrt{2}}$ mean?

Recall that every irrational number can be approximated by rational numbers to any degree of accuracy. We can approximate $\sqrt{2}$ by 1.4, 1.41, 1.414, or 1.4142, etc. Our choice of estimate depends only upon the accuracy we want in our work.

None of us can give an exact answer to the question "What is the decimal representation of the square root of 2?" (We can say that it is the number that, when multiplied by itself, gives 2, but this gives a property of the decimal rather than the decimal itself.) However, most of us feel that being able to approximate $\sqrt{2}$ as closely as we want gives us sufficient information about the number.

We do the same with a symbol such as $4^{\sqrt{2}}$. We have defined $4^{1.4}$ (since $4^{1.4} = 4^{14/10}$ and we know the meaning of rational exponents) and $4^{1.414}$. By replacing the irrational exponent, b, with rational numbers close to it we can evaluate a^b to any degree of accuracy. In practice, we will seldom use irrational exponents, but it is important to know that, if b is irrational, a^b defines a real number and we can approximate a^b as closely as we like by a^r where r is a rational approximation of b.

Our motivation for defining a^b has been the multiplication rule:

$$a^b \cdot a^c = a^{b+c}$$

It can be shown that this multiplication rule, shown as (1) below, and the other rules of arithmetic (2)–(5), hold for all exponentials. We state the next result without proof.

Let a, b, and c be any real numbers. If all terms on both sides of the equations below are defined, then

$$a^b \cdot a^c = a^{b+c} \tag{1}$$

$$a^b \div a^c = a^{b-c} \tag{2}$$

$$(ab)^c = a^c b^c \tag{3}$$

$$(a^b)^c = a^{bc} \tag{4}$$

$$a^{-b} = 1/a^b \tag{5}$$

You may wonder why we included the phrase "if all terms on both sides of the equations below are defined." It's to prevent anyone from using property (1) this way:

$$(-5)^{1/2} \cdot (-5)^{1/2} = (-5)^1 = -5$$

The terms on the left, $(-5)^{1/2}$, represent $\sqrt{-5}$. But a negative number has no square root, and the expression $(-5)^{1/2}$ is undefined. Similarly, $0^2 = 0$, but no one should use property (2) to think that $0^2 \div 0^2 = 0^0$, since 0^0 is undefined. (Since $0^2 = 0$, the left-hand side, $0 \div 0$, is also undefined.)

Excluding these unusual cases, our results state that all exponentials obey the arithmetic rules that exponentials with positive integer exponents obey.

Example 4

> We will use properties (1)–(5) to simplify the following expressions:
>
> A. $8^{-5} \cdot 2^{12}$ B. $x^{3/2} \div x^2$ C. $(a^{2/3})^4$
>
> A.   Since $8 = 2^3$, $8^{-5} = (2^3)^{-5} = 2^{-15}$.
>
> $$2^{-15} \cdot 2^{12} = 2^{-3} = \frac{1}{2^3} = \frac{1}{8}$$
>
> B. $x^{3/2} \div x^2 = x^{(3/2)-2} = x^{-(1/2)}$
> C. $(a^{2/3})^4 = a^{8/3}$

One result of our rules is that we may have different ways to evaluate the same expression. For example,

$$8^{2/3} \text{ is } (8^{1/3})^2 \text{ or } (8^2)^{1/3}$$

The first expression, $(8^{1/3})^2$, is $(\sqrt[3]{8})^2$. The second expression, $(8^2)^{1/3}$ is $(64)^{1/3} = \sqrt[3]{64}$. Not surprisingly, we get the same answer by using either method. In fact, $a^{b/c} = (\sqrt[c]{a})^b = \sqrt[c]{a^b}$ whenever both expressions are defined.

We have now defined a^b for all real numbers, b, and we have looked at some of the arithmetic properties these exponentials have.

Exercises 7.2

In Exercises 1–20, express the given number without using exponents.

1. $8^{2/3}$

2. $8^{-(2/3)}$

3. $25^{1/2}$

4. $25^{3/2}$

5. $\dfrac{1}{25^{1/2}}$

6. $\dfrac{1}{16^{-(1/2)}}$

7. $4^{-(3/2)}$

8. 225^0

9. $225^{-(1/2)}$

10. $64^{2/3}$

11. $\left(\dfrac{1}{4}\right)^{1/2}$

12. $\left(\dfrac{1}{8}\right)^{2/3}$

13. $\left(\dfrac{1}{8}\right)^{-(2/3)}$

14. $(0.1)^{-2}$

15. $(-32)^{1/5}$

16. $(-32)^{-(1/5)}$

17. $(1/3)^{-2}$

18. $32^{1.2}$

19. $16^{2.5}$

20. $16^{-(3/4)}$

In Exercises 21–26 express the given product or quotient in the form a^b.

21. $2^3 \cdot 8^4$

22. $3^2 \cdot 9$

23. $16^2 \cdot 4^3$

24. $3^{-3} \cdot 9^2$

25. $2^{20} \cdot \left(\dfrac{1}{8}\right)^6$

26. $5^7 \cdot 25^4$

27. Solve for x: $2^{3x+2} = 16^{x+1}$. (*Hint:* Write both sides of the equation as exponentials with the same base.)

28. Solve for x: $8^{x-1} = \dfrac{1}{2}$

29. a. Solve for x: $9^{2x-7} = 3^{x+1}$

   b. Could you use the same method to solve the equation $9^{2x-7} = 2^{x+1}$? Explain.

In Exercises 30–39, find the missing exponents.

30. $(2^3)^4 = 2^?$

31. $(2^3)^4 = 4^?$

32. $(2^3)^4 = 8^?$

33. $(3^2)^{-2} = 3^?$

34. $(8^4)^5 = 8^?$

35. $(8^4)^5 = 2^?$

36. $(5^7)^{-2} = 5^?$

37. $(5^7)^{-2} = \left(\dfrac{1}{5}\right)^?$

38. $(3^4)^2 = 3^?$

39. $(2^0)^5 = 7^?$

In Exercises 40–49, write the given expression using only positive exponents.

40. $x^2 y^{-3}$

41. $a^{-5} b^2$

42. $(xy)^{-2} z$

43. $s^2 t^{-5}$

44. $x^{-2} y^4 z^{-3}$

45. $a^5 b^{-3} c^{-4} d^3$

46. $(uv)^{-7}$

47. $(a^{-3} bc^2)^3$

48. $(x^{-2} y^{-2})^2$

49. $(xy^{-3})^2$

In Exercises 50–54, write the given expression in the form $a^{n_1}b^{n_2}c^{n_3}$.

50. $(a^2)(bc)^5$

51. $\dfrac{a}{b^2c}$

52. $\dfrac{1}{a^2b}$

53. $\dfrac{a^5}{a^{10}b^4}$

54. $\dfrac{(ab)^2}{(abc)^3}$

55. Assume that $2 = 10^{0.3010}$, $3 = 10^{0.4771}$, and $5 = 10^{0.6990}$. Use this information to express each of the following in the form 10^a.

a. 6	**b.** 15	**c.** 30
d. 8	**e.** 12	**f.** 75

56. Does $(-2)^{1/2} \cdot (-2)^{1/2} = (-2)^{1/2+1/2} = (-2)^1 = -2$? If not, why not?

SECTION 7.3

EXPONENTIAL FUNCTIONS AND THEIR GRAPHS

In Sections 7.1 and 7.2, we defined expressions of the form a^b for different real numbers, b. We now use these definitions to define a function.

Let a be a positive real number. The function, f, from **R** to **R**, defined by

$$f(x) = a^x$$

is called the **exponential function base a.**

Notice that the functions f, g, and h, defined by $f(x) = 2^x$, $g(x) = 3^x$, and $h(x) = 4^x$, are all exponential functions. Our definition actually defines a family of functions rather than any one particular function.

The graph of the exponential function base 2, $f(x) = 2^x$, is shown in Figure 7.3.1. We see that $(0, 1)$ lies on the graph since $2^0 = 1$. Actually, since $a^0 = 1$ for $a \neq 0$, $(0, 1)$ lies on the graph of each exponential function.

As x increases, 2^x increases. This means that, if $x_1 > x_2$, then $2^{x_1} > 2^{x_2}$. We saw in Section 7.1 that, if $x > 0$, then any increase in x results in a relatively large increase in 2^x. For example, starting at $(0, 1)$, if we move 10 units to the right, we must move 1023 units up to plot the point $(10, 2^{10})$.

To understand what is happening to the graph to the left of the origin, we will examine the values of 2^x for successively smaller negative values of x.

When $x = -2$, $2^x = 2^{-2} = \dfrac{1}{2^2} = \dfrac{1}{4}$.

When $x = -4$, $2^x = 2^{-4} = \dfrac{1}{2^4} = \dfrac{1}{16}$.

When $x = -10$, $2^x = 2^{-10} = \dfrac{1}{2^{10}} = \dfrac{1}{1024}$.

As x moves to the left, 2^x remains positive but gets closer and closer to zero.

Figure 7.3.1 $f(x) = 2^x$

x	-2	-1	0	1	2	3	4
2^x	$\frac{1}{4}$	$\frac{1}{2}$	1	2	4	8	16

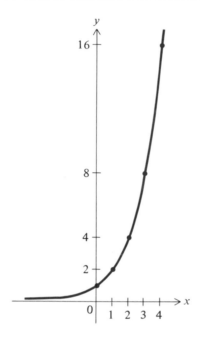

Equivalently, the graph gets closer and closer to the x-axis. This kind of behavior is described by saying that the graph of $f(x) = 2^x$ is **asymptotic** to the x-axis or, equivalently, that the x-axis is an **asymptote** of the graph.

Not all exponential functions increase as x increases. In Figure 7.3.2 we have graphed the function $f(x) = (\frac{1}{2})^x$.

This graph passes through the point $(0, 1)$ and is asymptotic to the x-axis, but because the base is $\frac{1}{2}$, the function decreases as x increases.

Figure 7.3.3 shows the graphs of $f(x) = 2^x$ and $g(x) = 3^x$ drawn on the same coordinate system. The graph of g is above the graph of f to the right of the origin and below it to the left, since if $x > 0$, $3^x > 2^x$, while if $x < 0$, $3^x < 2^x$. Both curves pass through $(0, 1)$.

Figure 7.3.4 shows the graphs of $f(x) = (\frac{1}{2})^x$ and $g(x) = (\frac{1}{3})^x$ drawn on the same coordinate system. If $x > 0$, $f(x) > g(x)$, and the graph of f is above the graph of g. If $x < 0$, $f(x) < g(x)$, and the graph of f is below the graph of g.

By examining the graphs in Figure 7.3.3 and 7.3.4, we notice that $f(x) = (\frac{1}{2})^x$ and $g(x) = 2^x$ are mirror images of each other with respect to the y-axis. The same holds true for the graphs of $(\frac{1}{3})^x$ and 3^x. In fact this is true for $f(x) = a^x$ and $g(x) = (1/a)^x$ for any real number a, $a > 0$. We ask you to prove this result in the exercises.

Figure 7.3.2 $f(x) = (\frac{1}{2})^x$

x	-4	-3	-2	-1	0	1	2
$(\frac{1}{2})^x$	16	8	4	2	1	$\frac{1}{2}$	$\frac{1}{4}$

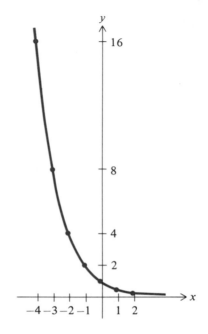

We saw in Sections 7.1 and 7.2 that if $a > 0$, a^x is defined for all real values of x. This is equivalent to saying that the domain of an exponential function is the set of all real numbers.

Also, we notice that a^x, $a > 0$, can never be zero or negative. The range of the exponential function is the set of positive real numbers rather than the entire set of real numbers. Notice that all the graphs we have sketched are above the x-axis. The y-coordinates, which are the values of the function, are all greater than zero.

We can summarize the important properties of the exponential function, $f(x) = a^x$, as follows:

A.   The domain consists of all real numbers; $D_f = (-\infty, \infty)$.
B.   The range consists of all positive real numbers; $R_f = (0, +\infty)$.
C.   The point $(0, 1)$ is on the graph of f.
D.   The point $(1, a)$ is on the graph of f.
E.   The graph of $y = a^x$ when $a > 1$ is pictured in Figure 7.3.5(a).
     Part (b) of Figure 7.3.5 shows the graph of $y = a^x$ when $0 < a < 1$.

Figure 7.3.3   **Figure 7.3.4**

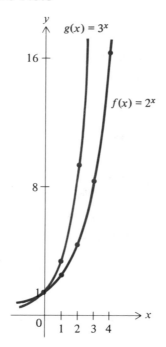

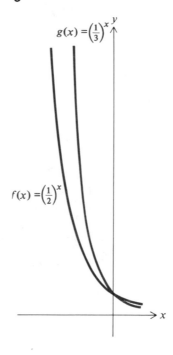

Figure 7.3.5

(a) If $a > 1$ **(b) If $0 < a < 1$**

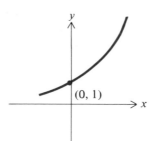

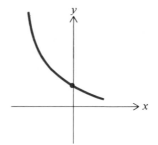

Example 1

Let $f(x) = 4^x$ and $g(x) = 2^x$. We can find expressions for $(f + g)(x)$, $(fg)(x)$, $\frac{f}{g}(x)$, $(f \circ g)(x)$, and $(g \circ f)(x)$ as follows:

$$(f + g)x = f(x) + g(x) = 4^x + 2^x$$

$$(fg)(x) = f(x)g(x) = 4^x 2^x = 2^{2x} 2^x = 2^{2x+x} = 2^{3x}$$

$$\left(\frac{f}{g}\right)(x) = \frac{f(x)}{g(x)} = \frac{4^x}{2^x} = \frac{2^{2x}}{2^x} = 2^x$$

$$(f \circ g)(x) = f(g(x)) = f(2^x) = 4^{(2^x)}$$

$$(g \circ f)(x) = g(f(x)) = g(4^x) = 2^{(4^x)}$$

An expression such as $2^{(4^x)}$ is sometimes written without parentheses using the following convention: $a^{b^c} = a^{(b^c)}$. Do not confuse the expression $(a^b)^c$ (which equals $a^{b \cdot c}$) with a^{b^c}. For example, $2^{3^2} = 2^9$ while $(2^3)^2 = 2^6$.

Example 2

Let $f(x) = 2^x$ and $g^x = \sin x$. We will find an expression for $(f \circ g)x$, the domain of $f \circ g$, and evaluate $(f \circ g)(\pi/2)$.

$$(f \circ g)(x) = f(g(x)) = f(\sin x) = 2^{\sin x}$$

The domain of $f \circ g$ is the set of real numbers.

$$(f \circ g)\left(\frac{\pi}{2}\right) = f\left(g\left(\frac{\pi}{2}\right)\right) = f\left(\sin \frac{\pi}{2}\right) = f(1) = 2^1 = 2$$

Exercises 7.3

1. This exercise requires a hand calculator. Draw the graph of $f(x) = 5^x$ for values of x between -1 and 1, plotting values of x at intervals of 0.1.
2. Sketch the graph of $g(x) = (\frac{1}{5})^x$ by plotting points.
3. Show that the graphs of $f(x) = 5^x$ and $g(x) = (\frac{1}{5})^x$ are mirror images of each other in the y-axis. (*Hint:* Show that $f(-x) = g(x)$.)
4. When is the following statement true? If x increases, a^x increases.
5. Sketch the graph of $f(x) = 3^{-x}$. Can you represent this function as $f(x) = a^x$?
6. On the same coordinate axes, sketch the graphs of $h(x) = 4^x$ and $j(x) = 5^x$. For what values of x is $h(x) > j(x)$?

In Exercises 7–10, graph the indicated function. (*Note:* You can sketch the graphs without plotting points by comparing the functions to $f(x) = 2^x$.)

7. $g_1(x) = 2^x + 2$ 8. $g_2(x) = 2^x - 2$
9. $g_3(x) = 2^{x+2}$ 10. $g_4(x) = 2^{x-2}$

11. We stated in the text that if $x_1 > x_2$, then $2^{x_1} > 2^{x_2}$. In fact, if $a > 1$, then $a^{x_1} > a^{x_2}$ whenever $x_1 > x_2$. Use this property to show that if $a^{x_1} = a^{x_2}$, then $x_1 = x_2$. (*Hint:* Assume $x_1 \neq x_2$.) Notice that this is the property that allows us to solve exponential equations using the method of Exercise 27 in Section 7.2.

12. Suppose we define the following function: $L(x) =$ the exponent to which 10 must be raised so that $10^{L(x)} = x$. For example, since $10^2 = 100$, $L(100) = 2$. Since $10^1 = 10$, $L(10) = 1$.

   a. Evaluate $L(1000)$.   **b.** Evaluate $L(1)$.
   c. Evaluate $L(1/10)$.   **d.** Evaluate $L(1/100)$.
   e. Find the domain of L.

13. Let $f(x) = 2^x$ and $g(x) = x^2 + 1$.

   a. Find $fg(x)$.   **b.** Find $\dfrac{f}{g}(x)$.   **c.** Find $f \circ g(x)$.

   d. Find $g \circ f(x)$.   **e.** Evaluate $f \circ g(1)$.   **f.** Evaluate $g \circ f(1)$.

14. Let $f(x) = 3^x$ and $g(x) = 1/x$.

   a. Find $fg(x)$.   **b.** Find $\dfrac{f}{g}(x)$.

   c. Find $f \circ g(x)$.   **d.** Find $g \circ f(x)$.

15. Let $f(x) = x$ and $g(x) = 2^x$.

   a. Find $fg(x)$.   **b.** Find $\dfrac{f}{g}(x)$.

   c. Find $f \circ g(x)$.   **d.** Find $g \circ f(x)$.

16. Throughout this section, we considered the function $f(x) = a^x$ only if $a > 0$.

   a. Graph the function $f(x) = 1^x$.
   b. What problem would you have trying to graph $f(x) = (-2)^x$? (*Hint:* The first problem is finding the domain.)

17. Consider the function $f(x) = 5 - 2^{-3x}$, $x \geq 0$.

   a. What is $f(0)$?
   b. As x increases, what happens to 2^{-3x}?
   c. As x increases, what happens to $f(x)$?
   d. Can $f(x)$ ever equal 5?
   e. Sketch the graph of f.

18. Sketch the graph of $f(x) = k - 2^{-cx}$, $c > 0$, $k > 0$.

19. Carbon-14 is a substance that decomposes as time passes. If there are initially 100 grams of carbon-14 in a sample, an equation that tells us the mass of carbon-14 remaining after t years is

$$M = 100 \cdot 2^{-(t/5568)}$$

   a. What mass of carbon-14 is left after 11,136 years?

b. How many years must pass for half the carbon-14 to decompose? (This value of t is called the **half-life**.)

20. A bacterial population increases exponentially. If the initial population is N, a formula for the population as a function of time is

$$P(t) = N \cdot 10^{kt}$$

where k is a constant and t is measured in hours.

a. Initially, a bacterial population is 250. After 2 hours, the population is 2,500,000. Use this information to find k and N.

b. What will the population be after 4 hours?

SECTION 7.4   **LOGARITHMS**

An important family of functions related to the exponential functions we studied in Sections 7.1, 7.2, and 7.3 is the family of logarithmic functions. In this section, we will discuss logarithms of numbers. After we have studied properties of logarithms we will define the logarithmic functions.

Definition

> Suppose $a > 0$ and $a \neq 1$. Then the logarithm of x, base a, is b if and only if $a^b = x$. (Notation: $\log_a x = b$.)

Our definition tells us that each time we write a statement about exponents, we can write an equivalent statement about logarithms. For example, $2^4 = 16$ is equivalent to $\log_2 16 = 4$ and $5^{-2} = \frac{1}{25}$ is equivalent to $\log_5 \frac{1}{25} = -2$.

As we study the properties of logarithms, always remember that *a logarithm is an exponent*. The logarithm of x, base a, is the exponent b such that $a^b = x$. Every statement about logarithms corresponds to a statement about exponents, and the properties of logarithms are properties of exponents.

Example 1

In this example, we will use the definition of logarithms to write logarithmic equivalents of statements about exponents.

A. $2^5 = 32 \Leftrightarrow \log_2 32 = 5$
B. $6^2 = 36 \Leftrightarrow \log_6 36 = 2$
C. $10^4 = 10,000 \Leftrightarrow \log_{10} 10,000 = 4$
D. $12^0 = 1 \Leftrightarrow \log_{12} 1 = 0$
E. $2^{-3} = \frac{1}{8} \Leftrightarrow \log_2 \frac{1}{8} = -3$
F. $16^{1/2} = 4 \Leftrightarrow \log_{16} 4 = \frac{1}{2}$

Example 2

We will use our definition of logarithm to evaluate

A.	$\log_5 25$	B.	$\log_{10} 1$	C.	$\log_3 27$
D.	$\log_2 64$	E.	$\log_2 2$	F.	$\log_8 4$

A. $\log_5 25 = 2$ since $5^2 = 25$
B. $\log_{10} 1 = 0$ since $10^0 = 1$
C. $\log_3 27 = 3$ since $3^3 = 27$
D. $\log_2 64 = 6$ since $2^6 = 64$
E. $\log_2 2 = 1$ since $2^1 = 2$
F. $\log_8 4 = \frac{2}{3}$ since $8^{2/3} = 4$

An important result of the definition of $\log_a x$ is that we cannot find $\log_a x$ if x is either zero or a negative number. For example, if $\log_2 0 = b$, then $2^b = 0$, but our work in Section 7.3 shows that there is no exponent satisfying this equation. Similarly, if $\log_2 (-8) = b$ then $2^b = -8$, but $2^b > 0$ for each real number b.

Exercises 7.4

In Exercises 1–10, write exponential statements equivalent to the given logarithmic statement.

1. $\log_{10} 1000 = 3$ **2.** $\log_{10} \left(\dfrac{1}{10} \right) = -1$

3. $\log_{25} 25 = 1$ **4.** $\log_7 49 = 2$

5. $\log_4 \left(\dfrac{1}{16} \right) = -2$ **6.** $\log_2 \dfrac{1}{8} = -3$

7. $\log_5 625 = 4$ **8.** $\log_{49} 7 = \dfrac{1}{2}$

9. $\log_6 216 = 3$ **10.** $\log_3 81 = 4$

In Exercises 11–20, write logarithmic statements equivalent to the given exponential statement.

11. $3^3 = 27$ **12.** $7^0 = 1$
13. $4^4 = 256$ **14.** $12^2 = 144$
15. $2^{-5} = \frac{1}{32}$ **16.** $6^2 = 36$
17. $8^{2/3} = 4$ **18.** $17^0 = 1$
19. $10^4 = 10,000$ **20.** $10^{-7} = 0.0000001$

In Exercises 21–30, evaluate the given logarithm.

21. $\log_4 16$ **22.** $\log_{10} 100,000$
23. $\log_{10} 0.01$ **24.** $\log_{23} 1$

25. $\log_{17} 1$ **26.** $\log_8 4$

27. $\log_{16} 8$ **28.** $\log_3 27$

29. $\log_{1/3} 27$ **30.** $\log_{12} 12$

31. Let $f(x) = 2^x$ and $g(x) = \log_2 x$. Show that, if (a, b) is a point on the graph of f, (b, a) is a point on the graph of g.

32. Assume that $\log_{10} 7 = 0.85$ and $\log_{10} 9 = 0.95$. (*Hint for (a):* Write $y = \log_{10} 7^2$, as an equivalent statement about exponents. Use the given information that $\log_{10} 7 = 0.85$.)

    a. Find $\log_{10} 7^2$ **b.** Find $\log_{10} 9^2$

    c. Find $\log_{10} 7 \cdot 9$ **d.** Find $\log_{10} 7^2 \cdot 9$

    e. Find $\log_{10} \dfrac{7}{9}$ **f.** Find $\log_{10} \dfrac{9}{7}$

33. The table of common logarithms at the end of the text gives the values of $\log_{10} x$ for $1 < x < 10$. Use the table to draw the graph of $\log_{10} x$ for values of x in $[1, 10]$.

34. Suppose we allowed $a = 1$ in the definition of $\log_a x$. Show that $f(x) = \log_1 x$ is not a function. For which values of x is $\log_1 x$ defined?

SECTION 7.5   **ALGEBRAIC PROPERTIES OF LOGARITHMS**

Suppose we are told that $\log_{10} 2 = 0.3010$ and $\log_{10} 6 = 0.7782$. Can we use this information to find $\log_{10} 12$? This is a special case of a more general question that we shall answer in this section: How is the logarithm of a product of two numbers related to the logarithm of the individual numbers? (That is, how is $\log_a xy$ related to $\log_a x$ and $\log_a y$?)

    Let us return to our example. Since $\log_{10} 2 = 0.3010$, $10^{0.3010} = 2$, and since $\log_{10} 6 = 0.7782$, $10^{0.7782} = 6$. Instead of writing

$$2 \times 6 = 12 \tag{1}$$

we can use the above results to write the equivalent statement

$$10^{0.3010} \times 10^{0.7782} = 12 \tag{2}$$

Our rules for exponents tell us that

$$10^{0.3010} \times 10^{0.7782} = 10^{0.3010 + 0.7782} = 10^{1.0792} \tag{3}$$

Comparing (2) and (3), we see that $12 = 10^{1.0792}$. We can write this exponential statement as the following statement about logarithms:

$$\log_{10} 12 = 1.0792$$

Since $1.0792 = 0.3010 + 0.7782 = \log_{10} 2 + \log_{10} 6$, we have

$$\log_{10} 12 = \log_{10} 2 + \log_{10} 6$$

Although 12 is the product of 2 and 6, $\log_{10} 12$ is the sum of $\log_{10} 2$ and $\log_{10} 6$! Actually, the result should not be surprising. Remember that a logarithm is an exponent, and when exponentials written with the same base are multiplied, their exponents are added. That is precisely what is happening here.

Exercise 35 of this section outlines a proof of the following rule:

$$\log_a xy = \log_a x + \log_a y \tag{4}$$

Although we stated rule (4) for the logarithm of a product of two factors, the rule generalizes to any number of factors. For example, $\log_a xyz = \log_a x + \log_a y + \log_a z$. Part (C) of Example 1 illustrates this idea.

Example 1

Suppose $\log_{10} 2 = 0.3010$ and $\log_{10} 3 = 0.4771$. Then, to find $\log_{10} 6$, $\log_{10} 4$, $\log_{10} 27$, and $\log_{10} 12$, we proceed as follows:

A. $\log_{10} 6 = \log_{10} (2 \cdot 3) = \log_{10} 2 + \log_{10} 3$
$$= 0.3010 + 0.4771 = 0.7781$$

B. $\log_{10} 4 = \log_{10} (2 \cdot 2) = \log_{10} 2 + \log_{10} 2$
$$= 2 \log_{10} 2 = 0.6020$$

C. $\log_{10} 27 = \log_{10} (3 \cdot 3 \cdot 3) = \log_{10} 3 + \log_{10} 3 + \log_{10} 3$
$$= 3 \log_{10} 3 = 1.4313$$

D. $\log_{10} 12 = \log_{10} (2 \cdot 2 \cdot 3) = \log_{10} 2 + \log_{10} 2 + \log_{10} 3$
$$= 0.3010 + 0.3010 + 0.4771$$
$$= 1.0791$$

Example 2

Suppose $\log_{10} 5 = 0.6990$. We can use this result to find $\log_{10} 50$, $\log_{10} 500$, and $\log_{10} (0.5)$.

A. $\log_{10} 50 = \log_{10} (10 \cdot 5) = \log_{10} 10 + \log_{10} 5$
$$= 1 + 0.6990 = 1.6990$$

B. $\log_{10} 500 = \log_{10} (100 \cdot 5) = \log_{10} 100 + \log_{10} 5$
$$= 2 + 0.6990 = 2.6990$$

C. $\log_{10} (0.5) = \log_{10} (\frac{1}{10} \cdot 5) = \log_{10} \frac{1}{10} + \log_{10} 5$
$$= -1 + 0.6990 = -0.3010$$

Parts (B) and (C) of Example 1 indicate how $\log_{10} x^b$ is related to $\log_{10} x$. We saw that $\log_{10} 2^2 = 2 \log_{10} 2$ and $\log_{10} 3^3 = 3 \log_{10} 3$.

In general,

$$\log_a x^b = b \log_a x \tag{5}$$

To prove (5), let

$$\log_a x = u$$

Then

$$x = a^u$$

and

$$x^b = (a^u)^b = a^{bu}$$

So

$$\log_a x^b = bu = b \log_a x$$

Example 3

> To illustrate how (5) is used, suppose $\log_{10} 2 = 0.3010$. Then
>
> A. $\log_{10} 16 = \log_{10} 2^4 = 4 \log_{10} 2 = 4(0.3010) = 1.2040.$
> B. $\log_{10} \sqrt[4]{2} = \log_{10} 2^{1/4} = \frac{1}{4} \log_{10} 2 = \frac{1}{4}(0.3010) = 0.0753.$

Frequently, both (4) and (5) are used together as in the next example.

Example 4

> Suppose $\log_{10} 2 = 0.3010$, $\log_{10} 3 = 0.4771$, $\log_{10} x = A$, $\log_{10} y = B$, and $\log_{10} z = C$. We will use this information to find $\log_{10} 18$ and $\log_{10} x^2 yz^3$.
>
> A. $\log_{10} 18 = \log_{10} (2 \cdot 3^2) = \log_{10} 2 + \log_{10} 3^2$
> $$= \log_{10} 2 + 2 \log_{10} 3$$
> $$= 0.3010 + 2(0.4771) = 1.2552$$
> B. $\log_{10} x^2 yz^3 = \log_{10} x^2 + \log_{10} y + \log_{10} z^3$
> $$= 2 \log_{10} x + \log_{10} y + 3 \log_{10} z$$
> $$= 2A + B + 3C$$

We now have rules (4) and (5) that help us evaluate $\log_a xy$ and $\log_a x^b$. We would like to derive a rule for evaluating $\log_a \frac{x}{y}$.

$$\log_a \frac{x}{y} = \log_a \left(x \cdot \frac{1}{y} \right)$$
$$= \log_a (x \cdot y^{-1})$$
$$= \log_a x + \log_a y^{-1} \qquad \text{(Rule 4)}$$
$$= \log_a x - \log_a y \qquad \text{(Rule 5)}$$

If we remember that a logarithm is an exponent, this rule reflects the following property of exponents: When dividing exponentials with the same base, subtract exponents.

Example 5

Suppose $\log_{10} 2 = 0.3010$ and $\log_{10} 3 = 0.4771$. Then

A. $\log_{10} \dfrac{1}{2} = \log_{10} 1 - \log_{10} 2$

$\qquad = 0 - 0.3010 = -0.3010$

B. $\log_{10} \left(\dfrac{\sqrt[3]{2}}{8} \right) = \log_{10} \sqrt[3]{2} - \log_{10} 8$

$\qquad\qquad = \log_{10} 2^{1/3} - \log_{10} 2^3$

$\qquad\qquad = \dfrac{1}{3} \log_{10} 2 - 3 \log_{10} 2 \doteq -0.8027$

C. Since $5 = \dfrac{30}{6} = \dfrac{3 \times 10}{2 \times 3}$

$\log_{10} 5 = \log_{10} \left(\dfrac{3 \times 10}{2 \times 3} \right)$

$\qquad = \log_{10} (3 \times 10) - \log_{10} (2 \times 3)$

$\qquad = \log_{10} 3 + \log_{10} 10 - (\log_{10} 2 + \log_{10} 3)$

$\qquad = 1.4771 - 0.7781 = 0.6990$

In this section we have obtained three results about logarithms. They will be used so frequently that we suggest you memorize them.

$$\log_a xy = \log_a x + \log_a y$$
$$\log_a x^b = b \log_a x$$
$$\log_a \frac{x}{y} = \log_a x - \log_a y$$

Logarithms base 10 are called **common logarithms**, and they are used so frequently that it has become an accepted convention (and one that we will adopt throughout the remainder of this text) to use the symbol log for $\log_{10}$. For example, log 2 is understood to be $\log_{10} 2$.

At the end of this text, there is a table of common logarithms. This table gives log x for values of x in $[1, 10]$, starting at 1 in increments of 0.01 unit. We shall see in the appendix to this section how to use such a table to compute (or at least closely approximate) log x for any positive number x.

Actually, it is less common now to use tables like the one in this text to compute logarithms than it once was, since most scientific calculators have a key that can be used to find the common logarithm of almost any number x, not just a number between 1 and 10.* In a sense, the log function is in the calculator: The calculator has been given a rule by which it can compute $\log x$ for any number x.

You will notice that tables for logarithms other than base 10 have not been included in this text.† The reason is that a single table of logarithms for any specified base (in our case, base 10) is sufficient to compute $\log_a x$ for any base and any positive number x. To illustrate, suppose we want to find $\log_3 8$. If we let $\log_3 8 = y$, then

$$\log_3 8 = y$$
$$\Leftrightarrow 3^y = 8$$

If numbers are equal, their logarithms are equal, and vice versa. So

$$3^y = 8$$
$$\Leftrightarrow \log (3^y) = \log 8$$
$$\Leftrightarrow y \log 3 = \log 8$$
$$\Leftrightarrow y = \frac{\log 8}{\log 3} = \frac{0.9031}{0.4771} = 1.8928$$

Notice that the logarithms in this last equation are common logarithms that we can find in our table. (Do not confuse $(\log 8)/(\log 3)$ with $\log \frac{8}{3} = \log 8 - \log 3$. In one case, we have the quotient of two logarithms; in the other, the logarithm of a quotient.)

Example 6

To find $\log_3 5$, let $\log_3 5 = A$, then

$$3^A = 5$$
$$\Leftrightarrow \log 3^A = \log 5$$
$$\Leftrightarrow A \log 3 = \log 5$$
$$\Leftrightarrow A = \frac{\log 5}{\log 3} = \frac{0.6990}{0.4771} = 1.4651$$

An equation such as $3^y = 8$, in which the variable is in the exponent, is called an **exponential equation.** You will surely see such equations while studying

---

*There are, of course, limits to the size of numbers that can be used in a calculator.

---

†A logarithm called a **natural logarithm** is used as frequently as the common logarithm. You will find tables for both common and natural logarithms in some texts, and scientific calculators frequently have both a natural and a common logarithm key.

calculus. Except for the simplest examples, solving exponential equations requires the use of logarithms.

Example 7

In this example, we will solve three exponential equations.

A. To solve $3^x = 81$, notice that $81 = 3^4$. Hence,

$$3^x = 81$$
$$\Leftrightarrow 3^x = 3^4$$
$$\Leftrightarrow x = 4$$

B. To solve $3^x = 80$, take the logarithm of both sides.

$$3^x = 80$$
$$\Leftrightarrow x \log 3 = \log 80$$
$$\Leftrightarrow x = \frac{\log 80}{\log 3} = \frac{1.9031}{0.4771} = 3.9889$$

C. To solve $3^{2x} = 4^{x-1}$ take the logarithm of both sides.

$$2x \log 3 = (x - 1) \log 4$$
$$\Leftrightarrow (2x)(0.4771) = (x - 1)(0.6021)$$
$$\Leftrightarrow 0.9542x = 0.6021x - 0.6021$$
$$\Leftrightarrow 0.3521x = -0.6021$$
$$\Leftrightarrow x = -\frac{0.6021}{0.3521} = -1.7100$$

Example 8

Given the function, p, defined by

$$p(t) = 1000(10^t)$$

A. What is the value of p when $t = 0$?
B. For what value of t will $p(t) = 2000$?

A. $p(0) = 1000(10^0) = 1000$.
B. Suppose t is the desired value. Then

$$2000 = 1000(10^t)$$
$$\Leftrightarrow 2 = 10^t$$
$$\Leftrightarrow \log 2 = t \log 10$$
$$\Leftrightarrow t = \frac{\log 2}{\log 10} = \frac{0.3010}{1} = 0.3010$$

Exercises 7.5   In Exercises 1–15, use the following information to evaluate each of the given logarithms.

$$\log 2 = 0.3010 \qquad \log 3 = 0.4771 \qquad \log 5 = 0.6990$$

1. $\log 16$	**2.** $\log 9$	**3.** $\log 6$	**4.** $\log 18$
5. $\log 75$	**6.** $\log 25$	**7.** $\log 0.006$	**8.** $\log \frac{1}{3}$
9. $\log \frac{1}{6}$	**10.** $\log \frac{3}{5}$	**11.** $\log 50$	**12.** $\log 36$
13. $\log 27$	**14.** $\log 250$	**15.** $\log 1.5$	

Express each of the expressions in Exercises 16–23 in terms of A, B, and C if $\log x = A$, $\log y = B$, and $\log z = C$.

16. $\log x^2 y$ 　　　　　　　　　　　　**17.** $\log 10\, yz$

18. $\log (xy/z)$ 　　　　　　　　　　　**19.** $\log (x^2/yz)$

20. $\log \sqrt[3]{xy}/z$ 　　　　　　　　　**21.** $\log \dfrac{(xy)^2}{z^3}$

22. $\log x^2 y^2 z^3$ 　　　　　　　　　　**23.** $\log (xyz^2)^3$

24. Find $\log_4 7$ 　　　　　　　　　　　**25.** Find $\log_3 2$

26. Find $\log_5 8$ 　　　　　　　　　　　**27.** Find $\log_7 12$

In Exercises 28–31, solve for x.

28. $3^{x-2} = 10^{x-1}$ 　　　　　　　　　**29.** $2^{x+1} = 16$

30. $2^{x^2} = 5$ 　　　　　　　　　　　　**31.** $x^2 \log 2 - \log 2 = 3x \log 5$

32. **a.** What is $\log 10^4$? 　　　　　　**b.** What is $\log_3 3^5$?
　　　c. What is $\log_7 7^9$?

33. **a.** What is $10^{\log 5}$? 　　　　　　**b.** What is $5^{\log_5 12}$?

34. **a.** What is $\log_k k^a$? 　　　　　　**b.** What is $k^{\log_k a}$?

35. This exercise outlines a proof of property (4), page 277. Let

$$\log_k a = x \quad \text{and} \quad \log_k b = y$$

　　a. Write equivalent exponential statements for

$$\log_k a = x \quad \text{and} \quad \log_k b = y$$

　　b. Use your answer to (a) to write ab as k raised to an exponent. $(ab = k^?)$

　　c. Use your answer to (b) to show that $\log_k ab = \log_k a + \log_k b$.

36. Problem 33 in Section 7.4 asked you to sketch a graph of $\log x$ on the interval $[1, 10]$. This problem asks you to sketch the graph of $\log x$ on the interval $[0.1, 100]$. If x is between 10 and 100, then $x = y \cdot 10$, where

$1 < y < 10$. (For example, $52 = 5.2 \cdot 10$.) Since $x = y \cdot 10$, $\log x = \log y + \log 10 = 1 + \log y$. Similarly, if $0.1 < x < 1$, then $x = y \cdot 10^{-1}$ where $1 < y < 10$. (For example, $0.52 = 5.2 \cdot 10^{-1}$.) Since $x = y \cdot 10^{-1}$, $\log x = \log y + 10^{-1} = \log y - 1$.

Use these facts and the table of common logarithms to complete the following table:

x	0.1	0.2	0.3	0.4	0.5	0.6	0.7	0.8	0.9	1	10	20	30	...	100
$\log x$															

Use this table to sketch the graph of $\log x$ on $[0.1, 100]$.

 37. A bacterial population's growth is described by the equation

$$P(t) = N \cdot 10^{kt}$$

where P is the population, t is the time in hours, and N and k are constants.

a. When $t = 0$, the population is 5000. Use this information to find N. (The population, when $t = 0$, is called the **initial population**.)

b. Two hours later the population has grown to 25,000. Find k.

c. Predict the population after 5 hours.

 38. The mass of a radioactive substance, in grams, is given by the formula

$$m = 100 \cdot 2^{-kt}$$

where t represents time in years.

a. How many grams of the substance are present initially (i.e., when $t = 0$)?

b. After 1 year, there are 90 grams of the substance left. Use this information to find k.

7.5 APPENDIX   **USING THE TABLE OF LOGARITHMS**

This section explains how to find the logarithm of a number using the table of common logarithms that appears at the end of this text. Actually, all the theory necessary to compute logarithms with the table has already been discussed in the text and in the exercises.

Any number can be written in the form

$$a \times 10^n$$

where a is a real number between 1 and 10, and n is an integer. For example,

$$850 = 8.5 \times 100 = 8.5 \times 10^2$$
$$2500 = 2.5 \times 1000 = 2.5 \times 10^3$$
$$31,650 = 3.165 \times 10,000 = 3.165 \times 10^4$$
$$0.05 = 5 \times \tfrac{1}{100} = 5 \times 10^{-2}$$

Writing numbers in this form is very common in scientific work, and when a number is written as $a \times 10^n$ we say that it is written in **scientific notation**.

If we try to write very large numbers, such as 16,000,000, or very small numbers, such as 0.0000164, in scientific notation, it may be difficult to decide the correct values for a and n, and the following rules may be helpful:

A.　Moving from the extreme left of the number, draw a vertical line to the right of the first nonzero digit.

B.　Let c be the number of digits between the line and the decimal point.

C.　If the decimal point lies to the right of the vertical line, the exponent of 10 is c. If the decimal point lies to the left of the vertical line, the exponent is $-c$.

D.　The number a is the number between 1 and 10 having the same digits as the original number, but with its decimal point at the vertical line.

We will illustrate the rules with the numbers 16,000,000 and 0.0000164:

$$1 \,|\, 6,000,000. = 1.6 \times 10^7$$

The vertical line is 7 places to the left of the decimal point and therefore the exponent of 10 is 7.

$$0.00001 \,|\, 64 = 1.64 \times 10^{-5}$$

The vertical line is 5 places to the right of the decimal point and therefore the exponent is -5.

Notice that when a number between 0 and 1 is written in scientific notation the exponent of 10 is a negative integer. When a number greater than 1 is written in scientific notation, the exponent is a non-negative integer.

Example 1

A.　$2120 = 2.12 \times 10^3$	B.　$324.5 = 3.245 \times 10^2$
C.　$10,500 = 1.05 \times 10^4$	D.　$0.00054 = 5.4 \times 10^{-4}$

To compute the common logarithm of a number, we use scientific notation and these two properties of logarithms.

$$\log xy = \log x + \log y$$
$$\log 10^n = n \log 10 = n$$

For example, suppose we wanted to find log 2120. Using scientific notation

$$2120 = 2.12 \times 10^3$$
$$\log 2120 = \log 2.12 + \log 10^3$$
$$\log 2120 = \log 2.12 + 3 \log 10$$
$$\log 2120 = \log 2.12 + 3 \tag{1}$$

The table of common logarithms lists the logarithms of all three-digit numbers between 1 and 10. To find $\log x$, locate the first two digits of x in the column to the left of the table and the third digit in the row above the table. Table 7.5.1 shows some of the entries of the row corresponding to 2.1. The logarithm of 2.12 is 0.3263. From equation (1) we see that the logarithm of 2120 is 3.3263.

Table 7.5.1

	0	1	2	3	4	5
2.1	0.3222	0.3243	0.3263	0.3284	0.3304	0.3324

The logarithm of a number x can be viewed as an integer plus a decimal between 0 and 1. (For example, log 2120 = 3.3263 = 3 + 0.3263.) The integer part is called the **characteristic** and the decimal part is called the **mantissa** of the logarithm. (The characteristic of log 2120 is 3 and the mantissa is 0.3263.)

Calculating a logarithm requires using the table to find the mantissa, or decimal part of the logarithm, and using scientific notation to find the characteristic, or integer part of the logarithm. (Notice that the characteristic is just the exponent of 10 when the number is written in scientific notation.) For example, log 215 = 2.3324. The mantissa, 0.3324, is found in the table. Since $215 = 2.15 \times 10^2$, the characteristic is 2.

Example 2

A. $\log 846 = \log (8.46 \times 10^2) = \log 8.46 + \log 10^2$
$$= 0.9274 + 2 = 2.9274$$

B. $\log 12{,}400 = \log (1.24 \times 10^4) = \log 1.24 + \log 10^4$
$$= 0.0934 + 4 = 4.0934$$

C. $\log 435{,}000 = \log (4.35 \times 10^5) = \log 4.35 + \log 10^5$
$$= 0.6385 + 5 = 5.6385$$

Example 3

A. $\log (0.5) = \log (5 \times 10^{-1})$

$$= \log 5 + \log 10^{-1}$$

$$= 0.6990 + (-1) = 0.6990 - 1 = -0.3010$$

Note that the characteristic of $\log 0.5$ is -1 and the mantissa is 0.6990.

    There are two errors commonly made here. The first is to think that the mantissa of $\log (0.5)$ is 0.3010, when it is actually 0.6990. The second error is to think that $0.6990 - 1$ is the same number as -1.6990.

B. $\log (0.0017) = \log (1.7 \times 10^{-3}) = \log 1.7 + \log 10^{-3}$

$$= 0.2304 - 3 = -2.7696$$

The characteristic of $\log (0.0017)$ is -3 and the mantissa is 0.2304.

C. $\log (0.00346) = \log (3.46 \times 10^{-3}) = \log 3.46 + \log 10^{-3}$

$$= 0.5391 - 3 = -2.4609$$

The characteristic of $\log (0.00346)$ is -3 and the mantissa is 0.5391.

    The process we have described above is reversible. If we are given a logarithm, y, we can use the table to find the number, x, whose logarithm is y. We say that we are finding the antilogarithm of y. For example, antilog $1.8451 = 70$, since $\log 70 = 1.8451$.

    To find antilog y:

A.   Express y in the form $c + n$, where n is an integer and $0 \le c < 1$.

B.   Use the table of common logarithms to find a number, a, such that $\log a = c$.

C.   Antilog $y = a \times 10^n$.

Example 4

To find antilog 3.8537:

A.   Express 3.8537 as $0.8537 + 3$ ($c = 0.8537$ and $n = 3$).

B.   From the table of common logarithms, we see that $\log (7.14) = 0.8537$.

C.   Antilog $3.8537 = 7.14 \times 10^3 = 7140$.

Example 5

> A. Find antilog 1.4232.
> B. Find antilog 0.6107 − 1.
> C. Find antilog −0.3010.
>
> Solution:
>
> A. $1.4232 = .4232 + 1$. Using the table of logarithms, we see that $\log 2.65 = 0.4232$.
>
>   antilog $1.4232 = 2.65 \times 10^1 = 26.5$
>
> B. The number 0.6107 is already in the necessary form. Using the table of logarithms, we see that $\log 4.08 = 0.6107$.
>
>   antilog $(0.6107 − 1) = 4.08 \times 10^{-1} = 0.408$
>
> C. Our first step is to write -0.3010 as $c + n$, where n is an integer and $0 \le c < 1$.
>
> $-0.3010 = -1 + 0.6990$
>
>   Therefore $n = -1$ and $c = 0.6990$. Using the table of logarithms we see that $\log 5 = 0.6990$.
>
>   antilog $-3010 = 5 \times 10^{-1} = 0.5$

In these examples, we have illustrated a method for finding antilog x for different values of x. It is important to realize that, when we find antilog x, we are finding 10^x. To understand why this is true, suppose $y =$ antilog x. This is equivalent to $\log y = x$, and writing this as an equivalent statement about exponentials, $y = 10^x$.

Finally, suppose we want to find log 5.655. Our table allows us to find log 5.65 and 5.66, but it has no entry for 5.655. We have two possible courses of action. The first is to find a better table. (This sounds facetious, but it isn't meant to be. When a table doesn't have a desired value, we frequently try to find a better table.) The other possibility is to argue that, since 5.655 is halfway between 5.65 and 5.66, log 5.655 should be halfway between log 5.65 and log 5.66. In fact, this argument is incorrect. The logarithm of 5.655 is not halfway between log 5.65 and log 5.66. However, when we estimate log 5.655 with the number halfway between log 5.65 and log 5.66, our error is small. This method is called **linear interpolation** and can be applied to any table.

Example 6

> We can use the method of linear interpolation to find log 1853. Using the table, we find $\log 1850 = 3.2672$ and $\log 1860 = 3.2695$. On a number line, we see that the distance between 1850 and 1853 is $\frac{3}{10}$ the distance

between 1850 and 1860. (See Figure 7.5.1.)

Figure 7.5.1

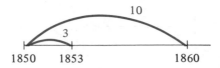

$$\text{1850} \qquad \text{1853} \qquad\qquad\qquad \text{1860}$$

We will assume that the same relationship holds for the corresponding logarithms.

$$\log 1853 - \log 1850 = \tfrac{3}{10} (\log 1860 - \log 1850)$$

$$\log 1853 - 3.2672 = \tfrac{3}{10} (3.2695 - 3.2672)$$

$$\log 1853 - 3.2672 = \tfrac{3}{10} (0.0023)$$

$$\log 1853 - 3.2672 = 0.00069$$

$$\log 1853 = 3.2679$$

Since the data we used to find log 1853 (the entries in the table of logarithms) are only accurate to four decimal places, we rounded our answer off to four decimal places.

Exercises 7.5
Appendix

Find the logarithms of the following numbers.

1. 842	**2.** 6080	**3.** 3070	**4.** 4250
5. 16,800	**6.** 0.00847	**7.** 0.0471	**8.** 8.21
9. 0.000315	**10.** 0.685		

Find antilogarithms of the following numbers.

11. 2.3617	**12.** 0.4548 − 1
13. 4.6693	**14.** 1.0374
15. 2.8007	**16.** 0.3181 − 3
17. −0.1524	**18.** −0.2204
19. 3.9791	**20.** 0.9881

21. Use linear interpolation to find log 225.4.

SECTION 7.6   **LOGARITHMIC FUNCTIONS AND THEIR GRAPHS**

In Section 7.4, we defined expressions of the form $\log_a x$ for different values of x. We now use these expressions to define the logarithmic functions.

Let a be a positive number not equal to 1. The function, f, with domain $(0, +\infty)$ defined by

$$f(x) = \log_a x$$

is called the **logarithmic function base a**.

We could graph the function $f(x) = \log_a x$ by plotting points. However, we prefer to use some of the theory we have already developed. In particular, we will use results about functions and their inverses.

Recall that a relationship between a function, f, and its inverse, f^{-1}, is that

$$y = f(x) \Leftrightarrow x = f^{-1}(y) \tag{1}$$

We have defined the logarithmic and exponential functions so that

$$y = \log_a x \Leftrightarrow x = a^y \tag{2}$$

Comparing (1) and (2) we see that the logarithmic function base a and the exponential function base a are inverses of each other.

We know that for any function, f, and its inverse, f^{-1}, $D_f = R_{f^{-1}}$ and $R_f = D_{f^{-1}}$. Since the domain of the exponential function base a is $(-\infty, \infty)$, this is also the range of the logarithmic function base a, and since the range of the exponential function is $(0, +\infty)$, this is also the domain of the logarithmic function base a. Also, we know that the graphs of $y = f(x)$ and $y = f^{-1}(x)$ are mirror images of each other with respect to the graph of $y = x$.

We will use this fact to sketch the graph of $y = \log x$. Figure 7.6.1 shows the graphs of $y = 10^x$, $y = x$, and $y = \log x$.

Figure 7.6.1

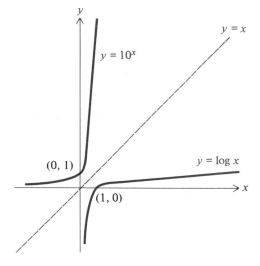

Notice that the graphs $y = 10^x$ and $y = \log x$ are mirror images of each other in the line $y = x$.

It appears from the graph of $y = \log x$ that, as x gets close to zero, $\log x$ decreases without bound while, as x gets larger, $\log x$ increases without bound.

To understand why this is true, notice that for all real numbers, k,

$$\log 10^k = k$$

Suppose $x = 1000 = 10^3$. Then $\log x = 3$. If $x = 1,000,000 = 10^6$, $\log x = 6$. If $x = 10^{12}$, $\log x = 12$ and, in general, we can make $\log x$ as large as we like. If we want $\log x$ to equal any large number, n, we choose $x = 10^n$.

Similarly, if $x = \frac{1}{10} = 10^{-1}$, $\log x = -1$. If $x = \frac{1}{100} = 10^{-2}$, $\log x = -2$. If $x = 10^{-6}$, $\log x = -6$. If $x = 10^{-12}$, $\log x = -12$, and, in general, we can make $\log x$ equal to any negative number. If we want $\log x$ to equal $-n$, $n > 0$, let $x = 10^{-n}$.

The point $(1, 0)$ is on the graph of $y = \log x$ since $\log 1 = 0$. More generally, since $\log_a 1 = 0$ for all positive a, $a \neq 1$, we see that $(1, 0)$ is on the graph of $y = \log_a x$.

To sketch the graphs of $y = \log_a x$ for different values of a, we could use the same approach that we used with $y = \log x$. We could draw the graph of $y = a^x$. The mirror image of this graph in the line $y = x$ is the graph of $y = \log_a x$. We could also sketch the graph by plotting points. We have plotted points for the function $f(x) = \log_2 x$ and sketched the graph in Figure 7.6.2.

Figure 7.6.2

x	$\frac{1}{64}$	$\frac{1}{32}$	$\frac{1}{16}$	$\frac{1}{8}$	$\frac{1}{4}$	$\frac{1}{2}$	1	2	4	8	16	32	64
$\log_2 x$	-6	-5	-4	-3	-2	-1	0	1	2	3	4	5	6

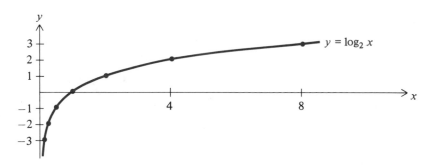

We will now illustrate a different method for sketching the graph of $y = \log_2 x$ which uses the relationship between $\log_2 x$ and $\log x$.

In Section 7.5, we saw that we could find the logarithm of a number with any base using the table of common logarithms. Suppose, for example, that x represents a number greater than 0, and we want to find $\log_2 x$. If $y = \log_2 x$, then

$$y = \log_2 x$$
$$\Leftrightarrow 2^y = x$$
$$\Leftrightarrow \log (2^y) = \log x$$

$$\Leftrightarrow y \log 2 = \log x$$

$$\Leftrightarrow y = \frac{\log x}{\log 2}$$

Since $y = \log_2 x$, we see

$$\log_2 x = \frac{\log x}{\log 2}$$

Since $\log 2 = 0.3010$,

$$\log_2 x = \frac{\log x}{0.3010} = \left(\frac{1}{0.3010}\right) \log x = 3.023 \, (\log x)$$

The function $f(x) = \log_2 x$ can also be written as $f(x) = 3.023 \, (\log x)$.

We know (from Section 4.6.3) the relationship between the graphs of $y = f(x)$ and $y = cf(x)$.

In Figure 7.6.3, we've drawn both $y = \log x$ and $y = \log_2 x$. Notice that each point on the graph of $y = \log_2 x$ is 3.023 times as far from the x-axis as the point on the graph of $y = \log x$ with the same x-coordinate.

Figure 7.6.3

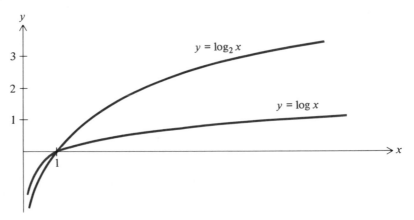

In general,

$$\log_a x = \frac{\log x}{\log a}$$

$$= \frac{1}{\log a} (\log x)$$

Since $1/(\log a)$ is a constant, any logarithmic function can be written as a multiple of the common logarithmic function, and this property helps us to sketch the graph of $y = \log_a x$.

Example 1

To sketch the graph of $y = \log_{100} x$, as shown in Figure 7.6.4, notice that

$$\log_{100} x = \frac{\log x}{\log 100} = \frac{\log x}{2} = \frac{1}{2} \log x$$

Figure 7.6.4

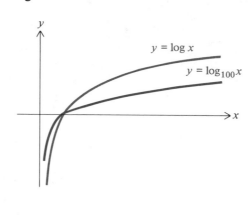

Exercises 7.6

1. **a.** On the same set of axes, draw the graphs of $y = \log_3 x$ and $y = 3^x$. Be sure to include negative values of x for the exponential curve and values of x between 0 and 1 for the logarithmic curve.
   b. What is the relationship between these graphs?

2. Sketch graphs of $y = \log x$ and $y = \log_{1/10} x$ on the same coordinate system. What is the relationship between the graphs?

In Exercises 3–8, graph the given functions. (*Hint:* You should be able to graph each function without plotting points by using the methods of Section 4.6.3.)

3. $g(x) = \log x + 3$

4. $g_2(x) = \log x - 3$

5. $g_3(x) = \log (x - 3)$

6. $g_4(x) = \log (x + 3)$

7. $g_5(x) = \log (x - 3) + 3$

8. $g_6(x) = 3 \log x$

9. Use the method of Example 1 to sketch the graphs of (a) and (b). (*Note:* $\log 5 = 0.6990$.)

   a. $y = \log_5 x$

   b. $y = \log_{1/5} x$

10. Use the method of Example 1 to sketch the graphs of (a) and (b). (*Note:* $\log 7 = 0.8451$.)

    a. $y = \log_7 x$

    b. $y = \log_{1/7} x$

11. For what value of x does $\log x = 1000$?

12. For what value of x does $\log x = 150$?

13. For what value of x does $\log x = -50$?
14. For what value of x does $\log x = -100$?
15. Sketch the graph of $y = \log x^2$.
16. Sketch the graph of $y = \log(-x)$.
17. Sketch the graph of $y = (\log x)^2$.
18. What is the domain of the function $f(x) = \log(\sin x)$?
19. Solve the equation $P = N \cdot 10^{kt}$, where N and k are constants, for t.
20. Solve the equation $m = 100 \cdot 2^{-kt}$, where k is a constant, for t.

SECTION 7.7

AN APPLICATION OF LOGARITHMS

In this section, we will discuss an important application of logarithms. Before considering this application, it will be convenient to generalize our definition of exponential function. We will consider functions defined by

$$f(x) = ca^{kx}$$

where k, a, and c are real numbers, $c \neq 0$, $a > 0$, and $a \neq 1$ to be exponential functions. When working with the exponential function defined by $f(x) = ca^{kx}$, the numbers c, a, and k are called the **parameters** of the function.

For example, the functions f, g, and h, defined by

$$f(x) = 3 \cdot 2^{4x} \qquad g(x) = 10 \cdot 5^{2x} \quad \text{and} \quad h(x) = \sqrt{2} \cdot 3^{1.4x}$$

are all exponential functions. The functions we studied earlier, functions defined by $f(x) = a^x$, also satisfy the above definition with $k = 1$ and $c = 1$.

Suppose we are working with the exponential function $f(x) = c \cdot 10^{kx}$, and we are told that $f(2) = 2000$ and $f(1) = 20$. We can use this information to find the values of c and k. Since $f(2) = 2000$, we can write

$$2000 = c \cdot 10^{2k} \tag{1}$$

Similarly, since $f(1) = 20$

$$20 = c \cdot 10^{k} \tag{2}$$

Since $20 = c \cdot 10^k$, we can divide the left side of equation (1) by 20 and the right side of equation (1) by $c \cdot 10^k$ to get

$$100 = 10^k \tag{3}$$

Therefore $k = 2$. Substituting $k = 2$ into equation (2), we have $20 = c \cdot 10^2$ and $c = \frac{1}{5}$. The exponential function f is defined by $f(x) = \frac{1}{5} \cdot 10^{2x}$.

Example 1

> Suppose the points $(2, 5)$ and $(4, 50)$ lie on the graph of $y = c \cdot 10^{kx}$. To find c and k we write
>
> $$50 = c \cdot 10^{4k} \tag{4}$$

$$5 = c \cdot 10^{2k} \tag{5}$$

and dividing the left and right sides of equation (4) by the corresponding sides of equation (5),

$$10 = 10^{2k} \tag{6}$$

$$\Leftrightarrow 2k = 1$$

$$\Leftrightarrow k = \tfrac{1}{2}$$

Substituting this result in equation (5)

$$5 = c \cdot 10^{2 \cdot \frac{1}{2}}$$

$$5 = c \cdot 10$$

$$c = \tfrac{1}{2}$$

Therefore, the points lie on the graph of $y = \tfrac{1}{2}(10^{(1/2)x})$.

In Example 1, we were told that the base of the exponential function was 10. The reason we assumed that $a = 10$ was that every exponential function can be written as an exponential function with base 10.

Suppose, for example, that we are working with the function $f(x) = 2^x$. Since $\log_{10} 2 = 0.3010$, $10^{0.3010} = 2$, and we can write $f(x) = 2^x$ as $f(x) = (10^{0.3010})^x = 10^{0.3010x}$, which is an exponential function with base 10.

Example 2

We can write the functions $f(x) = \tfrac{1}{4} \cdot 3^x$ and $g(x) = 4 \cdot 5^{2x}$ as exponential functions with base 10.
   Since $\log 3 = 0.4771$,

$$3 = 10^{0.4771}$$

and

$$f(x) = \tfrac{1}{4} \cdot 3^x = \tfrac{1}{4}(10^{0.4771})^x = \tfrac{1}{4}(10^{0.4771x})$$

Since $\log_{10} 5 = 0.6990$,

$$5 = 10^{0.6990}$$

and

$$g(x) = 4 \cdot (10^{0.6990})^{2x} = 4 \cdot 10^{1.398x}$$

In the remainder of this section, we will write exponential functions using base 10.

If we know that we are working with an exponential function, f, we have seen how to use the coordinates of two distinct points on the graph of f to find the parameters c and k. Frequently, however, we are faced with a much different question. Suppose we are given a collection of points. How can we decide if these points lie on the graph of an exponential function?

To illustrate how such a problem might arise, suppose a biologist wants to study the effect a nutrient has on bacterial growth. Nutrient is added to a bacterial culture and at various times, t, the biologist takes many samples, examines them under a microscope and computes the average number of bacteria per square millimeter on a slide.

The data are collected as shown in Table 7.7.1.

Table 7.7.1

t (in hours)	0	0.5	1	1.5	2	2.5	3
n (average number/mm^2	3	4.87	7.9	12.86	20.86	33.94	55.13

What is the relationship between t and n? A standard first step is to plot the data points. The result is shown in Figure 7.7.1.

Figure 7.7.1

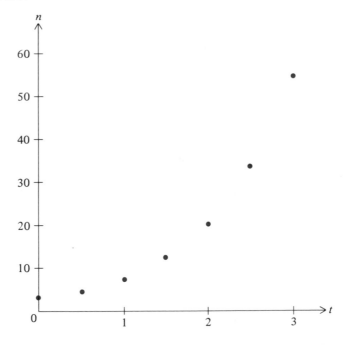

We might think, based on the shape of the graph suggested by the seven data points, that n is an exponential function of t. But this is difficult to observe visually. The six points plotted in Figure 7.7.2 determine a graph with roughly the same shape as the graph in Figure 7.7.1. However, you can verify that these points satisfy the equation $y = x^3 + 1$, which is not an exponential function.

Figure 7.7.2

x	0	0.5	1	1.5	2	2.5
y	1	1.125	2	3.75	9	15.625

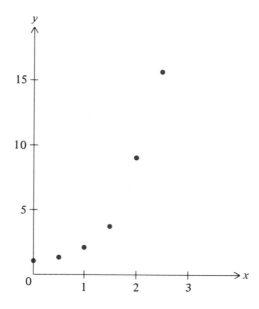

Visually, it is difficult to differentiate between graphs of exponential functions and graphs of other functions we have studied, based on only a few points.

Properties of logarithms give us a method to determine visually if points lie on the graph of an exponential function. We will discuss this method and then apply it to the biologist's data.

Suppose variables x and y are related exponentially; i.e., $y = c \cdot 10^{kx}$.

$$y = c \cdot 10^{kx}$$

$$\Leftrightarrow \log y = \log (c \cdot 10^{kx})$$

$$\Leftrightarrow \log y = \log c + \log 10^{kx}$$

$$\Leftrightarrow \log y = \log c + kx \log 10$$

$$\Leftrightarrow \log y = kx + \log c$$

Since c is a constant, $\log c$ is also a constant. Let $\log c = b$. Comparing the first equality, $y = c \cdot 10^{kx}$, and the last, $\log y = kx + b$, we see that y is an exponential function of x if and only if log y is a linear function of x.

This equivalence can be described in terms of graphs. Figure 7.7.3 shows the graph of $y = \frac{1}{5}10^x$. We get a familiar exponential curve.

Figure 7.7.3

x	-2	-1	0	1	2	3
y	$\frac{1}{500}$	$\frac{1}{50}$	$\frac{1}{5}$	2	20	200

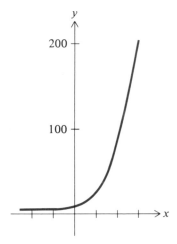

Suppose that, for each point (a, b) on the graph of $y = \frac{1}{5}10^x$, we plot the point $(a, \log b)$. We will use the same x-coordinate, but our new y-coordinate will be the logarithm of the original y-coordinate. The six points we used to sketch the graph in Figure 7.7.3 are transformed into the points listed in Table 7.7.2.

Table 7.7.2

x	-2	-1	0	1	2	3
$\log y$	-2.6990	-1.6990	-0.6990	0.3000	1.2010	2.3010

If we plot the points $(x, \log y)$ on a coordinate system, we get the graph given in Figure 7.7.4.

By plotting points of the form $(x, \log y)$, we have transformed an exponential graph into a linear one. This property gives us a method for determining if points lie on the graph of an exponential function.

If we have a collection of points, (x_1, y_1), (x_2, y_2), (x_3, y_3), . . . , and we want to see if they lie on the graph of an exponential function, we plot the points

Figure 7.7.4

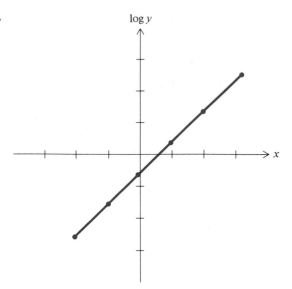

$(x_1, \log y_1)$ $(x_2, \log y_2)$, $(x_3, \log y_3)$ .... If these new points lie on a line, the original points lie on the graph of an exponential function. Otherwise, they do not.

We can apply this method to decide if the data points we examined earlier, collected by the biologist, lie on the graph of an exponential function.

The original data collected is summarized in Table 7.7.3.

Table 7.7.3

t	0	0.5	1	1.5	2	2.5	3
n	3	4.87	7.9	12.86	20.86	33.94	55.13

Next, for each point (t, n) we will find the point $(t, \log n)$, as in Table 7.7.4.

Table 7.7.4

t	0	0.5	1	1.5	2	2.5	3
$\log n$	0.4771	0.6875	0.8976	1.1092	1.3193	1.5307	1.7413

These points are plotted in Figure 7.7.5. Since the points $(t, \log n)$ lie on a line, the original data points lie on the graph of an exponential function.

Exponential functions occur frequently in natural phenomena. Population growth (even bacterial population) is often described using an exponential func-

Figure 7.7.5

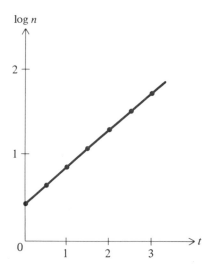

tion. Because exponential functions occur frequently, tests to see if data satisfy an exponential function are common and the method we have just described is used often. Because this method is used so often, a special type of graph paper is available to make the test easier.

Figure 7.7.6 shows points plotted on **semilog graph paper**. Points along the x-axis are evenly spaced. On the y-axis, however, the spacing is uneven. The distance from any point on the y-axis to the origin represents the logarithm of the coordinate of that point. If we consider the distance along the y-axis from 10 to the origin to be 1 unit, the distance from 100 to the origin is 2 units ($\log 100 = 2$) while the distance from 5 to the origin is approximately 0.6990 units ($\log 5 = 0.6990$). We say that the vertical scale is ruled **logarithmically**.

Plotting points (x, y) on semilog paper gives us the same graph we would get if we plotted points $(x, \log y)$ on regular graph paper. The points plotted on the semilog paper in Figure 7.7.6 are the points (t, n) of the biologist's data.

We will make one last comment about the examples we have studied. An experimenter will rarely (read: never) get the data that we have described. More likely, when a collection of data points (x_1, y_1), (x_2, y_2), (x_3, y_3), ... is graphed, it looks like Figure 7.7.7.

The corresponding graph of $(x_1, \log y_1)$, $(x_2, \log y_2)$ ... would look like Figure 7.7.8.

What should we think of this data? After taking logarithms, we did not find that our points were on a line, but they were "almost" on a line.

There are two possible explanations. The first is that the functional relationship is not exponential.

The second is that there was some error in the original data. Every experimenter knows that there is some error made when data are collected. There may be human error in reading a measurement or there may be an error in the measuring equipment.

Figure 7.7.6

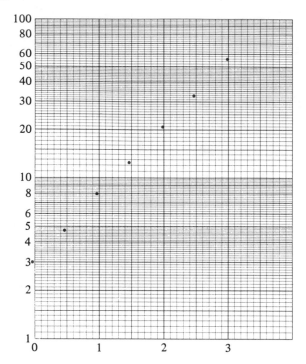

Figure 7.7.7

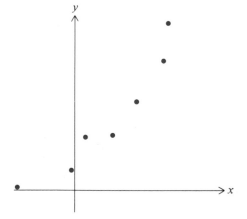

How likely is it that the relationship is linear? Mathematicians have developed sophisticated statistical tests to answer such questions. We point this out only to illustrate that problems in the real world are often more complicated than those in textbooks.

Figure 7.7.8

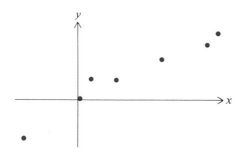

Even when data do not lie precisely on the graph of an exponential function, it is helpful to graph the points $(x, \log y)$. Notice that visually it is easier to recognize the almost linear relationship in Figure 7.7.8 than the almost exponential relationship in Figure 7.7.7.

Exercises 7.7

1. a. Rewrite the following using exponents:
        $$\log 2 = 0.3010$$
        $$\log 3 = 0.4771$$
        $$\log 5 = 0.6990$$

    b. Use your answer to (a) to rewrite each of the following in the form $k10^{cx}$:

    $$3 \cdot 2^{5x} \qquad 1.3 \cdot 3^{2x} \qquad 4 \cdot 5^{(1/2)x}$$

2. A biologist takes samples of bacterial cultures to determine how fast a bacterial population grows. At various times, different samples are studied and the number of cells per square millimeter are counted. He has the data listed in the following table.

Time (in hours)	0	1	2	3	4	5
Average number per mm²	5	7.92	12.56	19.9	31.55	50

Determine if the relationship between population and time is exponential. If it is, determine the parameters of the exponential function.

 3. a. The following table lists data for the function $y = 3x^2$.

x	1	2	3	4	5
y	3	12	27	48	75

Using this data, complete the following table and graph the points $(\log x, \log y)$ on a rectangular coordinate system.

$\log x$	
$\log y$	_____

b. The result in part (a) is general. Suppose $y = ax^n$. Show that the points $(\log x, \log y)$ lie on a line.

4. Use the result of Exercise 3 to show that the data in the following table satisfy an equation of the form $y = ax^n$.

x	1	2	3	4	5
y	4	16	36	64	100

5. Data were taken during an experiment and yielded the results shown in the following table. What type of functional relationship do you think exists between x and y? Take into account possible experimental errors in measurement.

x	1	2	3	4	5
y	0.2	0.8	1.5	3	6

6. a. The graph of $y = k \cdot 10^{cx}$ passes through the points $(2, 20)$ and $(4, 200)$. Find k and c.
    b. The graph of $y = ax^n$ passes through $(1, 4)$ and $(3, 36)$. Find a and n.

7. The graphs of $y = k \cdot 10^{cx}$ and $y = ax^n$ both pass through the points $(1, 10)$ and $(2, 80)$. Find k, c, a, and n.

Review Exercises
Chapter 7

In Exercises 1–15, evaluate each expression.

1. $\left(\dfrac{1}{16}\right)^{\frac{1}{2}}$ **2.** $\left(\dfrac{1}{16}\right)^{-\frac{1}{2}}$ **3.** $(16)^{-\frac{1}{2}}$ **4.** $(121)^{-\frac{3}{2}}$

5. $\left(\dfrac{1}{25}\right)^{\frac{1}{2}}$ **6.** $(3^2)^3$ **7.** $(17^0)^5$ **8.** $\log_9 27$

9. $\log_9 3$ **10.** $\log_3 27$ **11.** $\log_3 81$ **12.** $\log_6 216$

13. $\log_5 \frac{1}{5}$ **14.** $\log_{17} 1$ **15.** $\log_2 2^7$

In Exercises 16–19, solve for x.

16. $3^{2x-7} = 9$ **17.** $3^{2x-7} = 27$

18. $3^{2x-7} = 28$ **19.** $4^{x+3} = 8^{2x-4}$

20. For what value of x does $\log x = 7$?

21. Do the data in the following table satisfy an exponential function?

x	1	2	3	4	5
y	0.5	4	13.5	32	62.5

In Exercises 22–29, assume that $\log 2 = 0.3010$, $\log 3 = 0.4771$, and $\log 5 = 0.6990$. Use this information to evaluate the given logarithms.

22. $\log\left(\frac{1}{6}\right)$ **23.** $\log 150$ **24.** $\log (0.05)$

25. $\log 0.4$ **26.** $\log (1.2)$ **27.** $\log_3 5$

28. $\log 200$ **29.** $\log 81$

30. Find $\log_5 12$.

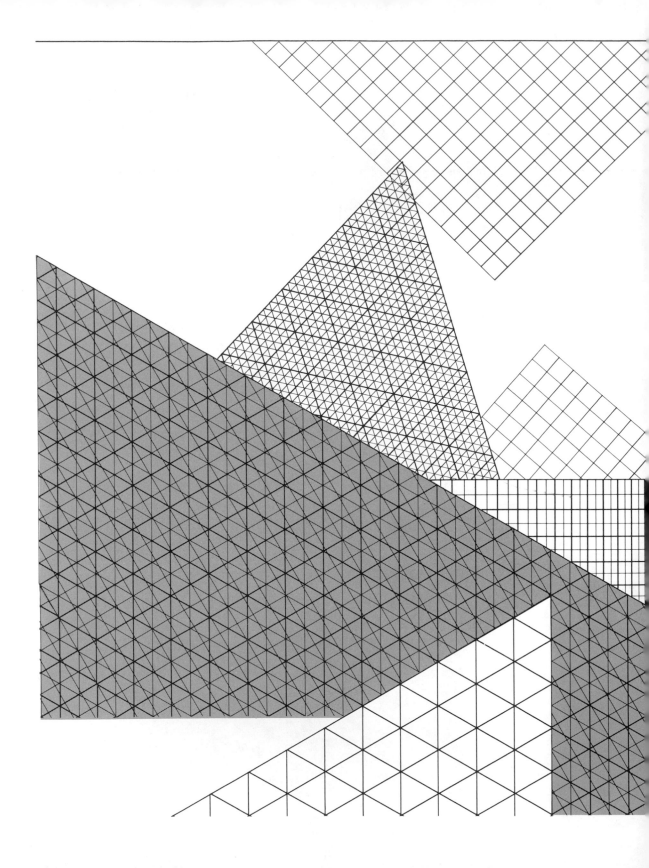

8 Inequalities and Absolute Value

SECTION 8.1

INTRODUCTION

Throughout the text we have studied problems of the following type: given a function f and a constant c, find the points in the domain of f at which f assumes the value c; that is, solve the equation

$$f(x) = c$$

Rather than finding values of x for which $f(x) = c$, we are often interested in finding values of x for which $f(x) > c$, $f(x) \geq c$, $f(x) < c$, or $f(x) \leq c$. Such expressions are called **inequalities**; and in this chapter, we present a systematic study of their solutions.

To show how inequalities may arise, we will consider an example from economics.

Suppose a manufacturer decides to manufacture automobiles. Manufacturing involves certain "set-up" costs: Before any cars are sold, a factory must be built and heavy machinery must be purchased. Assume that this set-up cost is $5 million.

Each automobile built has a cost attached to it in addition to the set-up costs. This is the cost of labor and raw materials. Suppose this cost is $1500 per car.

Finally, the manufacturer plans to sell each car for $4000. The manufacturer wants to know how many cars he must sell to realize a profit. He will realize a profit when his income exceeds his expenses. Suppose he sells x cars. His income is $4000x$ and his expenses are $5,000,000 + 1500x$. We are looking for values of x such that

$$4000x > 5,000,000 + 1500x$$

We will return to this example later in this chapter.

Our study of the solution of inequalities begins with linear inequalities.

Exercises 8.1

1. How realistic are our assumptions for the automobile manufacturer? In particular, do you think it will cost him $1500 to manufacture each car, whether he manufactures 1 or 100,000?
2. A clock manufacturer has set-up costs of $5000. The cost of labor and raw materials to manufacture the clocks is $20/clock, while the selling price is $30/clock.
   a. Find a function, f, such that $f(x)$ represents the manufacturer's profit when she sells x clocks.
   b. What is the profit when the manufacturer sells 1000 clocks?

   c.   What is the profit when the manufacturer sells 400 clocks? What is
        the significance of your answer?
   d.   What is the profit when the manufacturer sells 500 clocks? What is
        the significance of your answer?

SECTION 8.2   **LINEAR INEQUALITIES**

An **inequality** in (the variable) x is an expression that has one of the following
forms:

$$f(x) > g(x) \tag{1}$$

$$f(x) \geq g(x) \tag{2}$$

$$f(x) < g(x) \tag{3}$$

$$f(x) \leq g(x) \tag{4}$$

where f and g are functions. If f and g are both linear functions, the inequality
is called a **linear inequality**. Examples of linear inequalities are:

$$3x + 5 \leq 2x - 6 \qquad x + 4 < 10$$

and

$$2x + 17 > x + 1 \qquad x - 3 \geq 2x + 8$$

A **solution** of an inequality is any number that, when substituted for the
variable, makes the inequality a true statement.

If we consider the linear inequality

$$x + 4 < 10$$

we notice that 5 is a solution, since $5 + 4 < 10$, but 3 is also a solution, since
$3 + 4 < 10$. In fact, any number less than 6 is a solution of this inequality.

Solving an inequality means finding all solutions of the inequality. The solutions of the inequality $x + 4 < 10$ are the numbers, x, such that $x < 6$. We
will usually omit the phrase "the numbers, x, such that" and just write $x < 6$ to
represent the solutions. Another common method of representing the solutions
is to use a graph on a number line. The graph of the solutions of $x + 4 < 10$ is
shown in Figure 8.2.1.

Figure 8.2.1

We worked with the inequality $x + 4 < 10$ because we could solve this
inequality by inspection. Many linear inequalities are too complicated to solve
by inspection, and we would like to develop some general methods to solve these

inequalities. These methods are similar to those used to solve linear equations; and for this reason, it will be helpful to reconsider the solution of linear equations.

Below are the steps used to solve $4x + 5 = 2x + 7$:

$$4x + 5 = 2x + 7 \tag{5}$$

$$\Leftrightarrow 2x + 5 = 7 \tag{6}$$

$$\Leftrightarrow 2x = 2 \tag{7}$$

$$\Leftrightarrow x = 1 \tag{8}$$

All of the equations (5)–(8) are equivalent, meaning that they all have precisely the same solution. It is apparent that the only value of x that makes equation (8) a true statement is $x = 1$. Since 1 is the solution to (8) it is the only solution, as well, to (5), $4x + 5 = 2x + 7$.

To solve a linear equation, we transform it into equivalent equations until we get an equation whose solution is obvious. We know that the following procedures always transform equations into equivalent equations:

A.   Adding the same constant to both sides of an equation or subtracting the same constant from both sides of an equation.
B.   Multiplying or dividing both sides of an equation by the same nonzero constant.

We will solve linear inequalities by transforming them into equivalent inequalities (inequalities with precisely the same solutions) until we reach an inequality whose solution is obvious.

Although rules (A) and (B) always transform equations into equivalent equations, they do not always transform inequalities into equivalent inequalities. To understand how to transform inequalities into equivalent inequalities, it will be helpful to interpret addition and multiplication geometrically.

Let x be a number and suppose, as shown in Figure 8.2.2., that we have located it on the number line. If $c > 0$, $x + c$ is a point c units to the right of x while $x - c$ is a point c units to the left of x.

Figure 8.2.2

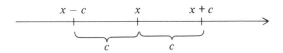

Geometrically, addition of a positive number shifts (sometimes we say "translates") a point to the right. Subtraction of a positive number shifts that point to the left.

If $c > 0$, the point cx is on the same side of the origin as x, but c times as far away. The point $3x$, for example, is three times as far from the origin as x, while $\frac{1}{3}x$ is only $\frac{1}{3}$ as far away. (See Figure 8.2.3.)

Figure 8.2.3

(a) $x > 0$

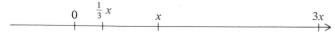

(b) $x < 0$

Multiplying x by -1 reflects the point in the origin; i.e., $(-1)x$ is symmetric to x with respect to the origin. The points x and $-x$ are equidistant from the origin and on opposite sides of the origin. (See Figure 8.2.4.)

Figure 8.2.4

(a) $x > 0$

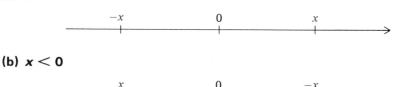

(b) $x < 0$

Multiplying x by an arbitrary negative number is equivalent to multiplying first by a positive number, and then multiplying the result by a negative one, as shown in the following example.

Example 1

Given x on the number line, we can locate $-3x$ and $-\frac{1}{2}x$ using the method indicated in Figure 8.2.5.

Figure 8.2.5

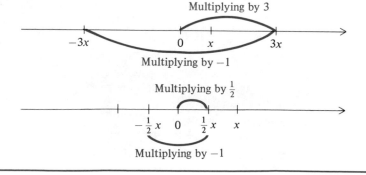

We have described the geometric effects of addition, subtraction, and multiplication. Since we can think of division as multiplication by reciprocals, we know the effect of division as well.

Example 2

Given x on the number line, we can locate $x \div (-4)$. We will separately consider in Figure 8.2.6 two cases, $x > 0$ and $x < 0$. We think of division by -4 as multiplication by $-\frac{1}{4}$.

Figure 8.2.6
(a) $x > 0$

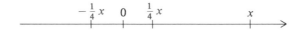

(b) $x < 0$

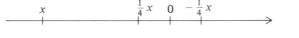

In Example 2, and in Figures 8.2.3 and 8.2.4, we included separate diagrams for $x > 0$ and $x < 0$. We did this to emphasize that we cannot assume that $3x$ represents a positive number or that $-x$ represents a negative number. If x represents a negative number, $3x$ is also negative and $-x$ is positive.

We now consider the effect addition and multiplication have on inequalities. Suppose $a < b$. Then a is to the left of b on the number line. If we add c to both a and b, we are shifting both a and b equal distances and the relative positions of $a + c$ and $b + c$ are the same as the relative positions of a and b. Equivalently,

$$\text{If} \quad a < b \quad \text{then} \quad a + b < b + c$$

We describe this behavior by saying that addition (or subtraction) preserves inequalities.

If $a < b$ and we multiply both a and b by a **positive** constant, c, geometrically, we are stretching or contracting distances to the origin. If a is to the left of b, ca is to the left of cb. Equivalently,

$$\text{If} \quad a < b \quad \text{and} \quad c > 0 \quad \text{then} \quad ca < cb$$

Now let us consider what would happen if $c < 0$, and we multiplied both a and b by c. As we see in Figure 8.2.7, multiplication by a negative involves reflection in the origin, and if a is to the left of b, ca is to the right of cb.

Figure 8.2.7

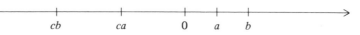

Equivalently,

> If $a < b$ and $c < 0$ then $ca > cb$

Multiplication (or division) by negative numbers reverses the sense of inequalities. It is this property that makes solving inequalities different from solving equations. Before considering examples, we will summarize our results. Although we state the results using $a < b$, we would get analogous results using other inequality signs.

Suppose a and b are any real numbers.

If c is any real number, then $a < b \Leftrightarrow a + c < b + c$ **(9)**

If $c > 0$, then $a < b \Leftrightarrow ca < cb$ **(10)**

If $c < 0$, then $a < b \Leftrightarrow ca > cb$ **(11)**

These rules allow us to solve inequalities with a method similar to the one for solving equations. We will illustrate this method in the next examples.

Example 3

To solve $8x + 4 < 3x + 14$, we proceed as follows:

$$8x + 4 < 3x + 14$$
$$\Leftrightarrow 5x + 4 < 14$$
$$\Leftrightarrow 5x < 10$$
$$\Leftrightarrow x < 2$$

The graph of the solution is shown in Figure 8.2.8.

Figure 8.2.8

$$2$$

Example 4

To solve $6 - 2x \leq 12$ we proceed as follows:

$$6 - 2x \leq 12$$
$$\Leftrightarrow -2x \leq 6$$
$$\Leftrightarrow x \geq -3$$

The graph of the solution is shown in Figure 8.2.9.

Notice that, when we divided both sides of the inequality by -2, we had to reverse the sense of the inequality. Also notice that we could have solved the same problem a different way.

$$6 - 2x \leq 12$$
$$\Leftrightarrow 6 < 12 + 2x$$
$$\Leftrightarrow -6 \leq 2x$$
$$\Leftrightarrow -3 \leq x$$

Figure 8.2.9

There is an interesting and useful way to think of solutions of inequalities using graphs.

The solutions of $3x + 6 > 0$ are the numbers x such that

$$x > -2$$

Figure 8.2.10 shows the graph of $y = 3x + 6$. Notice that, when $x > -2$, points on the graph are above the x-axis.

Figure 8.2.10   y = 3x + 6

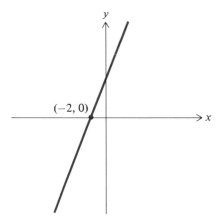

To understand why this is true, notice that, if x is the x-coordinate of a point of the graph, the y-coordinate is $3x + 6$. A number, x, that satisfies $3x + 6 > 0$ will make $y > 0$. But when $y > 0$, the point (x, y) is above the x-axis. Solving $3x + 6 > 0$ is equivalent to finding x-coordinates of points on the graph of $y = 3x + 6$ that are above the x-axis.

In general, solving $f(x) > 0$ is equivalent to finding values of x for which the graph $y = f(x)$ is above the x-axis. Similarly, solving $f(x) < 0$ is equivalent to finding values of x for which the graph $y = f(x)$ is below the x-axis.

Linear inequalities are relatively easy to solve algebraically, but this geometric argument is often very helpful for solving more complicated inequalities. We will use this argument in the following example and return to it in the next section when we solve quadratic inequalities.

Example 5

The graph of a function, f, is given in Figure 8.2.11. Find the values of x for which:

A. $f(x) > 0$ B. $f(x) < 0$ C. $f(x) = 0$

Figure 8.2.11

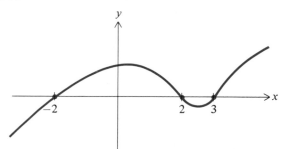

Solution:

A. $-2 < x < 2$ or $x > 3$
B. $x < -2$ or $2 < x < 3$
C. $x = -2$, $x = 2$, or $x = 3$

Exercises 8.2

In Exercises 1–15, solve each inequality and graph the solution.

1. $8x - 3 \le -16$
2. $6 + 2x \le 5$
3. $3x - 5 \ge 2x + 4$
4. $5 - 3x \le 11$
5. $-x - 5 > 7$
6. $\frac{1}{2}x + 4 \le 2x - 1$
7. $x - 8 \le 5x + 6$
8. $8 - \frac{1}{2}x \le \frac{1}{4}$
9. $0.1x + 2 \ge \frac{1}{2}x$
10. $3(2x - 6) \ge \frac{1}{3}(x - 4)$
11. $6(x - 1) \le 2(3x - \frac{1}{2})$
12. $x - 4 \le x - 5$
13. $17 \le 2(3x - 8)$
14. $x^2 + 8x + 5 \le (x + 2)(x - 7)$
15. $\dfrac{3x - 8}{5} \le 2$

16. Suppose $6/x < 12$. Is it always true that $6 < 12x$? When isn't it true?

17. Consider the following "solution" of $6/x < 12$.

$$\frac{6}{x} < 12$$

$$\Leftrightarrow 6 < 12x$$

$$\Leftrightarrow \frac{1}{2} < x$$

Notice that any negative number satisfies the original inequality, but our solution didn't find these solutions. Why not?

18. Solve the inequality $\dfrac{3}{x-1} < 6$. (*Hint:* Consider separately the cases when $x - 1 > 0$ and $x - 1 < 0$.)

19. The graph of a function, f, is shown in the figure on the right. Find all values of x for which:

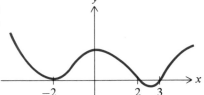

a. $f(x) > 0$
b. $f(x) < 0$
c. $f(x) = 0$

20. The graph of a function, g, is shown in the figure on the right. Find all values of x for which:

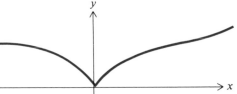

a. $g(x) > 0$
b. $g(x) < 0$
c. $g(x) = 0$

21. a. Draw the graph of $y = x^2 - 4$.
　　b. Use your sketch to find the solution of $x^2 - 4 \le 0$.

22. a. Draw the graph of $y = x^2 + 2x$.
　　b. Use your sketch to find the solutions of $x^2 + 2x > 0$.

23. Write statements analogous to (9), (10), and (11) when the inequality $a < b$ in these statements is replaced by

a. $a \le b$
b. $a > b$
c. $a \ge b$

24. Show that if $a > 0$, $b > 0$, and $a < b$, then $1/a > 1/b$.

25. An automobile manufacturer has set-up costs of $5,000,000. The cost of manufacturing each automobile is $1500 and each car is sold for $4000.

   a. Find a function, f, that represents the profit earned by the manufacturer when he sells x automobiles.
   b. How many automobiles must be sold for $f(x) = 0$? (This is called the break-even point.)
   c. How many automobiles must be sold for the manufacturer to realize a profit of at least $250,000?
   d. How many automobiles must be sold for the manufacturer to realize a profit of at least $750? (*Note:* Solving the equation $f(x) = 750$ will not give a correct answer.)

26. Fixed costs for a desk manufacturer are $58,800. Per-unit costs are $21/desk and the per-unit price is $35. What is the break-even point?

27. An oil dealer has 1000 gallons of gasoline in stock selling for 85¢/gallon. Next month the price rises to $1.00/gallon. If the dealer wants the total revenue from the 1000 gallons to be at least $900, what is the maximum number of gallons that can be sold this month?

28. $10,000 is to be invested. There are two possible investments: one returning 10 percent and the other, returning 8 percent. What is the largest amount that can be invested at 8 percent if the total return from both investments is to be at least 8.5 percent?

SECTION 8.3   **QUADRATIC INEQUALITIES**

A **quadratic inequality** is an inequality that is equivalent to one of the following:

$$ax^2 + bx + x < 0$$
$$ax^2 + bx + c \leq 0$$
$$ax^2 + bx + c > 0$$
$$ax^2 + bx + c \geq 0$$

where a, b, and c are any real numbers and $a \neq 0$. (If $a = 0$, we would have a linear inequality.)

Examples of quadratic inequalities are $x^2 + 5x + 6 < 0$ and $3t^2 - 7 \leq 0$. An inequality such as $2y + 5y^2 > y + y^2$ is also a quadratic inequality, since we can use the methods of the last section to rewrite it with zero on one side of the inequality. (We could subtract y and y^2 from both sides of the inequality to obtain $4y^2 + y > 0$.)

Solving quadratic inequalities uses many of the techniques studied in Chapter 2 for solving quadratic equations; and as in the case of quadratic equations, there are different methods of solution. We will discuss two of them in this section.

Suppose we want to solve the inequality

$$x^2 + 5x + 6 < 0$$

Since $x^2 + 5x + 6 = (x + 2)(x + 3)$,

$$x^2 + 5x + 6 < 0 \Leftrightarrow (x + 2)(x + 3) < 0$$

We know that *a product of two numbers is negative if the numbers have opposite signs*. In Figure 8.3.1, we have drawn number lines for each factor and have indicated when the factors are positive, negative, and zero.

Figure 8.3.1
(a) x + 2

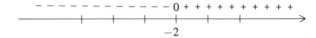

(b) x + 3

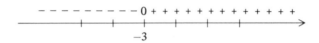

The first number line shows that $x + 2$ is zero when $x = -2$, $x + 2$ is positive when $x > -2$, and $x + 2$ is negative when $x < -2$. The second number line shows that $x + 3$ is zero when $x = -3$, $x + 3$ is positive when $x > -3$, and $x + 3$ is negative when $x < -3$.

When $-3 < x < -2$, $x + 2$ is negative and $x + 3$ is positive. Therefore, their product is negative. When x takes on any other value, the product is either positive or zero. The product is positive when $x > -2$, since both factors are positive when $x > -2$, and the product is positive when $x < -3$, since both factors are negative when $x < -3$. When $x = -3$ or $x = -2$, one of the factors is zero, which makes the product zero. We see that $x^2 + 5x + 6 < 0$ if and only if $-3 < x < -2$. The graph of the solution is given in Figure 8.3.2.

Figure 8.3.2 $-3 < x < -2$

Example 1

To solve $x^2 + 5x + 6 \leq 0$, we notice that the only difference between this problem and the problem solved above is that here we include the possibility that $x^2 + 5x + 6 = 0$. The solutions to this equation are $x = -3$ and $x = -2$. The solutions to the inequality are

$$-3 \leq x \leq -2$$

The graph of the solution is shown in Figure 8.3.3.

Figure 8.3.3

Example 2

$$x^2 - 7x + 6 \geq 0$$
$$\Leftrightarrow (x - 6)(x - 1) \geq 0 \qquad \text{(See Figure 8.3.4.)}$$

Figure 8.3.4
(a) $x - 6$

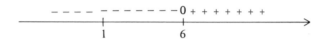

(b) $x - 1$

A product of two terms is positive if either both are positive or both are negative. The product is zero if either term is zero.

The solution is

$$x \leq 1 \quad \text{or} \quad x \geq 6 \qquad \text{(See Figure 8.3.5.)}$$

Figure 8.3.5

Although we have applied this method to quadratic inequalities where the left side is easily factored, the method can be used more generally. We make use of the following relationship between factors and roots studied in Chapter 2, Section 2.4.3: If a quadratic polynomial has roots r_1 and r_2, it has a factorization

$$a(x - r_1)(x - r_2)$$

Example 3

We will apply our method to the quadratic inequality

$$x^2 + 8x + 1 < 0$$

Using the quadratic formula, the roots of

$$x^2 + 8x + 1$$

are

$$\frac{-8 \pm \sqrt{64 - 4}}{2} = \frac{-8 \pm \sqrt{60}}{2} = \frac{-8 \pm 2\sqrt{15}}{2} = -4 \pm \sqrt{15}$$

(As an estimate, $\sqrt{15}$ is slightly less than 4, so $-4 + \sqrt{15}$ is between 0 and -1; $-4 - \sqrt{15}$ is approximately -8.)

Since we know the roots, we know how to factor the polynomial. (See Figure 8.3.6.)

$$x^2 + 8x + 1 = \left[x - (4 + \sqrt{15})\right]\left[x - (-4 - \sqrt{15})\right]$$

Figure 8.3.6

(a) $x - \left(-4 + \sqrt{15}\right)$

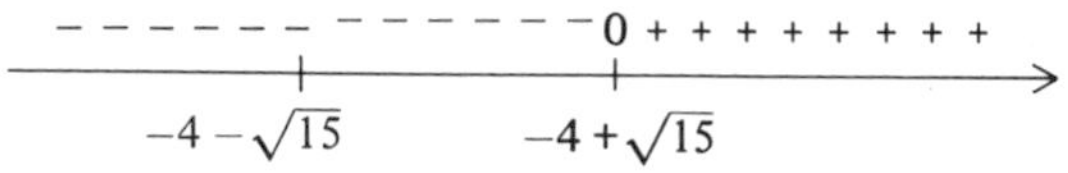

(b) $x - \left(-4 - \sqrt{15}\right)$

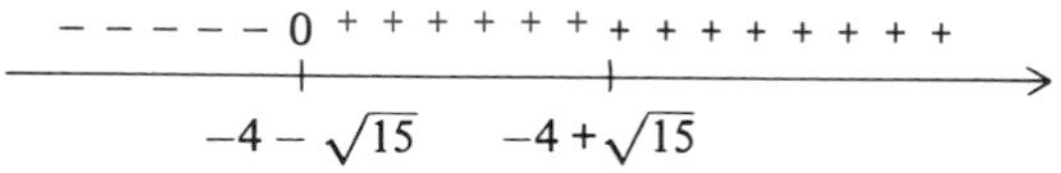

From our number lines, we see that the factors have different signs—and the product is negative—when x is between $-4 - \sqrt{15}$ and $-4 + \sqrt{15}$. The solution is $-4 - \sqrt{15} < x < -4 + \sqrt{15}$.

We now consider a second method for solving quadratic inequalities.

At the end of Section 8.2, we noted that the solutions of $f(x) > 0$ can be found by determining where the graph of $y = f(x)$ lies above the x-axis. Similarly, the solutions of $f(x) < 0$ can be found by determining when the graph of $y = f(x)$ lies below the x-axis. We will apply this method to the case where $f(x)$ is a quadratic polynomial. In order to solve the quadratic inequality $ax^2 + bx + c > 0$, we will sketch the graph $y = ax^2 + bx + c$ and notice where the graph is above the x-axis. Similarly, we will solve $ax^2 + bx + c < 0$ by seeing where the graph $y = ax^2 + bx + c$ is below the axis.

Since this method requires us to sketch graphs of quadratic polynomials, it will be helpful to recall some facts about these graphs.

First, the graph of a quadratic polynomial is a parabola.

Second, we can find the x-intercepts of the graph of $y = ax^2 + bx + c$ by solving the quadratic $ax^2 + bx + c = 0$.

Third, if $a > 0$, the graph of $y = ax^2 + bx + c$ opens upward. If $a < 0$, the graph opens downward.

Using these three facts, it is possible to sketch a quadratic polynomial very quickly.

Let us reconsider the first problem,

$$x^2 + 5x + 6 < 0$$

What will the graph $y = x^2 + 5x + 6$ look like? We know it is a parabola, since $x^2 + 5x + 6$ is a quadratic polynomial. The roots of $x^2 + 5x + 6$ are -2 and -3. (We could find the roots either by factoring or by using the quadratic formula.) This tells us that the graph $y = x^2 + 5x + 6$ crosses the x-axis when $x = -2$ and $x = -3$. Finally, since $a = 1$ (and $1 > 0$), the graph opens upward. The graph of $y = x^2 + 5x + 6$ must look like the graph in Figure 8.3.7, and the solution to $x^2 + 5x + 6 < 0$ is

$$-3 < x < -2$$

Figure 8.3.7

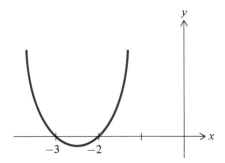

Example 4

Solve: $x^2 + 5x + 4 > 0$.

The roots of $x^2 + 5x + 4$ are -4 and -1. Since $a = 1$, the graph opens upward. (See Figure 8.3.8.) The graph is above the x-axis when $x > -1$ and when $x < -4$. The solution to $x^2 + 5x + 4 > 0$ is

$$x < -4 \quad \text{or} \quad x > -1$$

Figure 8.3.8

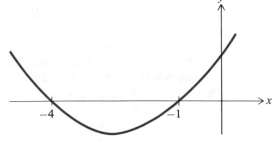

Example 5

Solve $-2x^2 + 10 < 0$.

The roots of $-2x^2 + 10$ are $\pm \sqrt{5}$. Since $a = -2 < 0$, the graph opens downward and the sketch of $f(x) = -2x^2 + 10$ is shown in Figure 8.3.9. The solution to $-2x^2 + 10 < 0$ is $x < -\sqrt{5}$ or $x > \sqrt{5}$.

Figure 8.3.9

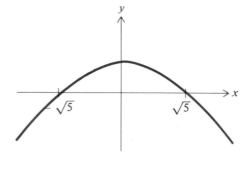

To compare the two methods of solution, we will reconsider a problem solved earlier.

Example 6

We will solve $x^2 + 8x + 1 < 0$ using our second method. (In Example 3, we solved it using the first method.) Using the quadratic formula, the roots of $x^2 + 8x + 1$ are

$$\frac{-8 \pm \sqrt{64 - 4}}{2}$$

Simplifying, the two roots are $-4 + \sqrt{15}$ and $-4 - \sqrt{15}$. Since $a > 0$, the graph opens upward. The graph of $f(x) = x^2 + 8x + 1$ is shown in Figure 8.3.10, and the solution to $x^2 + 8x + 1 < 0$ is

$$-4 - \sqrt{15} < x < -4 + \sqrt{15}$$

Figure 8.3.10

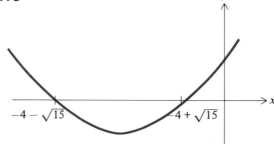

So far, all our examples have considered quadratic polynomials with two roots. This method also works if the polynomial has one root or no roots.

Example 7

Solve $x^2 - 10x + 25 < 0$.

Using the quadratic formula, we find that the only root is 5. Geometrically, this means that the graph touches the x-axis only once, when $x = 5$. Since $a > 0$, the parabola opens upward. Figure 8.3.11 shows the graph of $y = x^2 - 10x + 25$. (Note that $x^2 - 10x - 25 = 0$ when $x = 5$.) Since the graph is never below the x-axis, there are no values of x that make $x^2 - 10x + 25 < 0$.

Figure 8.3.11

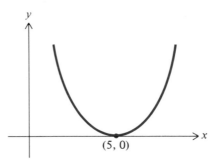

(5, 0)

Example 8

Solve:

$$-x^2 - 12x - 100 \geq 0$$

Using the quadratic formula, we find that $f(x) = -x^2 - 12x - 100$ has no roots. Geometrically, the graph, therefore, never touches the x-axis. Since $a = -1$, which is less than 0, the graph opens downward. (See Figure 8.3.12.) There are no values of x for which the graph is above the x-axis, so there are no solutions to $-x^2 - 12x - 100 \geq 0$.

Figure 8.3.12

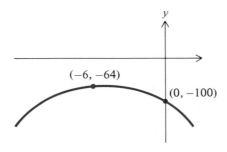

(−6, −64)

(0, −100)

Exercises 8.3   In problems 1–19 solve the given inequality and graph the solution.

1. $x^2 + 11x + 10 < 0$ 2. $16 - y^2 > 4$
3. $x^2 + 5 \geq 0$ 4. $t^2 + 2t + 1 \leq 0$
5. $z^2 - z > 0$ 6. $x^2 + 3x + 4 \leq 2x^2 - x - 8$
7. $x^2 + x - 1 < 0$ 8. $(y + 6)(y - 3)(y - 4) \leq 0$
9. $x^2 + 7x < 6$ 10. $t^2 - 1 \geq 3$
11. $2(x + 2)(x - 7) > 0$ 12. $r^2 + 3r + 20 < 0$
13. $20x^2 + 40x > 0$ 14. $(x + 3)(x^2 + 1) > 0$
15. $z^2 + 12z + 20 \leq 0$ 16. $r^2 + 4r + 4 \leq 0$
17. $x^2 - 4x + 3 \leq 0$ 18. $x(x + 1)(2x - 1) < 0$
19. $x^4 - 4x^2 + 3 \leq 0$
    (*Hint:* See Exercise 17.)

Solve problems 20–25 using the method of graphing. (Use the methods of Section 4.6.3 to sketch the graph.)

20. $\sin\left(x + \dfrac{\pi}{4}\right) \leq 0$ 21. $\log(x - 2) \geq 0$

22. $14^{x+1} < 0$ 23. $\tan\left(x - \dfrac{\pi}{4}\right) \leq 0$

24. $\sin x + 1 \leq 0$ 25. $3 \cos x < 0$

26. An open box is formed from a square sheet of cardboard by cutting a 3-inch square from each corner and then turning up the sides. Find the smallest dimensions of the square sheet of cardboard if the open box must have a volume of at least 27 cubic inches.
27. A printer wants to include an even margin on each 8 × 10-inch page. If the area of the printed matter must be at least 55.25 square inches, what is the largest allowable margin?
28. Suppose in problem 27 that the margin at each side of the page must be twice as wide as the margins at the top and bottom. If the printed matter must contain at least 32 square inches, what are the largest allowable margins?

SECTION 8.4   **ABSOLUTE VALUE FUNCTIONS**

When we first began studying functions in Chapter 4, among the examples we looked at were distance functions, functions that measured the distance between points on a line or between points in a plane. One reason we chose these functions as examples is that they are frequently used in mathematics. We are often interested in measuring the distance between points. Our next definition introduces a notation that we will often use when computing distances.

Definition

> If x is any real number, $|x|$ (read: **the absolute value of x**) is the distance from x to the origin.

Example 1

> A. $|5| = 5$ B. $|-5| = 5$ C. $|1 - 3| = |-2| = 2$
>
> D. $|3| = 3$ E. $|-3| = 3$ F. $|0| = 0$

Notice that $|x| \geq 0$ for all real numbers x. Also, (C) of Example 1 illustrates an important property of absolute value. The symbol $|1 - 3|$ can be thought of as the distance between -2 ($= 1 - 3$) and the origin, but we can also think of it as the distance between 1 and 3. In general, $|a - b|$ can be thought of either as the distance from the point $x = a - b$ to the origin or as the distance from a to b.

Example 2

> We can interpret $|5 - 3|$ as the distance from 2 to the origin or the distance from 5 to 3.
>
> We can interpret $|x - 1|$ as the distance from $x - 1$ to the origin or the distance from x to 1.

We now define a function using the absolute value. The function, f, from the real numbers into the non-negative real numbers, defined by $f(x) = |x|$, is called the **absolute value function**.

We have defined absolute value geometrically in terms of distance. It is also convenient to have an algebraic description of absolute value. There are two commonly used definitions:

$$|x| = \sqrt{x^2} \tag{1}$$

$$|x| = \begin{cases} x & \text{if} \quad x \geq 0 \\ -x & \text{if} \quad x < 0 \end{cases} \tag{2}$$

The two definitions are equivalent, and both are important.

Example 3

> We will apply definition (1) to find $|10|$ and $|-10|$.
>
> If $x = 10$ $|x| = \sqrt{10^2} = \sqrt{100} = 10$
>
> If $x = -10$ $|x| = \sqrt{(-10)^2} = \sqrt{100} = 10$

The second definition may seem complicated at first. It tells us that, to find $|x|$, we have to choose one of two instructions to follow. If $x \geq 0$, the instruction is $|x| = x$. If $x < 0$, the instruction is $|x| = -x$. For example, if $x = -3$ (which is < 0), $|-3| = -(-3) = 3$.

The graph (Figure 8.4.1) of the absolute value function, $f(x) = |x|$, can be sketched using Table 8.4.1.

Table 8.4.1

x	-3	-2	-1	0	1	2	3
$|x|$	3	2	1	0	1	2	3

Figure 8.4.1 $y = |x|$

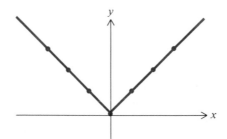

We will make two observations about the graph. The first is that the graph is symmetric with respect to the y-axis. To prove this, recall that the criterion for y-axis symmetry is that $f(-x) = f(x)$ for all x in D_f. In the case of the absolute value function, $f(x) = |x|$ and $f(-x) = |-x|$. But $|x| = |-x|$ for all real numbers x, so $f(x) = f(-x)$ for all x in D_f and the criterion for y-axis symmetry is met.

The second observation is that if we only look at the graph to the *right* of the y-axis, it looks just like the graph of $f(x) = x$ (Figure 8.4.2).

Figure 8.4.2 $y = |x|$ $x > 0$

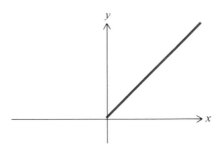

If we only look at the graph to the *left* of the y-axis, it looks just like the graph of $f(x) = -x$, shown in Figure 8.4.3.

Figure 8.4.3 $y = |x|$ $x < 0$

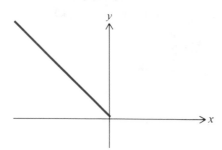

Now look at the second algebraic definition of $|x|$:

$$|x| = \begin{cases} x & \text{if } x \geq 0 \\ -x & \text{if } x < 0 \end{cases}$$

One way to view this definition (a very useful way) is that when $x \geq 0$, the absolute value function is the same as the function $g(x) = x$, while, when $x < 0$, the absolute value function is the same as the function $h(x) = -x$.

We can summarize our results about the absolute value function:

$$f(x) = |x| = \sqrt{x^2} = \begin{cases} x & \text{if } x \geq 0 \\ -x & \text{if } x < 0 \end{cases}$$

$$= \text{the distance from } x \text{ to the origin}$$

$$D_f = (-\infty, +\infty)$$

$$R_f = [0, +\infty)$$

The graph of f is shown in Figure 8.4.4.

Figure 8.4.4 $y = |x|$

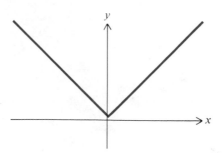

Exercises 8.4   Evaluate the expressions in Exercises 1–7.

1. $|5 - (-2)|$ 2. $|2 - (-5)|$ 3. $|-4 - 1|$

4. $|6 - 7|$ 5. $|3 - (-3)|$ 6. $|4 + (-2)|$

7. $|-1|$

8. Interpret each of the absolute values in Exercises 1–7 geometrically as a distance.

9. Graph the function $y = |x - 1|$. How does the graph of this function compare with the graph of $y = |x|$?

10. Graph the function $y = |x + 2|$.

11. Graph the function $y = |2x - 2|$. How does this graph compare with the graph of $y = |x - 1|$ in Exercise 9?

12. Graph the function $y = |x| - 1$.

13. Each of the following graphs looks like the graph of $y = |x|$, but displaced from the origin. Find an equation corresponding to each graph.

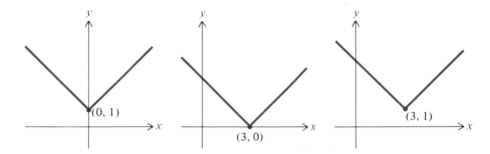

(*Hint:* The graph has been moved twice; first to the right, and then up.)

14. Graph the function $y = -|x|$.

15. a. Find numbers, x and y, such that

$$|x + y| = |x| + |y|$$

b. Find numbers, x and y, such that

$$|x + y| < |x| + |y|$$

c. What inequality compares the values of $|x + y|$ and $|x| + |y|$?

16. Which is larger, $|x - y|$ or $|x| - |y|$? Can they ever be equal?

17. Sketch the graph of the function f defined by $f(x) = |x^2|$.

18. Sketch the graph of the function f defined by $f(x) = |x^3|$.

19. Sketch the graph of the equation $y = |\log x|$.

20. Consider the function, f, defined by $f(x) = \log|x|$.

    a.   What is the domain of f?   **b.**   Sketch the graph of f.

21. Let f be the function defined by $f(x) = \dfrac{|x|}{x}$.

    a.   What is the domain of f?   **b.**   Find $f(x)$ if $x > 0$.
    c.   Find $f(x)$ if $x < 0$.   **d.**   Sketch the graph of f.

22. Sketch the graph of the equation $y = x + |x|$.

23. Sketch the graph of the equation $y = x - |x|$.

24. Sketch the graph of the equation $y = |x| - x$.

25. Prove that for each real number x, $|x| + x \ge 0$, using the definition of absolute value given in equation (2) of this section.

26. The graph of the function f is given below. Sketch the graph of the equation $y = |f(x)|$.

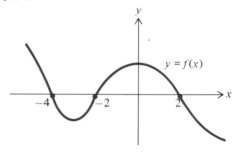

SECTION 8.5   **EQUATIONS, INEQUALITIES, AND ABSOLUTE VALUES**

In this section, we will consider equations involving absolute values. First, consider the equation $|x| = 6$.

Since $|x|$ is the distance from x to the origin, the solutions of $|x| = 6$ are those numbers six units from the origin. There are two solutions, $x = 6$ and $x = -6$.

In general, if $a \ge 0$, the solutions of $|x| = a$ are $x = a$ and $x = -a$. If $a < 0$, $|x| = a$ has no solutions since an absolute value cannot be negative. This argument generalizes as follows: If f is a function and $a \ge 0$, then x is a solution of $|f(x)| = a$ if and only if $f(x) = a$ or $f(x) = -a$. If $a < 0$, $|f(x)| = a$ has no solutions.

Example 1

To solve $|2x + 3| = 7$, we proceed as follows:

$$|2x + 3| = 7$$

$$\Leftrightarrow 2x + 3 = 7 \quad \text{or} \quad 2x + 3 = -7$$
$$\Leftrightarrow x = 2 \quad \text{or} \quad x = -5$$

So, the solutions of $|2x + 3| = 7$ are $x = 2$ and $x = -5$.

Solving the inequality $|x| < 6$ is equivalent to finding all points that are less than six units from the origin. The points on the number line that satisfy this condition are the numbers strictly between -6 and $+6$. Using the notation we developed in Chapter 1, $|x| < 6$ is equivalent to $-6 < x < 6$.

Recall that $-6 < x < 6$ indicates that two inequalities hold simultaneously: x is less than 6 and at the same time x is greater than -6. Both these conditions must be met for $|x|$ to be less than 6. If we only knew that $x < 6$, we could not be sure that $|x| < 6$. For example, $-100 < 6$, but $|-100| \not< 6$. Similarly, if we only knew that $-6 < x$, we could not be sure that $|x| < 6$. For example, $-6 < 100$, but $|100| \not< 6$.

Analogous arguments can be used to show that if $a > 0$, then $|x| < a$ if and only if $-a < x < a$. If $a \le 0$, then $|x| < a$ has no solution since $|x| \ge 0$ for each real number x.

This result can also be generalized: If f is a function and $a > 0$, then x is a solution of $|f(x)| < a$ if and only if $-a < f(x) < a$. If $a \le 0$, then $|f(x)| < a$ has no solution.

Example 2

To solve $|2x + 7| < 5$, we write

$$|2x + 7| < 5$$
$$\Leftrightarrow -5 < 2x + 7 < 5$$
$$\Leftrightarrow -12 < 2x < -2$$
$$\Leftrightarrow -6 < x < -1 \qquad \text{(Figure 8.5.1)}$$

Figure 8.5.1

Solving the inequality $|x| > 6$ is equivalent to finding all points more than six units from the origin. These are points either to the right of 6 or to the left of -6. So, $|x| > 6$ is equivalent to $x > 6$ or $x < -6$.

On the number line the solution of $|x| > 6$ consists of two disjoint sets of points as indicated in Figure 8.5.2. Analogous arguments can be used to show that if $a \ge 0$, $|x| > a$ if and only if $x > a$ or $x < -a$.

Figure 8.5.2

This result can be generalized to the following: If f is a function at $a \geq 0$, then x is a solution $|f(x)| > a$ if and only if $f(x) > a$ or $f(x) < -a$. If $a < 0$, the solutions of $|f(x)| > a$ are all numbers in the domain of f.

Example 3

To solve $|3x - 5| > 7$, we proceed as follows:

$$|3x - 5| > 7$$
$$\Leftrightarrow 3x - 5 > 7 \quad \text{or} \quad 3x - 5 < -7$$
$$\Leftrightarrow 3x > 12 \quad \text{or} \quad 3x < -2$$
$$\Leftrightarrow x > 4 \quad \text{or} \quad x < -\frac{2}{3} \qquad \text{(Figure 8.5.3)}$$

Figure 8.5.3

Example 4

The solution to $|x| > -1$ is all values of x since for any value of x, $|x| \geq 0$.

The solution to

$$\left| \frac{1}{x + 3} \right| > -2$$

is all values of $x \neq 3$, since the domain of $f(x) = 1/(x + 3)$ is all numbers not equal to -3.

In this section, we have solved equations and inequalities involving absolute values. The first step in each case is to write an equivalent problem that does not involve absolute values. This is done using the following rules:

Suppose f is a function.

If $a \geq 0$	$\|f(x)\| = a \Leftrightarrow f(x) = a \quad \text{or} \quad f(x) = -a$	(1)
If $a > 0$	$\|f(x)\| < a \Leftrightarrow -a < f(x) < a$	(2)
If $a \geq 0$	$\|f(x)\| > a \Leftrightarrow f(x) > a \quad \text{or} \quad f(x) < -a$	(3)

Example 5

> To solve $|2x - 3| \leq 5$, we proceed as follows:
>
> $$|2x - 3| \leq 5$$
> $$\Leftrightarrow -5 \leq 2x - 3 \leq 5$$
> $$\Leftrightarrow -2 \leq 2x \leq 8$$
> $$\Leftrightarrow -1 \leq x \leq 4$$

Example 6

> To solve $|7 - x| \geq 4$, we proceed as follows:
>
> $$|7 - x| \geq 4$$
> $$\Leftrightarrow 7 - x \geq 4 \quad \text{or} \quad 7 - x \leq -4$$
> $$\Leftrightarrow -x \geq -3 \quad \text{or} \quad -x \leq -11$$
> $$\Leftrightarrow x \leq 3 \quad \text{or} \quad x \geq 11$$

Example 7

> To solve $|\log x| \leq 2$, we find the numbers in the domain of log that satisfy the inequality.
>
> $$|\log x| \leq 2$$
> $$\Leftrightarrow -2 \leq \log x \leq 2$$
> $$\Leftrightarrow 10^{-2} \leq x \leq 10^{2}$$

Example 8

> To solve $\log |x| < 3$, we proceed:
>
> $$\log |x| < 3$$
> $$\Leftrightarrow |x| < 10^{3} \quad \text{and} \quad x \neq 0 \qquad (\text{since } \log 0 \text{ is not defined})$$
> $$\Leftrightarrow -10^{3} < x < 10^{3} \quad \text{and} \quad x \neq 0$$

Exercises 8.5

Solve each of the following inequalities or equations. In each case, graph the solution.

1. $|8x + 3| = 5$
2. $|8x + 3| = -5$
3. $|x - 3| < 4$
4. $|2x - 1| \geq 5$
5. $|x - 15| < 1$
6. $|3 - x| < 5$
7. $|3y - 15| > 0$
8. $|8x + 4| \geq 4$
9. $3 + |4x - 1| < 5$
10. $|2x - 3| > -4$

11. $6 < |2x - 10|$

12. $|3 + 2x| \leq 1$

13. $|3(2x + 4) + x| \leq 5$

14. $\left| \dfrac{x}{3} \right| > 5$

15. $|3(x - 1)| < 6$

16. $|x^2 - 2| \geq 1$

17. $|y^2 - 3| > 1$

18. $\left| \dfrac{3}{x^2 - 4} \right| > -2$

19. $|\log x| > -1$

20. $|\log x| \leq 1$

21. $|\sin x| < \frac{1}{2}$

Review Exercises
Chapter 8

In Exercises 1–20, solve each inequality.

1. $8x - 4 \leq 2x + 2$

2. $3x + 1 \geq x - 5$

3. $x^2 + 8x + 7 > 0$

4. $x^2 + 3 > 0$

5. $2(x + 4) \leq 3(2x + 9) - (4x + 19)$

6. $-\frac{1}{2}x \leq 6$

7. $\dfrac{2x - 4}{5} \leq 10$

8. $3x^2 + 6x + 3 \leq 0$

9. $x^2 - 16 \leq 0$

10. $(x + 4)(x - 3)(x + 2) \leq 0$

11. $x^2 + 4x \leq 2x^2 + 8x + 5$

12. $|x + 4| \geq 7$

13. $|2x - 5| \leq 5$

14. $|3x + 5| < 10$

15. $|x - 3| > 12$

16. $15 \leq 2(3x - 8)$

17. $\dfrac{3}{x} < 5$

18. $(x + 3)(x^2 + 5) > 0$

19. $(x^2 + 5)(x^2 + 7) > 0$

20. $x^2 + 5x + 1 \leq 0$

In Exercises 21–23, solve each equation.

21. $|3x + 5| = 2$ **22.** $|x + 5| = -1$ **23.** $|2x - 6| = 0$

In Exercises 24–28, sketch the graph of the given equations.

24. $y = |\cos x|$ **25.** $y = \cos |x|$ **26.** $y = |2^x|$

27. $y = 2^{|x|}$ **28.** $y = |x^3|$

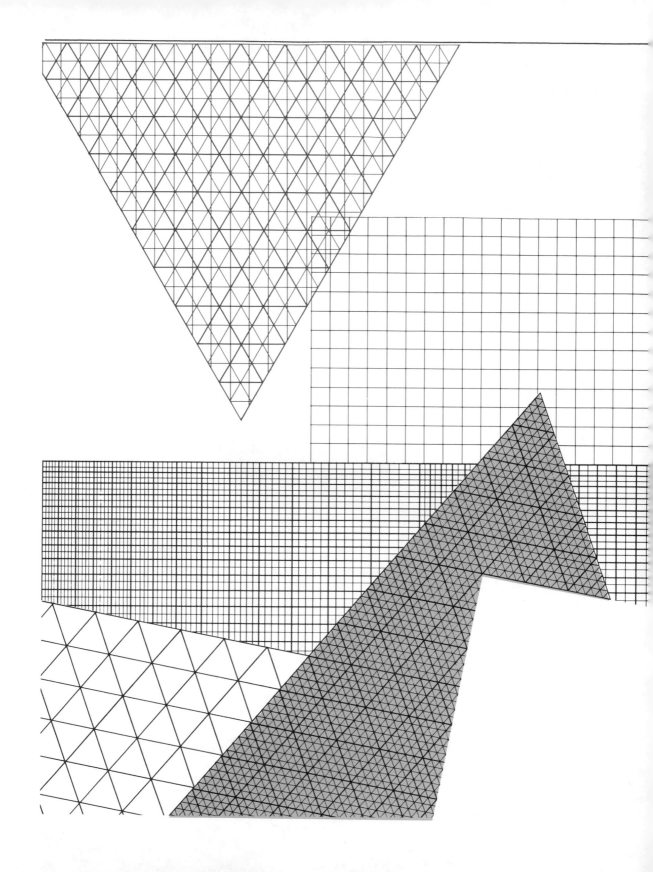

9 Some Applications and a First Look at Limits

SECTION 9.1 **INTRODUCTION**

In this text, we have discussed many properties of functions—intercepts, symmetry, periodicity, zeros, and others—in an attempt to analyze properties that functions have in common. In this chapter, we will study another property of functions, **limits**. This property is in many ways more subtle than the others, but it lies at the heart of calculus and, as such, is an appropriate conclusion for this text.

Suppose you were asked to describe the function $f(x) = 1/x$. We choose this function because we have used it frequently and its graph (Figure 9.1.1) is probably familiar to you.

Figure 9.1.1 $f(x) = 1/x$

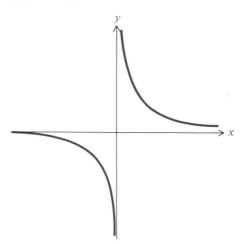

We may describe this function by saying that its graph has no intercepts. The graph has no x-intercepts since $f(x)$ is never zero and it has no y-intercepts since the function is undefined when $x = 0$. Also, the graph of f is symmetric with respect to the origin. But how can we describe its behavior near zero?

We might say that f is very large if x is near zero and positive, and very small if x is near zero and negative. (By very small, we mean $f(x)$ is negative and $|f(x)|$ is large.) Although this is true, this description does not distinguish f from the function whose graph appears in Figure 9.1.2.

Figure 9.1.2

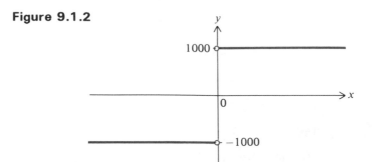

We could say that f increases as x approaches zero from the right and decreases as x approaches zero from the left. Again, this inadequately describes the behavior of f near zero. Figure 9.1.3 shows the graph of a function that increases to one side of zero and decreases to the other.

Figure 9.1.3

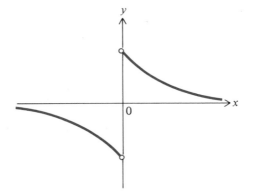

The property that distinguishes the function $f(x) = 1/x$ from the functions graphed in Figures 9.1.2 and 9.1.3 is that we can make $f(x)$ "as large as we want" by requiring x to be positive and "sufficiently near" zero and "as small as we want" by requiring x to be negative and "sufficiently near" zero. We shall make the phrases "as large as we like" and "sufficiently near" more precise.

No matter what number, K, we name, however large, we can find a positive number a, such that if $0 < x < a$ then $f(x) > K$. If $K = 1000$, let $a = \frac{1}{1000}$. If $0 < x < \frac{1}{1000}$, then $f(x) > 1000$. If $K = 1,000,000$, let $a = \frac{1}{1,000,000}$. If $0 < x < \frac{1}{1,000,000}$, then $f(x) > 1,000,000$. A similar argument holds for any large number, K, not just for 1000 and 1,000,000.

The property we have just discussed—the behavior of $f(x)$ as x "approaches" a number—is a very useful one in mathematics; and in this chapter, we will consider a number of applications of this property. We can describe the behavior of $f(x) = 1/x$ near zero as follows.

> As x approaches zero from the right, $f(x)$ increases without bound; i.e., $f(x)$ gets as large as we like. As x approaches zero from the left, $f(x)$ decreases without bound; i.e., $f(x)$ gets as small as we like.

There is a convenient notation for this behavior. Using the abbreviation "lim," for limit, we write

$$\lim_{x \to 0^+} f(x) = +\infty$$

(read: the **limit** of $f(x)$ as x approaches zero from the right is positive infinity) and

$$\lim_{x \to 0^-} f(x) = -\infty$$

(read: the limit of $f(x)$ as x approaches zero from the left is negative infinity).

The symbols $+\infty$ and $-\infty$ are not numbers. We cannot add, subtract, multiply, or divide with them. When used as above, they indicate that f increases without bound ($+\infty$) or decreases without bound ($-\infty$).

Let us consider another use of the limit notation using the same function, $f(x) = 1/x$. What happens to the values of f as x gets larger?

When $x = 100$, $f(x) = \frac{1}{100}$. When $x = 1000$, $f(x) = \frac{1}{1000}$. When $x = 1,000,000$, $f(x) = \frac{1}{1,000,000}$. We can say that as x increases, $f(x)$ decreases. Although this is true, we can say the same about the function $g(x) = -x^3$, whose graph looks much different from the graph of f. (See Figure 9.1.4.) One thing that distinguishes f from g is that as x gets larger, $f(x)$ gets "as close to zero as we like," while this is not true of g.

Figure 9.1.4

(a) $f(x) = 1/x$

(b) $g(x) = -x^3$

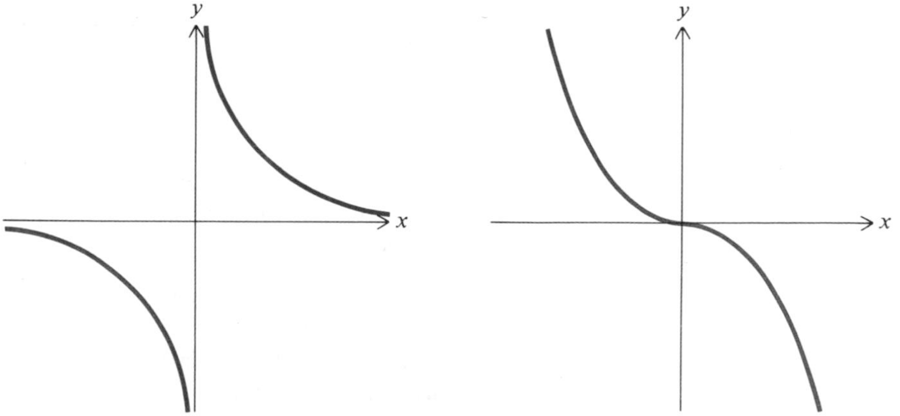

Using the limit notation, we can express this fact as follows:

$$\lim_{x \to +\infty} f(x) = 0$$

Many subtle ideas are contained in the limit notation. One is illustrated above. The function $f(x)$ is never zero. Its limit, as $x \to +\infty$, *is* zero. When we talk about the limit of a function, we are not necessarily talking about a value the function achieves. A second subtle idea is illustrated in the first limit statement we wrote:

$$\lim_{x \to 0^+} f(x) = +\infty$$

We cannot, in this case, ask "What happens when $x = 0$?" When $x = 0$, the function f is undefined, and $f(0)$ makes no sense. We *can* ask "What happens to $f(x)$ as x approaches zero?"

In this section, we have tried to show how the concepts of limits are helpful when describing the behavior of certain functions. We have also introduced the notation for limits. You will see it used again throughout this chapter.

Exercises 9.1

1. Consider the function $f(x) = 2 - 1/x$.
   a. Evaluate $f(1), f(2), f(5), f(10), f(1000)$.
   b. As x gets larger, does $f(x)$ get larger or smaller?
   c. Is there a number, x, such that $f(x) = 2$?
   d. What is $\lim_{x \to +\infty} f(x)$?
   e. Does the phrase "$f(x)$ gets larger as x increases" mean the same thing as "f increases without bound as x increases"?

2. Consider the function $g(x) = 1/x^2$.
   a. Evaluate $g(x)$ when $x = 0.1, 0.01, 0.001$, and 0.0001.
   b. What is $\lim_{x \to 0^+} g(x)$?
   c. Evaluate $g(x)$ when $x = -0.1, -0.01, -0.001$, and -0.0001.
   d. What is $\lim_{x \to 0^-} g(x)$?

3. Consider the function $h(x) = (1/x) - 3$.
   a. Evaluate this function when $x = 1, 2, 5, 10$, and 1000.
   b. As x gets larger, does $h(x)$ get larger or smaller?
   c. What is $\lim_{x \to +\infty} h(x)$?
   d. Does the phrase "$h(x)$ gets smaller" mean the same thing as "h decreases without bound"?
   e. How is the graph of $h(x)$ related to the graph of $f(x) = 1/x$? (Use the methods of Chapter 4.)

4. Sketch graphs of function F, G, and H, which have the following properties:

    a. $\quad \lim\limits_{x \to +\infty} F(x) = 4$ **b.** $\quad \lim\limits_{x \to +\infty} G(x) = -2$

    c. $\quad \lim\limits_{x \to 2} H(x) = +\infty$ (*Hint:* Try to shift the graph of the function of Exercise 2.)

5. Sketch the graph of $y = \tan x$.

    a. Use limit notation to describe the behavior of the graph near $x = \pi/2$.

    b. Use limit notation to describe the behavior of the graph near $x = -\pi/2$.

6. Sketch the graph of $y = \log x$. Use limit notation to describe the behavior of the graph near $x = 0$.

7. Sketch the graph $y = 3^x$. Use limit notation to discuss the behavior of the graph as x gets smaller without bound; i.e., x is negative and gets increasingly large in absolute value.

8. Sketch the graph $y = 3^x + 2$. Use limit notation to discuss the behavior of the graph as x gets smaller without bound.

9. Consider the function

$$f(x) = \begin{cases} x^2 & x \geq 0 \\ x + 2 & x < 0 \end{cases}$$

whose graph is sketched on the right.

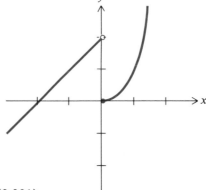

    a. Compute $f(1)$, $f(0.1)$, $f(0.01)$, $f(0.001)$.

    b. What is $\lim\limits_{x \to 0^+} f(x)$?

    c. Compute $f(-1)$, $f(-0.1)$, $f(-0.01)$, $f(-0.001)$.

    d. What is $\lim\limits_{x \to 0^-} f(x)$?

10. Consider the function, f, defined by $f(x) = x^2 + 1$.

    a. Evaluate $f(1)$, $f(0.1)$, $f(0.01)$, $f(0.001)$.

    b. Evaluate $f(-1)$, $f(-0.1)$, $f(-0.01)$, $f(-0.001)$.

    c. What is $\lim\limits_{x \to 0} f(x)$? (*Note:* If $\lim\limits_{x \to 0^+} f(x) = \lim\limits_{x \to 0^-} f(x)$, we can represent the common limit by $\lim\limits_{x \to 0} f(x)$.)

11. Consider the function

$$f(x) = \begin{cases} x^2 & x \neq 0 \\ 2 & x = 0 \end{cases}$$

the graph of which is sketched here.

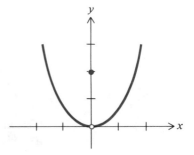

a. Compute $f(1)$, $f(0.1)$, $f(0.01)$, and $f(0.001)$.
b. Compute $f(-1)$, $f(-0.1)$, $f(-0.01)$, and $f(-0.001)$.
c. What is $\lim_{x \to 0} f(x)$?

12. Consider the function g, defined by

$$g(x) = \begin{cases} x^2 & x \neq 0 \\ 3 & x = 0 \end{cases}$$

a. What is $\lim_{x \to 0} g(x)$?
b. Would your answer to (a) be different if $g(0)$ were a number other than 3?

SECTION 9.2

LEARNING CURVES

Psychologists frequently use functions to describe patterns of behavior. As an example, we will investigate relationships that psychologists believe describe how much we learn about a subject as a function of the time spent studying.

Suppose you are one of a large group given a precalculus test that has a perfect score of 100. Before taking the test, you are asked to notice how many minutes you studied for the test.

The person administering this test is interested in the relationship between times spent studying and scores on tests, and would like to find a function, f, such that $f(x)$ represents the average score obtained by those people who studied x minutes. For example,

$f(10) =$ the average score obtained by people who studied 10 minutes.

$f(30) =$ the average score obtained by people who studied half an hour.

If we found such a function, we would certainly expect it to have the following two properties.

A. $f(x)$ must never be greater than 100, since 100 is the highest possible score.

B. f must be an increasing function (i.e., if $x_2 > x_1$ then $f(x_2) > f(x_1)$) since, although any particular student may do worse after additional studying, we expect the average of a large group to improve.

Using experimental and statistical techniques, psychologists have concluded that a function, f, describing the score after x minutes of study is given by an equation of the form

$$f(x) = c - a \cdot 10^{-kx}$$

where c, a, and k are positive constants. Functions of this form satisfy (A) and (B) above. We will illustrate this with an example. Consider the function $f(x) = 100 - 30 \cdot 10^{-0.003x}$. We have computed some values of f below, rounded off to the nearest tenth.

$f(0) = 70.0$

$f(30) = 75.6$

$f(60) = 80.2$

$f(120) = 86.9$

$f(240) = 94.3$

It may be easier to analyze the function if we rewrite it without negative exponents:

$$f(x) = 100 - \frac{30}{10^{0.003x}}$$

From our work with exponential functions, we know that $10^{0.003x}$ will always be positive. Because of this, $30/10^{0.003x}$ will also always be positive, and $f(x) = 100 - 30/10^{0.003x}$ will always be less than 100. Therefore f satisfies property (A).

We can also show that f satisfies property (B); i.e., if $x_2 > x_1$, then $f(x_2) > f(x_1)$.

Suppose $x_2 > x_1$. Then

$$10^{0.003x_2} > 10^{0.003x_1} > 0$$

Hence,

$$\frac{30}{10^{0.003x_2}} < \frac{30}{10^{0.003x_1}}$$

$$\frac{-30}{10^{0.003x_2}} > \frac{-30}{10^{0.003x_1}}$$

and so,

$$100 - \frac{30}{10^{0.003x_2}} > 100 - \frac{30}{10^{0.003x_1}}$$

The expression on the left of the last inequality is $f(x_2)$; the expression on the right is $f(x_1)$. So, if $x_2 > x_1$, $f(x_2) > f(x_1)$, and, therefore, f is an increasing function.

Another property of the function we consider in some detail is the behavior of this function as x increases. Looking at the values of f we computed earlier it seems that as x increases, $f(x)$ gets closer to 100. When $x = 60$, for example, $f(x)$ is within 20 units of 100. When $x = 240$, $f(x)$ is within 6 units of 100. How close does $f(x)$ get to 100?

We claim that by choosing x large enough, we can make $f(x)$ as close to 100 as we like. Using the notation of the previous section,

$$\lim_{x \to +\infty} f(x) = 100$$

Before we prove this result, we will first illustrate the proof with an example. We will show that by choosing x large enough we can make $f(x)$ come within 0.1 unit of 100. More precisely, we will find a number, x_0, such that if $x \geq x_0$, $|f(x) - 100| < 0.1$.

Recall from our work with absolute values that

$$|f(x) - 100| < 0.1$$
$$\Leftrightarrow 99.9 < f(x) < 100.1$$

Since we saw earlier that $f(x) < 100$ for all x, $99.9 < f(x) < 100.1$ is equivalent to $99.9 < f(x)$, which is equivalent to each of the following:

$$99.9 < 100 - \frac{30}{10^{0.003x}}$$

$$-0.1 < -\frac{30}{10^{0.003x}}$$

$$10^{0.003x} > \frac{30}{0.1} = 300$$

$$0.003x > \log 300 \qquad \text{(taking logs of both sides)}$$

$$x > \frac{\log 300}{0.003} = 825.7$$

If we choose $x > 825.7$, $f(x) > 99.9$, and $f(x)$ will be within 0.1 of 100. The number 825.7 is the x_0 we are trying to find.

In the above example, we used 0.1 and found an x_0 such that $f(x)$ would be within 0.1 units of 100 if $x > x_0$ ($x_0 = 825.7$). How large would x have to be if we wanted $f(x)$ to be within 0.01 units of 100? We could solve this problem, but we prefer to solve the more general problem where 0.01 is replaced by any small number u.

The steps, using u instead of 0.1, are these.

$$|f(x) - 100| < u$$
$$\Leftrightarrow 100 - u < f(x) < 100 + u$$
$$\Leftrightarrow 100 - u < f(x)$$
$$\Leftrightarrow 100 - u < 100 - \frac{30}{10^{0.003x}}$$
$$\Leftrightarrow -u < \frac{-30}{10^{0.003x}}$$
$$\Leftrightarrow 10^{0.003x} > \frac{30}{u}$$
$$\Leftrightarrow 0.003x > \log\left(\frac{30}{u}\right)$$
$$\Leftrightarrow x > \frac{\log(30/u)}{0.003}$$

No matter what small positive number u we choose, if

$$x > \frac{\log(30/u)}{0.003}$$

then $f(x)$ is within u units of 100. In general, then, our choice for x_0 is

$$\frac{\log(30/u)}{0.003}$$

It is possible to interpret this behavior geometrically. In Figure 9.2.1, we have drawn coordinate axes, drawn the line with equation $y = 100$, and "surrounded" it by two other lines; the lines with equations $y = 100 + u$ and $y = 100 - u$, each of them u units away from $y = 100$. No matter how small

Figure 9.2.1

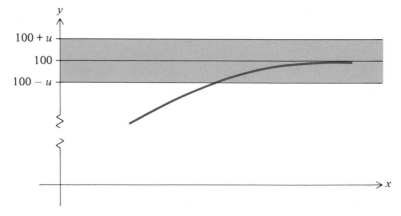

we make u, the graph of f will be inside the band between the graphs of $y = 100 - u$ and $y = 100 + u$ for all x greater than some x_0, which is determined by the size of u (i.e., by how wide the band is).

We began this section by describing an experiment in learning. Because of their effectiveness in describing learning, graphs of functions of the form

$$f(x) = c - a \cdot 10^{-kx} \quad \text{with } a, c, \text{ and } k > 0$$

are called **learning curves**.

Exercises 9.2

Exercises 1–4 refer to the function $f(x) = 100 - 30 \cdot 10^{-0.003x}$, which describes test scores after x minutes of study.

1. What is $f(0)$?
2. Does your answer to Exercise 1 seem reasonable? (Recall that $f(0)$ is the average score after 0 minute study.)
3. As x increases $f(x)$ approaches 100. Does this result seem reasonable?
4. Suppose the test were not a precalculus test (a subject we presumably know something about) but a test in a subject we know nothing about. Would $f(x) = 100 - 30 \cdot 10^{-0.003x}$ still give a good description of the results of a test? (Think about $f(0)$.)
5. Which of the following two graphs do you think better represents a learning curve if we know nothing about the subject when we start studying, as in Exercise 4? Explain.

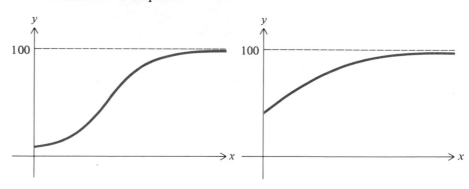

SECTION 9.3

THE FUNCTION (sin x)/x

An important concept in calculus is the determination of rates of change of functional values. For example, the height of a body falling near the surface of the earth can be given as a function of time. The rate of change of height is the velocity of that body and the rate of change of velocity is the acceleration of that body. Most functions used in scientific work can be expressed as sums, products, quotients, and compositions of functions we have studied in this text, and finding the rates of change of the values of functions, in general, can be reduced to finding the rates of change of these more elementary functions.

The function f defined by

$$f(x) = \frac{\sin x}{x}$$

is used to determine rates of change of trigonometric functions. Because of the importance of this function in the calculus, this section will be devoted to a study of f. Actually a knowledge of f is useful in many scientific investigations. We shall see in the next section how this function can be used to determine the area of a circle.

Since the domain of the sine function is $(-\infty, +\infty)$, we see that

$$D_f = \text{all real numbers } x \neq 0$$

Also,

$$f(-x) = \frac{\sin(-x)}{-x} = \frac{-\sin x}{-x} = \frac{\sin x}{x} = f(x)$$

so we see that the graph of f is symmetric with respect to the y-axis.

Because of this symmetry, we will concentrate on f and its graph when $x > 0$ and draw inferences about the behavior of f on $(-\infty, 0)$ from our knowledge of symmetry.

To get an understanding of the behavior of f on $\left(0, \frac{\pi}{2}\right]$, we begin by sketching the graph of f via the point-plotting method. It is especially important for future applications to investigate f when x is close to 0.

In Figure 9.3.1 you will find a sketch of a graph using the values of f on $\left(0, \frac{\pi}{2}\right]$. Table 9.3.1 correspondingly lists these values. Using the graph and table

Figure 9.3.1

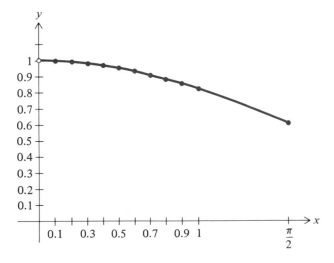

try to predict

$$\lim_{x \to 0} \frac{\sin x}{x}$$

(i.e., determine what number $(\sin x)/x$ will be near when x is near but not equal to 0.)

Table 9.3.1

x	$\dfrac{\sin x}{x}$	x	$\dfrac{\sin x}{x}$	x	$\dfrac{\sin x}{x}$
$\dfrac{\pi}{2}$	0.63661977	0.1	0.99833417	0.009	0.9998650
1	.84147098	.09	.99865055	.008	.99998932
0.9	.87036323	.08	.99893367	.007	.99999183
.8	.89669511	.07	.99918353	.006	.99999398
.7	.92031098	.06	.99940011	.005	.99999582
.6	.94107079	.05	.99958338	.004	.99999733
.5	.95885108	.04	.99973335	.003	.99999847
.4	.9735486	.03	.99985001	.002	.99999930
.3	.98506736	.02	.99993333	.001	.99999980
.2	.99334665	.01	.99998333	.0001	1 (accurate to 8 decimals)

The information given by Figure 9.3.1 and Table 9.3.1, together with symmetry, indicate that

$$\lim_{x \to 0} \frac{\sin x}{x} = 1$$

Although the work above might convince some people that $(\sin x)/x$ approaches 1 when x approaches 0, we can provide a more rigorous argument based on our understanding of the definitions of the trigonometric functions and a more obvious limiting statement,

$$\lim_{x \to 0} (1 - x) = 1$$

Consider an angle of measure x (we may assume $0 < x < \pi/2$) in standard position as depicted in Figure 9.3.2. Recall that the length of the arc of the unit circle subtended by this angle is x units, and the point P (the intersection of the terminal ray of the angle with the unit circle) has coordinates $(\cos x, \sin x)$.

Construct a line through P perpendicular to the horizontal axis and let R denote the point of intersection. Then construct a line through Q (i.e., the point $(1, 0)$) perpendicular to the horizontal axis and let S denote the point of intersection of this line with the line determined by OP.

Figure 9.3.2

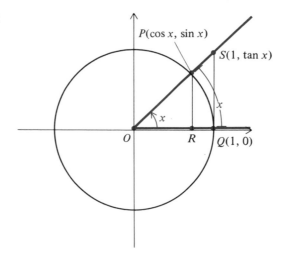

Now

$$d(P, R) \leq \text{length of arc } PQ \leq d(S, Q) \qquad \text{(See Exercise 17.)} \qquad (1)$$

Since $d(P, R)$ is the y-coordinate of the point P, we see

$$d(P, R) = \sin x$$

As noted above, length of arc $PQ = x$ and since

$$\tan x = \frac{d(S, Q)}{d(O, Q)} = \frac{d(S, Q)}{1}$$

$$d(S, Q) = \tan x = \frac{\sin x}{\cos x}$$

Thus, (1) is equivalent to

$$\sin x \leq x \leq \frac{\sin x}{\cos x}$$

Since we are assuming $0 < x < \pi/2$, $\sin x > 0$ and $\cos x > 0$, and so this sequence of inequalities is equivalent to

$$1 \leq \frac{x}{\sin x} \leq \frac{1}{\cos x}$$

Recall that if $a > 0$, $b > 0$, and $a < b$ then $1/b < 1/a$. We can apply this to the above inequality to obtain

$$\cos x \leq \frac{\sin x}{x} \leq 1 \qquad (2)$$

Now, $\cos x = d(O, R)$ so $d(R, Q) = 1 - \cos x$. Since the line segment RQ is a leg, and the line segment PQ the hypotenuse of the right triangle PQR, it follows that $d(R, Q) < d(P, Q)$. But the chord PQ is not longer than the arc PQ and so $d(R, Q) \leq$ length of arc $PQ = x$. That is,

$$1 - \cos x \leq x$$

This inequality is equivalent to

$$1 - x \leq \cos x \tag{3}$$

Inequalities (2) and (3) imply that

$$1 - x \leq \frac{\sin x}{x} \leq 1 \tag{4}$$

As x approaches 0, $1 - x$ approaches 1. Inequality (4) tells us that $(\sin x)/x$ is between $1 - x$ and 1. Therefore, as x approaches 0, $(\sin x)/x$ also approaches 1; that is,

$$\lim_{x \to 0^+} \frac{\sin x}{x} = 1$$

Because of the symmetry of the graphs of f, we also know that

$$\lim_{x \to 0^-} \frac{\sin x}{x} = 1$$

Hence

$$\lim_{x \to 0} \frac{\sin x}{x} = 1$$

Example 1

A.   If x approaches 0, then $2x$ also approaches 0, so

$$\lim_{x \to 0} \frac{\sin 2x}{2x} = 1$$

B.   We could generalize the above example as follows: If $c \neq 0$ is a constant and x approaches 0, then cx also approaches 0. So

$$\lim_{x \to 0} \frac{\sin cx}{cx} = 1$$

Example 2

To determine

$$\lim_{x \to 0} \frac{\sin cx}{x} \qquad c \neq 0$$

a constant, write

$$\frac{\sin cx}{x} = c \frac{\sin cx}{cx}$$

Since $(\sin cx)/cx$ approaches 1 as x approaches 0, it follows that $c \cdot (\sin cx)/cx$ approaches $c \cdot 1 = c$ as x approaches 0. So

$$\lim_{x \to 0} \frac{\sin cx}{x} = c$$

Suppose the measure of an angle is $x°$ (i.e., x degrees, not x radians) and we looked at the quotient $(\sin x°)/x$. The next example shows that

$$\lim_{x \to 0} \frac{\sin x°}{x} \neq 1$$

Example 3

$$\lim_{x \to 0} \frac{\sin x°}{x} = \lim_{x \to 0} \frac{\sin \left(\dfrac{\pi}{180} x\right)}{x} = \frac{\pi}{180} \doteq 0.01745$$

Note that this example shows that

$$\lim_{x \to 0} \frac{\sin x}{x} \neq 1$$

if $\sin x$ is computed as if x were the degree measurement of an angle. We pointed out in Section 6.3 that radian measurement has several advantages over degree measurement. Using radian measurement instead of degree measurement not only gives the important result

$$\lim_{x \to 0} \frac{\sin x}{x} = 1$$

but also simplifies the calculations of the rate of change of the sine function.

Example 4

$$\lim_{x \to 0} \frac{\cos x - 1}{x} = \lim_{x \to 0} \frac{\cos [2(\tfrac{1}{2} x)] - 1}{x}$$

$$= \lim_{x \to 0} \frac{1 - 2 \sin^2 \tfrac{1}{2} x - 1}{x} = \lim_{x \to 0} -2 \frac{\sin^2 \tfrac{1}{2} x}{x}$$

$$= \lim_{x \to 0} (-2) \sin \tfrac{1}{2} x \left(\frac{\sin \tfrac{1}{2} x}{x}\right)$$

As x approaches 0, $\sin \tfrac{1}{2} x$ approaches 0, and $(\sin \tfrac{1}{2} x)/x$ approaches $\tfrac{1}{2}$. Thus, as x approaches 0,

$$(-2)(\sin \tfrac{1}{2} x)\left(\frac{\sin \tfrac{1}{2} x}{x}\right)$$

approaches

$$(-2)(0)(\tfrac{1}{2}) = 0$$

So

$$\lim_{x \to 0} \frac{\cos x - 1}{x} = 0$$

Example 5

Since

$$\lim_{x \to 0} \frac{\sin x}{x} = 1$$

we know that when x is near 0,

$$\frac{\sin x}{x} \doteq 1 \quad \text{and} \quad \sin x \doteq x$$

So when x is sufficiently small, we may replace $\sin x$ in any function we are considering with x. This simplification is used frequently in scientific work.

It is also important to understand the behavior of $(\sin x)/x$ when x is large in absolute value. For example, functions similar to $(\sin x)/x$ are useful when studying mechanical vibrations like the motion of a spring whose vibrations are being damped by a shock absorber.

To begin our analysis of f on $\left[\frac{\pi}{2}, +\infty\right)$ recall that for $x > 0$

$$\sin x = 0 \quad \text{if} \quad x = 0, \pi, 2\pi, 3\pi, \ldots$$

Hence,

$$\frac{\sin x}{x} = 0 \quad \text{if} \quad x = \pi, 2\pi, 3\pi, \ldots$$

Also, note that if $x = \frac{\pi}{2} + 2k\pi$, where $k = 0, 1, 2, \ldots$, then $\sin x = 1$ so $(\sin x)/x = 1/x$. Hence,

$$f(x) = \frac{1}{x} \quad \text{if} \quad x = \frac{\pi}{2}, \frac{5\pi}{2}, \ldots, \frac{\pi}{2} + 2n\pi, \ldots$$

Moreover, if

$$x = \frac{3\pi}{2} + 2k\pi \qquad k = 0, 1, 2, \ldots$$

then $\sin x = -1$, so

$$f(x) = -\frac{1}{x} \quad \text{if} \quad x = \frac{3\pi}{2}, \frac{7\pi}{2}, \ldots, \frac{3\pi}{2} + 2n\pi, \ldots$$

This analysis tells us that the points $\left(\frac{\pi}{2} + 2k\pi,\ f\left(\frac{\pi}{2} + 2k\pi\right)\right)$ on the graph of f lie on the graph of $g(x) = 1/x$. The points $\left(\frac{3\pi}{2} + 2k\pi,\ f\left(\frac{3\pi}{2} + 2k\pi\right)\right)$ lie on the graph of the function $h(x) = -1/x$.

Furthermore, since

$$-1 \leq \sin x \leq 1$$

we see that for $x > 0$

$$-\frac{1}{x} \leq \frac{\sin x}{x} \leq \frac{1}{x}$$

Thus, the graph of f on $(0, +\infty)$ must lie between the graphs of $g(x) = 1/x$ and $h(x) = -1/x$.

Figure 9.3.3 shows the graph of f on the intervals $\left[\frac{\pi}{2}, +\infty\right)$ and $\left(-\infty, -\frac{\pi}{2}\right]$. This graph suggests that when x increases without bound or decreases without bound, $f(x)$ approaches zero. We use the limit statements

$$\lim_{x \to +\infty} f(x) = 0 \quad \text{and} \quad \lim_{x \to -\infty} f(x) = 0$$

to describe these phenomena.

Figure 9.3.3

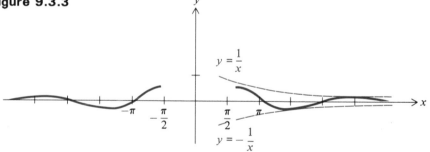

Since

$$\lim_{x \to +\infty} \frac{1}{x} = 0 \quad \text{and} \quad \lim_{x \to +\infty} \frac{-1}{x} = 0$$

we can draw the conclusion that

$$\lim_{x \to +\infty} f(x) = 0$$

from the fact that

$$-\frac{1}{x} \leq \frac{\sin x}{x} \leq \frac{1}{x} \quad \text{if} \quad x > 0$$

Indeed, if $1/x$ and $-(1/x)$ are near zero, then any number between $1/x$ and $-1/x$ will be near zero.

Exercises 9.3   In Exercises 1–13, evaluate the given limit.

1. $\lim\limits_{x \to 0} \dfrac{\sin x}{3x}$

2. $\lim\limits_{x \to 0} \dfrac{\sin 3x}{x}$

3. $\lim\limits_{x \to 0} \dfrac{\sin \pi x}{x}$

4. $\lim\limits_{x \to 0} \dfrac{\sin x}{-5x}$

5. $\lim\limits_{x \to 0} \dfrac{\sin \frac{1}{2} x}{3x}$

6. $\lim\limits_{x \to 0} \dfrac{\sin x^2}{x^2}$

7. $\lim\limits_{x \to 0} \dfrac{\sin x^2}{x}$

8. $\lim\limits_{x \to 0} \dfrac{1 - \cos 3x}{3x}$

   $\left(Hint: \dfrac{\sin x^2}{x} = x \dfrac{\sin x^2}{x^2}\right)$

9. $\lim\limits_{x \to 0} \dfrac{1 - \cos 3x}{x}$

10. $\lim\limits_{x \to 0} \dfrac{1 - \cos x}{3x}$

11. $\lim\limits_{x \to 0} \dfrac{\cos x - 1}{x}$

12. $\lim\limits_{x \to 0} \dfrac{1 - \cos 7x}{2x}$

13. $\lim\limits_{x \to 0} \dfrac{1 - \cos 2x}{-3x}$

14. Consider the function y defined by

$$g(x) = \frac{1 - \cos x}{x}$$

   a. What is D_g?
   b. Is the graph of g symmetric with respect to the vertical axis? Origin?
   c. For what values in D_g is $g(x) = 0$?
   d. Show that $0 \le 1 - \cos x \le 2$ for all x.
   e. Use part (d) to show that if $x > 0$, then $0 \le g(x) \le 2/x$.
   f. For what values of $x > 0$ is $g(x) = 2/x$?
   g. What is $\lim\limits_{x \to +\infty} g(x)$?
    h. Sketch a graph of g.

 15. Evaluate

$$\frac{\sin 1°}{1}, \frac{\sin 0.5°}{0.5}, \frac{\sin 0.1°}{0.1}, \quad \text{and} \quad \frac{\sin 0.01°}{0.01}$$

   and compare your answers with the result obtained in Example 3 of this section.

16. In this exercise, we will evaluate

$$\lim\limits_{x \to 0} \frac{\sin (\theta + x) - \sin \theta}{x}$$

   (θ is a fixed number.)

a. Using the identity

$$\sin(A + B) = \sin A \cos B + \sin B \cos A$$

show

$$\sin(\theta + x) - \sin \theta = \sin \theta (\cos x - 1) + \sin x \cos \theta$$

b. Show that

$$\frac{\sin(\theta + x) - \sin \theta}{x} = \sin \theta \frac{\cos x - 1}{x} + \cos \theta \frac{\sin x}{x}$$

c. Use the fact that

$$\lim_{x \to 0} \left(\sin \theta \frac{\cos x - 1}{x} + \cos \theta \frac{\sin x}{x} \right)$$

$$= \lim_{x \to 0} \sin \theta \frac{\cos x - 1}{x} + \lim_{x \to 0} \cos \theta \frac{\sin x}{x}$$

to find

$$\lim_{x \to 0} \frac{\sin(\theta + x) - \sin \theta}{x}$$

17. Verify statement (1) in Section 9.3. (*Hints:* To verify $d(P, R) \leq x$ (see Figure 9.3.1) show that $d(P, R) \leq d(P, Q)$ and that $d(P, Q) \leq x$. To verify that $x \leq d(S, Q)$ use the area of circular sector OPQ to be less than or equal to the area of triangle OSQ.)

SECTION 9.4   AREA OF A CIRCLE

The computation of the area of a circle is one of the oldest problems in mathematics. Clay tablets dating from approximately 3000 B.C. indicate that Babylonian mathematicians used the formula $A = 3r^2$. The Rhind papyrus, our source for much ancient Egyptian mathematics, uses a formula equivalent to $A = 3.16r^2$. In this section, we will show how to obtain the correct formula, $A = \pi r^2$, using limits.

Our method for obtaining a formula for the area of a circle is motivated by the diagrams in Figure 9.4.1.

We have inscribed three regular polygons in a circle. (A regular polygon is one whose sides and angles are congruent.) As the number of sides increases, the areas of the inscribed polygons give us increasingly better estimates for the area of the circle. We will use the areas of these regular polygons to find the area of the circle.

Figure 9.4.2 shows part of a regular polygon with n sides inscribed in a circle of radius r. We will derive a formula for the area of this polygon. Since the

Figure 9.4.1

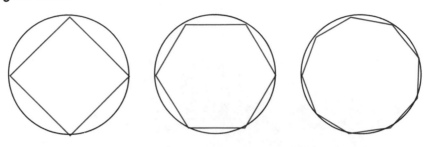

Figure 9.4.2

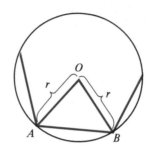

Figure 9.4.3

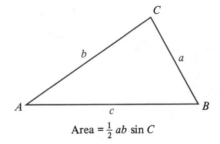

$$\text{Area} = \tfrac{1}{2}\,ab\,\sin C$$

central angle of a circle is 2π radians, angle AOB measures $2\pi/n$ radians, where n is the number of sides in the polygon. In Chapter 6, we derived the following formula for the area of a triangle: Area $= (\tfrac{1}{2}\,ab)(\sin C)$. (See Figure 9.4.3.)

Applying this formula to triangle AOB of Figure 9.4.2, we have

$$\text{Area of } AOB = (\tfrac{1}{2}r \cdot r)\left(\sin \frac{2\pi}{n}\right)$$

$$= (\tfrac{1}{2}r^2)\left(\sin \frac{2\pi}{n}\right)$$

Triangle AOB was formed by using the center of the circle as a vertex and one side of the polygon as a base. If we did this for each side of the polygon, we would have n congruent triangles, and the area of the polygon would be

$$\text{Area of polygon} = n(\tfrac{1}{2}r^2)\left(\sin\frac{2\pi}{n}\right)$$

$$= (\tfrac{1}{2}n)(r^2)\left(\sin\frac{2\pi}{n}\right)$$

We can use the area of an inscribed regular polygon to estimate the area of the circle. In Table 9.4.1, $A(n)$ represents the area of the inscribed regular polygon with n sides.

Notice that as n increases $A(n)$ seems to approach πr^2. This should not be surprising, since we know that the area of the circle is πr^2 or approximately $3.1416r^2$. As the inscribed polygons are constructed with more and more sides, their areas approach the area of the circle. The area of an inscribed polygon with 1000 sides differs from the area of the circle by less than $(1/10,000)r^2$!

To find the area of the circle, we must find the limit of $A(n)$ as n, the number of sides, increases without bound. Using limit notation we are looking for

$$\lim_{n \to +\infty} A(n).$$

We can use the result obtained in the last section,

$$\lim_{x \to 0} \frac{\sin x}{x} = 1$$

to evaluate this limit and obtain the exact formula for the area of a circle.

We know that

$$A(n) = (\tfrac{1}{2}nr^2)\left(\sin\frac{2\pi}{n}\right)$$

We can rewrite the formula this way:

$$A(n) = (\tfrac{1}{2}nr^2)\left(\frac{2\pi}{n}\right)\left(\frac{\sin\left(\frac{2\pi}{n}\right)}{\left(\frac{2\pi}{n}\right)}\right) = \pi r^2 \frac{\sin\left(\frac{2\pi}{n}\right)}{\frac{2\pi}{n}}$$

Table 9.4.1

n = Number of Sides of the Regular Inscribed Polygons	$A(n) = (\tfrac{1}{2}nr^2)\left(\sin\dfrac{2\pi}{n}\right)$
3	$1.299r^2$
5	$2.1398r^2$
10	$2.9389r^2$
15	$3.0505r^2$
20	$3.0902r^2$
25	$3.6086r^2$
30	$3.1187r^2$
100	$3.1395r^2$
1000	$3.14157r^2$

(All we have done is multiply by $2\pi/n$ and divide by $2\pi/n$.) As n increases $2\pi/n$ approaches 0. Using the result from the previous section we get

$$\lim_{n \to +\infty} \frac{\sin(2\pi/n)}{(2\pi/n)} = 1$$

Since

$$A(n) = (\pi r^2) \frac{\sin(2\pi/n)}{(2\pi/n)}$$

the area of the circle is

$$\lim_{n \to +\infty} A(n) = \lim_{n \to +\infty} (\pi r^2) \frac{\sin(2\pi/n)}{(2\pi/n)} = \pi r^2$$

Exercises 9.4

Instead of approximating the area of a circle with regular inscribed polygons as we did in the text, we could approximate the area with regular circumscribed polygons, as shown in Figures (a), (b), and (c) below.

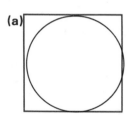

(a)

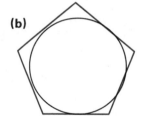

(b)

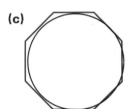

(c)

The exercises outline an alternative method to find the area of a circle using circumscribed regular polygons. All statements refer to diagram (d). AB is one side of a regular, n-sided polygon, circumscribing a circle with radius r.

Prove each of the following statements.

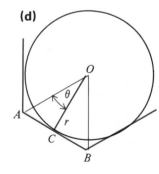

(d)

1. $AC = r \tan \theta$.
2. If AB is one side of a polygon of n sides, then angle AOB has measure $2\pi/n$.
3. $\theta = \pi/n$.
4. The area of $\triangle AOB$ is $\frac{1}{2}r \cdot 2r \tan\theta = r^2 \tan\theta$.
5. The area of $\triangle AOB = r^2 \dfrac{\sin\theta}{\cos\theta} = r^2 \dfrac{\sin(\pi/n)}{\cos(\pi/n)}$.

6. The area of the circumscribed polygon of n sides is

$$(nr^2)\left(\sin\frac{\pi}{n}\right)\left(\frac{1}{\cos(\pi/n)}\right)$$

7. The area of the circumscribed polygon of n sides, $A(n)$, is

$$A(n) = \left(\frac{\pi}{n}\right)(nr^2)\left(\frac{\sin(\pi/n)}{(\pi/n)}\right)\left(\frac{1}{\cos(\pi/n)}\right)$$

$$A(n) = \pi r^2\left(\frac{\sin(\pi/n)}{(\pi/n)}\right)\left(\frac{1}{\cos(\pi/n)}\right)$$

We have shown in this section that

$$\lim_{n\to+\infty}\frac{\sin(\pi/n)}{(\pi/n)} = 1$$

Assume that

$$\lim_{n\to+\infty}\frac{1}{\cos(\pi/n)} = 1$$

(We have not proved this result, but it should not be surprising.) As n gets larger without bound, π/n approaches 0. As π/n approaches 0, $\cos(\pi/n)$ approaches 1, and

$$\frac{1}{\cos(\pi/n)}$$

approaches 1.

8. $\lim\limits_{n\to+\infty} A(n) = \pi r^2$.

SECTION 9.5   **APPROXIMATING ZEROS OF FUNCTIONS**

Throughout this text, we have discussed how to find the points where a function has value 0 (i.e., the zeros of a function). If f is a polynomial function, finding zeros is equivalent to finding the roots of the polynomial defining f. In the case where f is linear or quadratic, we developed formulas (e.g., the quadratic formula) that gave us the roots. However, for polynomials of degree 3 or greater, we have no definitive method for determining roots. (There exist formulas that can be used to compute the roots of cubic or quartic polynomials, but we have chosen not to discuss them because of their complexity.) For example, finding the roots of $x^3 + x^2 + x - 1$ by the methods we have discussed in this text is nearly impossible.

Finding zeros of a function is very important in scientific work, and several algorithmic techniques have been developed to approximate the zeros of a function that cannot be found exactly. This section will illustrate one of these algorithms.

Suppose f is a function defined on the closed interval $[a, b]$ and f has the following properties:

A. $f(a) > 0$ and $f(b) < 0$ or $f(a) < 0$ and $f(b) > 0$.
B.   You can draw the graph over $[a, b]$ without "lifting your pencil off the paper."

We have a special name for functions satisfying property (B). Such functions are said to be continuous. Figure 9.5.1 shows the graph of a function that is continuous on $[1, 3]$. Figure 9.5.2 shows the graph of a function that is not continuous on $[1, 3]$. It will be helpful to note on what intervals $[a, b]$ the functions we have studied are continuous.

Figure 9.5.1   **A Function that is Continuous on $[1, 3]$**

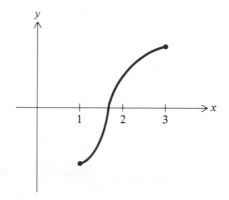

Figure 9.5.2   **A Function that is not Continuous on $[1, 3]$**

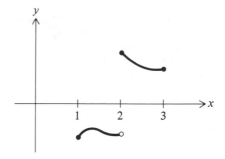

Each polynomial function is continuous on every interval $[a, b]$.

Each rational function

$$R(x) = \frac{P(x)}{Q(x)}$$

with $P(x)$ and $Q(x)$ polynomials and $Q(x)$ not the zero polynomial, is continuous on a closed interval $[a, b]$ if and only if there is no x in $[a, b]$ such that $Q(x) = 0$.

Each exponential function is continuous on every interval $[a, b]$.

Each logarithmic function, $f(x) = \log_c x$, is continuous on $[a, b]$ whenever $a > 0$.

The functions sine and cosine are continuous on every interval $[a, b]$.

The tangent function is continuous on an interval $[a, b]$ if and only if there is no number of the form $(\pi/2) + k\pi$, $k = 0, \pm 1, \ldots$ in $[a, b]$.

Suppose that f satisfies $f(a) < 0$ and $f(b) > 0$. Then the point $(a, f(a))$ lies below the x-axis and $(b, f(b))$ lies above the x-axis. (See Figure 9.5.1.) Since f satisfies property (B) we should be able to draw the graph of f starting at $(a, f(a))$ and ending at $(b, f(b))$ without "lifting our pencil off the paper." Thus there must be some point, c, in $[a, b]$ where the graph crosses the horizontal axis. The point c is a zero of f, i.e., $f(c) = 0$. We can summarize this as follows: If f is a function satisfying (A) and (B) on $[a, b]$, then f has a zero in $[a, b]$.

Note that f must satisfy both (A) and (B) in order for us to conclude that f has a zero in $[a, b]$. Figure 9.5.2 is the graph of a function satisfying (A) but not (B) on $[1, 3]$. Figure 9.5.3 is the graph of a function satisfying (B) but not (A). In both examples, f has no zeros in $[1, 3]$. (Also, note that just because a function f fails to satisfy either (A) or (B) on $[a, b]$ this does not necessarily mean f has no zeros on $[a, b]$.) (See Exercises 9 and 10.)

Figure 9.5.3

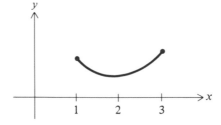

Let f be a function satisfying (A) and (B) above on the interval $[a_0, b_0]$. Without loss of generality, we may assume that $f(a_0) < 0$ and $f(b_0) > 0$. We know f has a zero between a_0 and b_0. In an attempt to locate a zero of f, consider the point $(a_0 + b_0)/2$, the point half-way between a_0 and b_0. If

$$f\left(\frac{a_0 + b_0}{2}\right) = 0$$

then we have located a zero of f. If

$$f\left(\frac{a_0 + b_0}{2}\right) \neq 0$$

then either

$$f\left(\frac{a_0 + b_0}{2}\right) < 0 \quad \text{or} \quad f\left(\frac{a_0 + b_0}{2}\right) > 0$$

If

$$f\left(\frac{a_0 + b_0}{2}\right) < 0$$

then f satisfies properties (A) and (B) on

$$\left[\frac{a_0 + b_0}{2}, b_0\right] \quad \text{(See Figure 9.5.4)}$$

and f has a zero between

$$\frac{a_0 + b_0}{2} \quad \text{and} \quad b_0$$

Figure 9.5.4

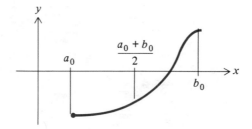

If

$$f\left(\frac{a_0 + b_0}{2}\right) > 0$$

then f satisfies (A) and (B) on

$$\left[a_0, \frac{a_0 + b_0}{2}\right] \quad \text{(See Figure 9.5.5)}$$

and f has a zero between

$$a_0 \quad \text{and} \quad \frac{a_0 + b_0}{2}$$

Figure 9.5.5

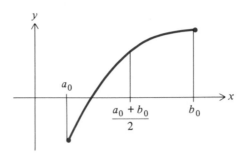

If

$$f\left(\frac{a_0 + b_0}{2}\right) < 0$$

let

$$a_1 = \frac{a_0 + b_0}{2} \quad \text{and} \quad b_1 = b_0$$

If

$$f\left(\frac{a_0 + b_0}{2}\right) > 0$$

let

$$a_1 = a_0 \quad \text{and} \quad b_1 = \frac{a_0 + b_0}{2}$$

In either case f satisfies (A) and (B) on $[a_1, b_1]$ and so f has a root between a_1 and b_1. Note that if c is a zero of f in $[a_1, b_1]$ then using b_1 as an approximation to c has a maximum error of $b_1 - a_1 = \frac{1}{2}(b_0 - a_0)$.

Now repeat the above process replacing $[a_0, b_0]$ by $[a_1, b_1]$ to obtain an interval $[a_2, b_2]$ such that

$$a_2 = \frac{a_1 + b_1}{2} \qquad b_2 = b_1 \quad \text{if} \quad f\left(\frac{a_1 + b_1}{2}\right) < 0$$

and

$$a_2 = a_1 \qquad b_2 = \frac{a_1 + b_1}{2} \quad \text{if} \quad f\left(\frac{a_1 + b_1}{2}\right) > 0$$

Then f satisfies (A) and (B) on $[a_2, b_2]$ and has a zero, c, between a_2 and b_2. If b_2 is used as an approximation to c, then the maximum error in this approximation is

$$b_2 - a_2 = \tfrac{1}{2}(b_1 - a_1) = (\tfrac{1}{2})^2(b_0 - a_0)$$

Figure 9.5.6 illustrates how the intervals $[a_1, b_1]$ and $[a_2, b_2]$ might be found.

Figure 9.5.6

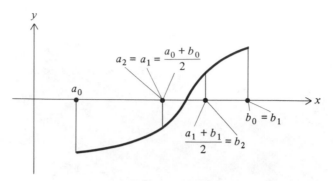

If we continue this process, we can construct a *sequence* of intervals $[a_1, b_1]$, $[a_2, b_2]$, $[a_3, b_3]$, ... having the following properties:

1. f satisfies (A) and (B) on each $[a_i, b_i]$ and thus has a zero between a_i and b_i.

2. $b_1 - a_1 = \frac{1}{2}(b_0 - a_0)$

   $b_2 - a_2 = \frac{1}{2}(b_1 - a_1) = (\frac{1}{2})^2(b_0 - a_0)$

   $b_3 - a_3 = \frac{1}{2}(b_2 - a_2) = (\frac{1}{2})^3(b_0 - a_0)$

   $$\vdots$$

   $b_n - a_n = \frac{1}{2}(b_{n-1} - a_{n-1}) = (\frac{1}{2})^n(b_0 - a_0)$

3.   If b_n, $n = 1, 2, \ldots$ is used as an approximation to a zero of f in $[a_0, b_0]$, then the maximum error in this approximation is

   $$b_n - a_n = (\tfrac{1}{2})^n(b_0 - a_0)$$

Figure 9.5.7 shows the graph of the function $f(x) = (b_0 - a_0)(\frac{1}{2})^x$. The graph indicates that $\lim_{x \to +\infty} (b_0 - a_0)(\frac{1}{2})^x = 0$. Thus, as n takes on successively larger integral values, $(\frac{1}{2})^n(b_0 - a_0)$ approaches 0 and b_n becomes a better approximation to a zero, c, of f. That is, $\lim_{n \to +\infty} b_n = c$.

Figure 9.5.7

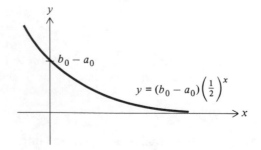

Example 1

To illustrate this technique, consider the function f defined by

$$f(x) = x^3 + x^2 + x - 1$$

Since $f(0) = -1$ and $f(1) = 2$, f has a zero between $a_0 = 0$ and $b_0 = 1$. Now

$$\frac{a_0 + b_0}{2} = \frac{1 + 0}{2} = \frac{1}{2} \quad \text{and} \quad f\left(\frac{1}{2}\right) = -\frac{1}{8}$$

Since $f(\frac{1}{2}) < 0$, f has a zero in $[\frac{1}{2}, 1]$, and we take $a_1 = \frac{1}{2}$ and $b_1 = 1$. If we use $b_1 = 1$ as an approximation to a zero of f, the maximum error in this approximation is $\frac{1}{2}(b_0 - a_0) = \frac{1}{2}$.

We now repeat the above process on $[a_1, b_1] = [\frac{1}{2}, 1]$.

$$\frac{a_1 + b_1}{2} = \frac{\frac{1}{2} + 1}{2} = \frac{3}{4}$$

$$f\left(\frac{3}{4}\right) = \frac{45}{64} = 0.70$$

Since $f(\frac{3}{4}) > 0$, we set $a_2 = a_1 = \frac{1}{2}$ and $b_2 = (a_1 + b_1)/2 = \frac{3}{4}$. If we use $b_2 = \frac{3}{4}$ to approximate a zero of f, then the maximum error in this approximation is $b_2 - a_2 = \frac{1}{4} = (\frac{1}{2})^2(b_0 - a_0)$. Figure 9.5.8 illustrates how a_1, b_1, a_2, and b_2 are found, and Table 9.5.1 gives the sequence of a_n's and b_n's found using iterations of this technique.

Figure 9.5.8

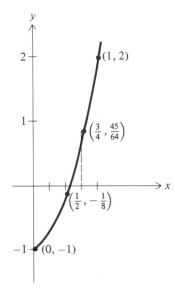

Table 9.5.1

n	a_n	b_n	$\dfrac{a_n + b_n}{2}$	$f\left(\dfrac{a_n + b_n}{2}\right)$	$b_n - a_n$
0	0	1	0.5	−0.12500	1
1	0.5	1	.75	.73438	0.5
2	.5	0.75	.625	.25977	.25
3	.5	.625	.5625	.05688	.125
4	.5	.5625	.53125	− .03659	.0625
5	.53125	.5625	.564875	.00950	.03125
6	.53125	.564875	.5390625	− .01370	.015625
7	.5390625	.564875	.54296875	− .00214	.0078125
8	.54296875	.564875	.544921875	.00367	.003906250
9	.54296875	.544921875	.543945313	.00076	.001953125
10	.54296875	.543945313	.543457032	− .00069	.000976563

If we wish to obtain a certain degree of accuracy in approximating a zero of f by this technique, we can tell how many iterations we must use to obtain this accuracy. For example, if we wanted to have our approximation, b_n, accurate to within two decimal points, then we want the error in approximation to be less than 0.005. Since $b_n - a_n$ is an upper bound on the error in approximating the zero of f, we will obtain the desired degree of accuracy if we choose n so that $b_n - a_n < 0.005$. If we examine the above table we see that $b_6 = 5.64875$ gives us the desired accuracy. A more general approach to finding the number, n, of iterations needed is given by the following argument.

Since

$$b_n - a_n = (\tfrac{1}{2})^n (b_0 - a_0)$$

we wish to choose n so that

$$(\tfrac{1}{2})^n (b_0 - a_0) < 0.005$$

We can find n as follows:

$$(\tfrac{1}{2})^n (b_0 - a_0) < 0.005$$

$$\Leftrightarrow (\tfrac{1}{2})^n < \frac{0.005}{b_0 - a_0}$$

$$\Leftrightarrow n \log (\tfrac{1}{2}) < \log \left(\frac{0.005}{b_0 - a_0} \right)$$

$$\Leftrightarrow n > \frac{\log \left(\dfrac{0.005}{b_0 - a_0} \right)}{\log (\tfrac{1}{2})} \qquad \text{(Recall } \log (\tfrac{1}{2}) < 0.)$$

Thus, if we choose n to be the smallest integer greater than

$$\frac{\log\left(\frac{0.005}{b_0 - a_0}\right)}{\log\left(\frac{1}{2}\right)}$$

then b_n will approximate a zero of f to within 0.005 unit.

In our example $a_0 = 0$, $b_0 = 1$. Hence, we will get two-decimal-place accuracy if

$$n > \frac{\log\left(0.005\right)}{\log\left(\frac{1}{2}\right)} \doteq 7.64$$

Hence, $b_8 = 0.564875$ approximates a zero of $f(x) = x^3 + x^2 + x - 1$ accurate to 0.005 unit.

Example 2

Suppose we wanted to solve the equation $\frac{1}{20}x = \log x$. Finding x where $\frac{1}{20}x = \log x$ is equivalent to finding x such that $\frac{1}{20}x - \log x = 0$ or to finding a zero of the function f defined by

$$f(x) = \frac{1}{20}x - \log x$$

Note that

$$f(1) = \frac{1}{20} - 0 = \frac{1}{20} > 0 \quad \text{and}$$
$$f(10) = \frac{1}{20}(10) - \log 10 = \frac{1}{2} - 1 = -\frac{1}{2} < 0$$

We can then use our iterative technique to approximate a zero of f between 1 and 10 by starting with $a_0 = 1$ and $b_0 = 10$. If we wish to obtain two-decimal-place accuracy, we choose n so that

$$b_n - a_n = \left(\tfrac{1}{2}\right)^n (b_0 - a_0) < 0.005$$

Now,

$$\left(\tfrac{1}{2}\right)^n (b_0 - a_0) < 0.005$$
$$\Leftrightarrow 9\left(\tfrac{1}{2}\right)^n < 0.005$$
$$\Leftrightarrow \left(\tfrac{1}{2}\right)^n < \frac{0.005}{9}$$
$$\Leftrightarrow n \log\left(\tfrac{1}{2}\right) < \log\left(\frac{0.005}{9}\right)$$
$$\Leftrightarrow n > \frac{\log\left(0.005/9\right)}{\log\left(\frac{1}{2}\right)} \doteq 10.81$$

Hence, we choose $n = 11$ to obtain the desired degree of accuracy. Table 9.5.2 describes this iterative procedure.

Table 9.5.2

n	a_n	b_n	$\dfrac{a_n + b_n}{2}$	$f\left(\dfrac{a_n + b_n}{2}\right)$
0	1	10	5.5	−0.47
1	1	5.5	3.25	−0.3493
2	1	3.25	2.125	−0.22111
3	1	2.125	1.5625	−0.11570
4	1	1.5625	1.28125	−0.04357
5	1	1.28125	1.140625	−0.00011
6	1	1.140625	1.0703125	0.02400
7	1.0703125	1.140625	1.10546875	0.011726
8	1.10546875	1.140625	1.123046875	0.00575
9	1.123046875	1.140625	1.131835938	0.00281
10	1.131835938	1.140625	1.136230469	0.00135
11	1.136230469	1.140625	1.138427735	0.000616

It should be clear from our two examples that implementing this algorithm involves considerable computation and requires a hand calculator or computer to be practical.

When the above algorithm is used to locate a zero, c, of a function it may seem that we have not found c exactly. However, since we can approximate c, with b_n, to any desired degree of accuracy (provided we are willing to spend the time computing enough b_n's) we can explicitly give any digit in the decimal expansion of c.

Practically speaking, it is seldom necessary to know c exactly. When c is used in calculations, it is usually sufficient to have a truncated decimal representation of c that gives us the degree of accuracy we desire. For example, if we use π in a calculation, we may find that the approximation $\frac{22}{7}$ is sufficient for our need. Others may require 3.1416 as an approximation to π, while others, desiring a higher degree of accuracy, may use 3.141592654. Each of these numbers, $\frac{22}{7}$, 3.1416, and 3.141592654, is only an approximation to π. Nevertheless, since methods have been devised that, theoretically, specify each digit in the decimal representation of π, we have the feeling we know π exactly.

Exercises 9.5

1. **a.** Use the quadratic formula to find the roots of

$$x^2 + 7x - 3$$

b. If $f(x) = x^2 + 7x - 3$, then $f(0) = -3$ and $f(1) = 5$. Use the iterative technique of this section to approximate the root of $x^2 + 7x - 3$ in $[0, 1]$ to two-decimal-place accuracy.

2. a. Use the quadratic formula to find the roots of

$$-2x^2 + x + 4$$

  b. If $f(x) = -2x^2 + x + 4$, then $f(1) = 3$ and $f(2) = -2$. Use the iterative technique of this section to approximate the root of $-2x^2 + x + 4$ in $[1, 2]$ to two-decimal-place accuracy.

3. Approximate a zero of $f(x) = x^3 - 7x^2 + x - 3$ to three-decimal-place accuracy.

4. Approximate a zero of $f(x) = x^3 + x^2 - x - 2$ to three-decimal-place accuracy.

5. Approximate a zero of $f(x) = 10^x - x - 2$ to two-decimal-place accuracy.

6. Approximate a zero of $f(x) = \log x - \frac{1}{30}x$ to three-decimal-place accuracy.

7. Approximate a zero of $f(x) = \cos x - x$ to four-decimal-place accuracy.

8. Approximate the zero of $f(x) = \tan x - x - 1$ that lies in $(0, \pi/2)$ to two-decimal-place accuracy.

9. Sketch the graph of a function f such that f is continuous on $[1, 3]$, $f(1) > 0$, $f(3) > 0$, and f has at least one zero in $[1, 3]$.

10. Sketch the graph of a function f such that $f(1) < 0$, $f(3) > 0$, f is **not** continuous on $[1, 3]$, and f has at least one zero in $[1, 3]$.

Review Exercises
Chapter 9

1. Let $f(x) = 6 + \dfrac{2}{x^2}$. Find:

    a. $\lim\limits_{x \to +\infty} f(x)$ b. $\lim\limits_{x \to 0} f(x)$

2. Find $\lim\limits_{x \to +\infty} \dfrac{2x^2 + x + 1}{x^2}$.   (*Hint:* Divide.)

3. a. Find $\lim\limits_{x \to 2} \dfrac{x^2 - 4}{x - 2}$.

    b. How do the functions $f(x) = \dfrac{x^2 - 4}{x - 2}$ and $g(x) = x + 2$ differ?

4. Find $\lim\limits_{x \to 3} \dfrac{x^2 - x - 6}{x^2 - 9}$.

5. Find a function, f, such that $\lim\limits_{x \to +\infty} f(x) = 3$.

6. Let $f(x) = x^2 + 8$.

   a. Find $f(1.1), f(1.01)$, and $f(1.001)$.

   b. What is $\lim_{x \to 1} f(x)$?

   c. How does $\lim_{x \to 1} f(x)$ compare with $f(1)$?

7. Let $f(x) = \begin{cases} x^2 & x \neq 1 \\ 5 & x = 1 \end{cases}$

   a. Sketch the graph of f.
   b. Find $f(1.1), f(1.01)$, and $f(1.001)$.
   c. What is $\lim_{x \to 1} f(x)$?
   d. How does $\lim_{x \to 1} f(x)$ compare with $f(1)$?

8. What is the relationship between horizontal asymptotes and limits?
9. What is the relationship between vertical asymptotes and limits?

In Exercises 10–15, find the given limit.

10. $\lim_{x \to 0} \dfrac{\sin(-3x)}{3x}$.

11. $\lim_{x \to 0} \dfrac{\sin x}{6x}$.

12. $\lim_{x \to 0} \dfrac{\sin 2x}{x\sqrt{3}}$.

13. $\lim_{x \to 0} \dfrac{\sin x}{\pi x}$.

14. $\lim_{x \to 0} \dfrac{1 - \cos 2x}{2x}$.

15. $\lim_{x \to 0} \dfrac{1 - \cos x}{3x}$.

16. Use the identities $\sin(-A) = -\sin A$, $\cos(-A) = \cos A$, $\cos(A) = \sin\left(\frac{\pi}{2} - A\right)$, $\sin A = \cos\left(\frac{\pi}{2} - A\right)$, and the result of Exercise 16 of Section 9.3

   (i.e., $\lim_{x \to 0} \dfrac{\sin(\theta + x) - \sin\theta}{x} = \cos\theta$)

   to prove that

   $$\lim_{x \to 0} \frac{\cos(\theta + x) - \cos\theta}{x} = -\sin\theta$$

   for each real number θ.

17. Suppose an inscribed regular polygon with 10 sides were used to estimate the area of a circle with radius $r = 1$. What will the error be?
18. A square is circumscribed around a circle of radius 4 while another square is inscribed in it. If we use the average of the areas of these two squares to estimate the area of the circle, what will the error be?

In Exercises 19–26, use the iterative technique of Section 9.5 to find an approximate zero of the given function on the indicated interval. Give three-decimal-place accuracy.

19. $f(x) = x^2 + 3x - 5$ $[0, 2]$
20. $f(x) = 3x^2 - 2x - 2$ $[-1, 1]$
21. $f(x) = x^5 + x^3 + 3$ $[-2, 0]$
22. $f(x) = x^4 - 3x + 1$ $[0, 1]$
23. $f(x) = 10^x + x^2 - 4$ $[0, 1]$
24. $f(x) = \log x - \frac{1}{20} x$ $[1, 10]$

25. $f(x) = \cos x - \sin x$ $\left[0, \dfrac{\pi}{2}\right]$

26. $f(x) = \tan x + x - 1$ $\left[\dfrac{2\pi}{3}, \dfrac{4\pi}{3}\right]$

27. Use the iterative technique of Section 9.5 to approximate $\sqrt{3}$ accurate to three decimal places. (*Hint:* Approximate a zero of $f(x) = x^2 - 3$ on $[1, 2]$.)

28. Use the iterative technique of Section 9.5 to approximate $\sqrt{5}$ accurate to three decimal places.

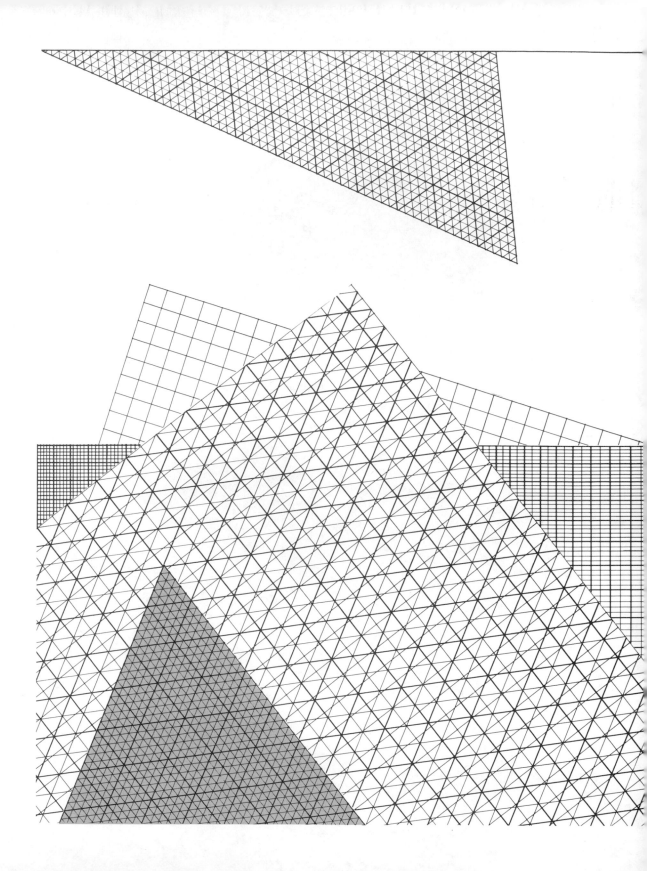

Appendix
Mathematical
Induction

Many interesting results in mathematics describe properties of the natural numbers, 1, 2, 3, 4 .... Among these results is the following, which you will very likely use in calculus:

The sum of the first n natural numbers is $n(n + 1)/2$; i.e.,

$$1 + 2 + 3 + 4 + 5 + \cdots + n = \frac{n(n + 1)}{2} \tag{1}$$

We can use formula (1) to find the sum of the first four natural numbers by letting $n = 4$: $1 + 2 + 3 + 4 = 4(5)/2 = 10$. The sum of the first seven natural numbers is

$$1 + 2 + 3 + 4 + 5 + 6 + 7 = \frac{7(8)}{2} = 28$$

and the sum of the first one hundred natural numbers is

$$1 + 2 + 3 + 4 + \cdots + 98 + 99 + 100 = \frac{100(101)}{2} = 5050$$

This appendix discusses how results such as (1) may be proved.

We can always check that (1) is true for any *particular* value of n. To prove that formula (1) is true when $n = 4$, for example, we could add the numbers 1, 2, 3, and 4, and see if the result is $4(5)/2$. Similarly, we could check to see if the formula is true when $n = 7$ or when $n = 100$. However, there are an infinite number of natural numbers, and we can never use this method to check that formula (1) is true for all of them. A common method used to prove that properties hold for all natural numbers is called **mathematical induction**.

Principle of Mathematical Induction

Let $P(n)$ be a proposition about the natural number n. Suppose

A. $P(1)$ is true.
B. For any natural number, n, if $P(n)$ is true then $P(n + 1)$ is also true.

Then $P(n)$ is true for all natural numbers.

If we have a proposition, $P(n)$, about the natural number n, we can sometimes use mathematical induction to prove that the proposition is true for all values of n. Formula (1), for example, is a proposition about the natural number, n. We saw that $P(4)$ is true by substituting $n = 4$. We will soon use mathematical induction to show that formula (1) is correct for all natural numbers, n.

Every proof using mathematical induction has two parts. The first part is to prove that $P(1)$ is true. This part usually consists of substituting $n = 1$ in the proposition $P(n)$ and noticing if the result is correct. The second part is usually more difficult. In this part we *assume* that $P(n)$ is true, and use this assumption to *prove* that $P(n + 1)$ is true.

Intuitively, (A) guarantees that the proposition is true for $n = 1$. Applying (B), if the proposition is true for $n = 1$, it must be true for $n = 2$. Applying (B) again, since the proposition is true for $n = 2$, it is true for $n = 3$. Continuing in this way, we can show that the proposition is true for any natural number.

We will use mathematical induction to prove (1), $1 + 2 + 3 + \cdots + n = n(n + 1)/2$. Let $P(n)$ be the statement, the sum $1 + 2 + 3 + \cdots + n = n(n + 1)/2$. We would like to show that $P(n)$ is true for all natural numbers, n. The proof by induction is in two parts:

A. *P(1) is true.* We must show that formula (1) is true when $n = 1$. If we substitute $n = 1$ in (1), we get

$$1 = \frac{1(2)}{2}$$

which is true.

B. *For any n, if P(n) is true, then P(n + 1) is also true.* We assume that $P(n)$ is true. (This is called the **induction hypothesis**.) $P(n)$ is the proposition that

$$1 + 2 + 3 + \cdots + n = \frac{n(n + 1)}{2} \tag{2}$$

We now try to show that $P(n + 1)$ is true. $P(n + 1)$ is the proposition that

$$1 + 2 + 3 + \cdots + n + (n + 1) = \frac{(n + 1)(n + 2)}{2} \tag{3}$$

(Notice that the right-hand side of (3) is obtained from the right-hand side of (2) by replacing n by $n + 1$.) Now

$$1 + 2 + 3 + \cdots + n + (n + 1)$$
$$= (1 + 2 + 3 + \cdots + n) + (n + 1) \tag{4}$$

By the induction hypothesis, (4) is equivalent to

$$1 + 2 + 3 + \cdots + n + (n + 1) = \frac{n(n + 1)}{2} + (n + 1) \tag{5}$$

$$\Leftrightarrow 1 + 2 + 3 + \cdots + n + (n + 1) = \frac{n(n + 1)}{2} + \frac{2(n + 1)}{2} \tag{6}$$

$$\Leftrightarrow 1 + 2 + 3 + \cdots + n + (n + 1) = \frac{(n + 1)(n + 2)}{2} \tag{7}$$

But this last equation is $P(n + 1)$. By assuming that $P(n)$ is true, we have shown that $P(n + 1)$ is true and our proof is complete.

As a second application of mathematical induction, we will prove that the product of any three consecutive natural numbers is divisible by three.

Let $P(n)$ be the proposition that $n(n + 1)(n + 2)$ is divisible by 3.

A.   *P(1) is true.* $P(1)$ is the proposition that $1(2)(3)$ is divisible by 3, which is clearly a true statement.

B.   *If P(n) is true, then P(n + 1) is true.* $P(n + 1)$ is the proposition that $(n + 1)(n + 2)(n + 3)$ is divisible by 3. Note that

$$(n + 1)(n + 2)(n + 3) = [(n + 1)(n + 2)](n + 3)$$
$$= n(n + 1)(n + 2) + 3(n + 1)(n + 2)$$

If we assume that $P(n)$ is true (the induction hypothesis), then $n(n + 1)(n + 2)$ is divisible by 3. The product $3(n + 1)(n + 2)$ is divisible by 3 since 3 is a factor. Since the sum of numbers divisible by 3 is a number divisible by 3, $(n + 1)(n + 2)(n + 3) = n(n + 1)(n + 2) + 3(n + 1)(n + 2)$ is divisible by 3 and $P(n + 1)$ is true.

Although mathematical induction is a powerful tool, we cannot use it to discover any formulas. Rather, we use induction to verify that formulas we suspect are true are, in fact, true.

As you may have observed in the two problems we solved, it is usually easy to check $P(1)$. The difficult part of an induction proof is usually proving that if $P(n)$ is true, $P(n + 1)$ is also true. We suggest that, when starting this part of the induction proof, you write out $P(n + 1)$.

Appendix Exercises

In each exercise, use a proof by induction.

1.   Prove that the sum of the first n odd integers is n^2. (*Hint:* The nth odd integer is $2n - 1$. Let $P(n)$ be the proposition that $1 + 3 + 5 + \cdots + (2n - 1) = n^2$.)

2.   Prove that $1^2 + 2^2 + 3^2 + \cdots + n^2 = \dfrac{n(n + 1)(2n + 1)}{6}$ for all natural numbers, n.

3.   Prove that the product of any four consecutive natural numbers is divisible by 4.

4.   Prove that $4 + 8 + 12 + 16 + \cdots + 4n = 2n(n + 1)$.

5.   Prove that

$$\frac{1}{1 \cdot 2} + \frac{1}{2 \cdot 3} + \frac{1}{3 \cdot 4} + \cdots + \frac{1}{n(n + 1)} = \frac{n}{n + 1}$$

6.   For any natural number, n, the symbol $n!$ (read: n factorial) means $n(n - 1)(n - 2) \cdots (1)$.   For   example, $3! = 3 \cdot 2 \cdot 1 = 6$, $4! =$

$4 \cdot 3 \cdot 2 \cdot 1 = 24$, and $5! = 5 \cdot 4 \cdot 3 \cdot 2 \cdot 1 = 120$. Prove that $n^n \geq n!$ for all natural numbers n.

7. Prove that for all natural numbers, n, if $x > y > 0$, $x^n > y^n$.

8. A geometric series is a series of terms a, ar, ar^2, ar^3, $\ldots$, ar^n, $\ldots$, where each term is obtained from the preceding term by multiplication by a number, r, called the **common ratio**. Prove the formula for the sum of the first n terms of a geometric series:

$$a + ar + ar^2 + \cdots + ar^{n-1} = \frac{a - ar^n}{1 - r}$$

9. Prove the formula for the sum of the first n terms of an arithmetic series:

$$a + (a + d) + (a + 2d) + \cdots + [a + (n - 1)d] =$$
$$n\left[a + \frac{(n - 1)d}{2}\right]$$

10. Prove that

$$1 + 4 + 7 + \cdots + (3n - 2) = \frac{n(3n - 1)}{2}$$

11. Prove that if n is a natural number and $x \neq y$ then $x^n - y^n$ is divisible by $x - y$. (*Hint:* Note that $x^{n+1} - y^{n+1} = x^{n+1} - xy^n + xy^n - y^{n+1}$.)

Table 1
Trigonometric
Functions

Degrees	Radians	Sine	Cosine	Tangent
0	0.000	0.000	1.000	0.000
1	.017	.017	1.000	.017
2	.035	.035	.999	.035
3	.052	.052	.999	.052
4	.070	.070	.998	.070
5	.087	.087	.996	.087
6	.105	.105	.995	.105
7	.122	.122	.993	.123
8	.140	.139	.990	.141
9	.157	.156	.988	.158
10	.175	.174	.985	.176
11	.192	.191	.982	.194
12	.209	.208	.978	.213
13	.227	.225	.974	.231
14	.244	.242	.970	.249
15	.262	.259	.966	.268
16	.279	.276	.961	.287
17	.297	.292	.956	.306
18	.314	.309	.951	.325
19	.332	.326	.946	.344
20	.349	.342	.940	.364
21	.367	.358	.934	.384
22	.384	.375	.927	.404
23	.401	.391	.921	.424
24	.419	.407	.914	.445
25	.436	.423	.906	.466
26	.454	.438	.899	.488
27	.471	.454	.891	.510
28	.489	.469	.883	.532
29	.506	.485	.875	.554
30	.524	.500	.866	.577
31	.541	.515	.857	.601
32	.559	.530	.848	.625
33	.576	.545	.839	.649
34	.593	.559	.829	.675
35	.611	.574	.819	.700
36	.628	.588	.809	.727

Degrees	Radians	Sine	Cosine	Tangent
37	.646	.602	.799	.754
38	.663	.616	.788	.781
39	.681	.629	.777	.810
40	.698	.643	.766	.839
41	.716	.656	.755	.869
42	.733	.669	.743	.900
43	.750	.682	.731	.933
44	.768	.695	.719	.966
45	.785	.707	.707	1.000
46	.803	.719	.695	1.036
47	.820	.731	.682	1.072
48	.838	.743	.669	1.111
49	.855	.755	.656	1.150
50	.873	.766	.643	1.192
51	.890	.777	.629	1.235
52	.908	.788	.616	1.280
53	.925	.799	.602	1.327
54	.942	.809	.588	1.376
55	.960	.819	.574	1.428
56	.977	.829	.559	1.483
57	.995	.839	.545	1.540
58	1.012	.848	.530	1.600
59	1.030	.857	.515	1.664
60	1.047	.866	.500	1.732
61	1.065	.875	.485	1.804
62	1.082	.883	.469	1.881
63	1.100	.891	.454	1.963
64	1.117	.899	.438	2.050
65	1.134	.906	.423	2.145
66	1.152	.914	.407	2.246
67	1.169	.921	.391	2.356
68	1.187	.927	.375	2.475
69	1.204	.934	.358	2.605
70	1.222	.940	.342	2.747
71	1.239	.946	.326	2.904
72	1.257	.951	.309	3.078
73	1.274	.956	.292	3.271
74	1.292	.961	.276	3.487
75	1.309	.966	.259	3.732
76	1.326	.970	.242	4.011
77	1.344	.974	.225	4.331
78	1.361	.978	.208	4.705
79	1.379	.982	.191	5.145

Degrees	Radians	Sine	Cosine	Tangent
80	1.396	.985	.174	5.671
81	1.414	.988	.156	6.314
82	1.431	.990	.139	7.115
83	1.449	.993	.122	8.144
84	1.466	.995	.105	9.514
85	1.484	.996	.087	11.430
86	1.501	.998	.070	14.301
87	1.518	.999	.052	19.081
88	1.536	.999	.035	28.636
89	1.553	1.000	.017	57.290

Table 2
Common Logarithms

	0	1	2	3	4	5	6	7	8	9
10	0.0000	0.0043	0.0086	0.0128	0.0170	0.0212	0.0253	0.0294	0.0334	0.0374
11	.0414	.0453	.0492	.0531	.0569	.0607	.0645	.0682	.0719	.0755
12	.0792	.0828	.0864	.0899	.0934	.0969	.1004	.1038	.1072	.1106
13	.1139	.1173	.1206	.1239	.1271	.1303	.1335	.1367	.1399	.1430
14	.1461	.1492	.1523	.1553	.1584	.1614	.1644	.1673	.1703	.1732
15	.1761	.1790	.1818	.1847	.1875	.1903	.1931	.1959	.1987	.2014
16	.2041	.2068	.2095	.2122	.2148	.2175	.2201	.2227	.2253	.2279
17	.2304	.2330	.2355	.2380	.2405	.2430	.2455	.2480	.2504	.2529
18	.2553	.2577	.2601	.2625	.2648	.2672	.2695	.2718	.2742	.2765
19	.2788	.2810	.2833	.2856	.2878	.2900	.2923	.2945	.2967	.2989
20	.3010	.3032	.3054	.3075	.3096	.3118	.3139	.3160	.3181	.3201
21	.3222	.3243	.3263	.3284	.3304	.3324	.3345	.3365	.3385	.3404
22	.3424	.3444	.3464	.3483	.3502	.3522	.3541	.3560	.3579	.3598
23	.3617	.3636	.3655	.3674	.3692	.3711	.3729	.3747	.3766	.3784
24	.3802	.3820	.3838	.3856	.3874	.3892	.3909	.3927	.3945	.3962
25	.3979	.3997	.4014	.4031	.4048	.4065	.4082	.4099	.4116	.4133
26	.4150	.4166	.4183	.4200	.4216	.4232	.4249	.4265	.4281	.4298
27	.4314	.4330	.4346	.4362	.4377	.4393	.4409	.4425	.4440	.4456
28	.4472	.4487	.4502	.4518	.4533	.4548	.4564	.4579	.4594	.4609
29	.4624	.4639	.4654	.4669	.4683	.4698	.4713	.4728	.4742	.4757
30	.4771	.4786	.4800	.4814	.4829	.4843	.4857	.4871	.4885	.4900
31	.4914	.4928	.4942	.4955	.4969	.4983	.4997	.5011	.5024	.5038
32	.5051	.5065	.5079	.5092	.5105	.5119	.5132	.5145	.5159	.5172
33	.5185	.5198	.5211	.5224	.5237	.5250	.5263	.5276	.5289	.5302
34	.5315	.5328	.5340	.5353	.5366	.5378	.5391	.5403	.5416	.5428
35	.5441	.5453	.5465	.5478	.5490	.5502	.5514	.5527	.5539	.5551
36	.5563	.5575	.5587	.5599	.5611	.5623	.5635	.5647	.5658	.5670
37	.5682	.5694	.5705	.5717	.5729	.5740	.5752	.5763	.5775	.5786
38	.5798	.5809	.5821	.5832	.5843	.5855	.5866	.5877	.5888	.5899
39	.5911	.5922	.5933	.5944	.5955	.5966	.5977	.5988	.5999	.6010
40	.6021	.6031	.6042	.6053	.6064	.6075	.6085	.6096	.6107	.6117
41	.6128	.6138	.6149	.6159	.6170	.6180	.6191	.6201	.6212	.6222
42	.6232	.6243	.6253	.6263	.6274	.6284	.6294	.6304	.6314	.6325
43	.6335	.6345	.6355	.6365	.6375	.6385	.6395	.6405	.6415	.6425
44	.6434	.6444	.6454	.6464	.6474	.6484	.6493	.6503	.6513	.6522
45	.6532	.6542	.6551	.6561	.6571	.6580	.6590	.6599	.6609	.6618
46	.6628	.6637	.6646	.6656	.6665	.6674	.6684	.6693	.6702	.6712

	0	1	2	3	4	5	6	7	8	9
47	.6721	.6730	.6739	.6749	.6758	.6767	.6776	.6785	.6794	.6803
48	.6812	.6821	.6830	.6839	.6848	.6857	.6866	.6875	.6884	.6893
49	.6902	.6911	.6920	.6928	.6937	.6946	.6955	.6964	.6972	.6981
50	.6990	.6998	.7007	.7016	.7024	.7033	.7041	.7050	.7059	.7067
51	.7076	.7084	.7093	.7101	.7110	.7118	.7126	.7135	.7143	.7152
52	.7160	.7168	.7177	.7185	.7193	.7202	.7210	.7218	.7226	.7234
53	.7243	.7251	.7259	.7267	.7275	.7283	.7292	.7300	.7308	.7316
54	.7324	.7332	.7340	.7348	.7356	.7364	.7372	.7380	.7388	.7396
55	.7404	.7411	.7419	.7427	.7435	.7443	.7451	.7458	.7466	.7474
56	.7482	.7490	.7497	.7505	.7513	.7520	.7528	.7536	.7543	.7551
57	.7559	.7566	.7574	.7581	.7589	.7597	.7604	.7612	.7619	.7627
58	.7634	.7642	.7649	.7657	.7664	.7671	.7679	.7686	.7694	.7701
59	.7708	.7716	.7723	.7730	.7738	.7745	.7752	.7760	.7767	.7774
60	.7781	.7789	.7796	.7803	.7810	.7817	.7825	.7832	.7839	.7846
61	.7853	.7860	.7867	.7875	.7882	.7889	.7896	.7903	.7910	.7917
62	.7924	.7931	.7938	.7945	.7952	.7959	.7966	.7973	.7980	.7986
63	.7993	.8000	.8007	.8014	.8021	.8028	.8034	.8041	.8048	.8055
64	.8062	.8069	.8075	.8082	.8089	.8096	.8102	.8109	.8116	.8122
65	.8129	.8136	.8142	.8149	.8156	.8162	.8169	.8176	.8182	.8189
66	.8195	.8202	.8208	.8215	.8222	.8228	.8235	.8241	.8248	.8254
67	.8261	.8267	.8274	.8280	.8287	.8293	.8299	.8306	.8312	.8319
68	.8325	.8331	.8338	.8344	.8350	.8357	.8363	.8369	.8376	.8382
69	.8388	.8395	.8401	.8407	.8414	.8420	.8426	.8432	.8438	.8445
70	.8451	.8457	.8463	.8469	.8476	.8482	.8488	.8494	.8500	.8506
71	.8512	.8519	.8525	.8531	.8537	.8543	.8549	.8555	.8561	.8567
72	.8573	.8579	.8585	.8591	.8597	.8603	.8609	.8615	.8621	.8627
73	.8633	.8639	.8645	.8651	.8657	.8663	.8669	.8675	.8680	.8686
74	.8692	.8698	.8704	.8710	.8716	.8721	.8727	.8733	.8739	.8745
75	.8751	.8756	.8762	.8768	.8774	.8779	.8785	.8791	.8797	.8802
76	.8808	.8814	.8819	.8825	.8831	.8837	.8842	.8848	.8854	.8859
77	.8865	.8870	.8876	.8882	.8887	.8893	.8899	.8904	.8910	.8915
78	.8921	.8926	.8932	.8938	.8943	.8949	.8954	.8960	.8965	.8971
79	.8976	.8982	.8987	.8993	.8998	.9004	.9009	.9015	.9020	.9025
80	.9031	.9036	.9042	.9047	.9052	.9058	.9063	.9069	.9074	.9079
81	.9085	.9090	.9095	.9101	.9106	.9111	.9117	.9122	.9127	.9133
82	.9138	.9143	.9149	.9154	.9159	.9164	.9170	.9175	.9180	.9185
83	.9191	.9196	.9201	.9206	.9212	.9217	.9222	.9227	.9232	.9238
84	.9243	.9248	.9253	.9258	.9263	.9268	.9274	.9279	.9284	.9289
85	.9294	.9299	.9304	.9309	.9314	.9320	.9325	.9330	.9335	.9340
86	.9345	.9350	.9355	.9360	.9365	.9370	.9375	.9380	.9385	.9390
87	.9395	.9400	.9405	.9410	.9415	.9420	.9425	.9430	.9435	.9440
88	.9445	.9450	.9455	.9460	.9464	.9469	.9474	.9479	.9484	.9489
89	.9494	.9499	.9504	.9508	.9513	.9518	.9523	.9528	.9533	.9538

	0	1	2	3	4	5	6	7	8	9
90	.9542	.9547	.9552	.9557	.9562	.9566	.9571	.9576	.9581	.9586
91	.9590	.9595	.9600	.9605	.9609	.9614	.9619	.9624	.9628	.9633
92	.9638	.9643	.9647	.9652	.9657	.9661	.9666	.9671	.9675	.9680
93	.9685	.9689	.9694	.9699	.9703	.9708	.9713	.9717	.9722	.9727
94	.9731	.9736	.9740	.9745	.9750	.9754	.9759	.9763	.9768	.9773
95	.9777	.9782	.9786	.9791	.9795	.9800	.9804	.9809	.9814	.9818
96	.9823	.9827	.9832	.9836	.9841	.9845	.9850	.9854	.9859	.9863
97	.9868	.9872	.9877	.9881	.9885	.9890	.9894	.9899	.9903	.9908
98	.9912	.9917	.9921	.9925	.9930	.9934	.9939	.9943	.9947	.9952
99	.9956	.9961	.9965	.9969	.9974	.9978	.9983	.9987	.9991	.9996

Answers to Odd-Numbered Exercises

Exercise Set 1.2

1. T **3.** T **5.** T **7.** T **9.** F **11.** T **13.** $2 < 5$
15. $-2 < 5$ **17.** $0 > -1$ **19.** $\pi^2 > \pi$ **21.** $(-1/2)^2 > (-1/4)^2$

23.

25.

27. No real numbers satisfy the inequalities

29.

31.

33.

35. a. $\frac{1}{2} < 1$ since $1 - \frac{1}{2} = \frac{1}{2} > 0$ **37.** $x - 2 \geq y - 2$
  b. $-\frac{1}{2} > -\frac{3}{2}$ since $-\frac{1}{2} - (-\frac{3}{2}) = 1 > 0$
  c. $5 \leq 7$ since $7 - 5 = 2 > 0$
39. $2x < 2y, \ -2x > -2y$

Exercise Set 1.3

1.

3.

5.

7.

9. $(-1, +\infty)$ **11.** $(1, 3)$ **13.** $(2, 3]$ **15.** $(0, 5)$
17. $(-2, 2)$ **19.** $[2, 5)$ **21.** $(-\infty, -2)$ **23.** Not an interval
25. Not an interval **27.** $(3, +\infty)$ **29.** $(-\infty, 3]$
31. $(-\infty, +\infty)$

Exercise Set 1.4

1. $-1, 0, 1, 2, 3, 4, 5$ **3.** $-2, -1, 0, 1, 2$ **5.** $-3, -2$ **7.** None
9. $-3, -2, -1, 0, 1$ **11.** $\ldots, -4, -3, -2, -1, 0, 1, 2$ **13.** 1
15. $1, 2, 3, 4, 5$ **17.** $1, 2$ **19.** 1 **21.** 1 **23.** $1, 2, 3$
25. None **27.** $1, 2, 3, 4$ **29.** $1, 2, 3, 4, 5$
31. $0.571428\,\overline{571428} \ldots$
33. a. 1.505 **b.** 1.5005 **c.** 1.50005 **d.** 1.5 **e.** 3.5
35. No **37.** $\sqrt{2} \cdot \sqrt{8} = 4$

**Chapter 1
Review Exercises**

1. $-2 < -1$ **3.** $1 > -2$ **5.** $\sqrt{2} > 1$ **7.** $\pi > 3.14$
9. $1/(2^2) > 1/(3^2)$

11.
$[5, +\infty)$

13.
$[2, 5)$

15.
$[-2, 0]$

17.
$[-3, 3]$

19.
$[3, +\infty)$

21.
$(0, 3)$

Not an Interval

Not an Interval

23. ——|———|———|———|———|———◇——▷ $(4, +\infty)$ **25.** ——|———|———|———◆———◆———◆———◆———|——▷   **27.** ——◆———◆———◆———◆———◆———◆———◆——▷
$\quad\quad$ 0 $\;$ 1 $\;$ 2 $\;$ 3 $\;$ 4 $\quad\quad\quad\quad\quad\quad\quad$ -4 $\;$ -2 $\quad$ 0 $\;$ 1 $\;$ 2 $\;$ 3 $\;$ 4 $\quad\quad\quad\quad\quad$ -2 -1 $\;$ 0 $\;$ 1 $\;$ 2 $\;$ 3 $\;$ 4

-3 $\;$ -1

29. ——|———|———◆———|———|———|———|———|———|——▷ $[-2, 4]$
$\quad\quad$ -4 -3 -2 -1 $\;$ 0 $\;$ 1 $\;$ 2 $\;$ 3 $\;$ 4

31. $0.285714\,\overline{285714}\ldots$ **33.** $1.444\overline{4}\ldots$ **35.** $2517/1000$ **37.** No
39. Yes   **41.** Yes   **43.** No   **45.** Yes, -1
47. $2 + c < 3 + c$ **49.** $2c < 3c$

Exercise Set 2.2

1. a. Polynomial, degree 5, coefficient of x^5 is 1, coefficient of x^4 is 1, coefficient of x^3 is 0, coefficient of x^2 is 0, coefficient of x is 2, and the constant coefficient is -3.
b. Not a polynomial
c. Polynomial, degree 599, coefficients of x^{599}, x^{100}, and x^6 are all 1; constant coefficient is -7000, and all other coefficients are 0.
d. Not a polynomial
3. 12, 6, 2, 0, 0, 2   **5.** 0, -1, 0, 3   **7.** 3, 7, 35
9. $a = -3, b = 0, c = 4$
11. $a = 1; b = c = d = 0; e = -1$ **13.** No
15. $P(x) = x^3; P(2) = 8; P(4) = 64$

Exercise Set 2.3

1. $P(x) + Q(x) = 4x - 1; P(x) - Q(x) = -2x + 7;$
$P(x) \cdot Q(x) = 3x^2 + 5x - 12$
3. $P(x) + Q(x) = x^5 + x^3 + 1, P(x) - Q(x) = -x^5 - x^3 - 1, P(x) \cdot Q(x) = 0$
5. $P(x) + Q(x) = x^4 + x^3 + x^2 + 2x, P(x) - Q(x) = x^4 + x^3 + x^2 + 2,$
$P(x) \cdot Q(x) = x^5 - 1$
7. $P(x) + Q(x) = 3x^2 + 2x + 2, P(x) - Q(x) = -x^2 + 4x - 6,$
$P(x) \cdot Q(x) = 2x^4 + 5x^3 - 3x^2 + 14x - 8$
9. $P(x) + Q(x) = x^{40} + 3x^2 - x - 1, P(x) - Q(x) = x^{40} - 3x^2 + 3x + 3,$
$P(x) \cdot Q(x) = 3x^{42} - 2x^{41} - 2x^{40} + 3x^3 + x^2 - 4x - 2$
11. a. $6x^2 - 7x - 3$ **b.** $x^2 + x - 20$ **c.** $-6x^2 + 8x - 2$
d. $-14x^2 + 6x + 8$ **e.** $3x^2 - 34x + 80$
f. $\sqrt{2}x^2 + (\pi\sqrt{2} - 1)x - \pi$
15. 0   **17.** 0
19. b. $P(x) = x^5 + ax^4 + a^2x^3 + a^3x^2 + a^4x + a^5$
c. $P(x) = x^{n-1} + ax^{n-2} + a^2x^{n-3} + \cdots + a^{n-2}x + a^{n-1}$
23. $P(x) = 4x$ **25.** $P(x) = 10 + 0.1x$

Exercise Set 2.4.2

1. $3/2$ **3.** $-3/5$ **5.** 12   **7.** 60   **9.** $-5\sqrt{2}/2$ **11.** 2
13. $4/5$ **17.** 1 and -7 **19.** -8 **21.** No roots
23. No roots   **25.** $x^2 - 9x + \dfrac{81}{4} = \left(x - \dfrac{9}{2}\right)^2$

27. $x^2 - \pi x + \dfrac{\pi^2}{4} = \left(x - \dfrac{\pi}{2}\right)^2$ **29.** $x^2 + x + 1/4 = (x + 1/2)^2$

31. $x^2 - 12x + 36 = (x - 6)^2$

33. $P(x) = 5000 + 5x$; the cost of 10,000 bolts $= P(10) = \$5050$; the cost of 20,000 bolts $= P(20) = \$5100$; 200,000 bolts can be produced for \$6000; 400,000 bolts can be produced for \$7000.

35. **a.** $x^2 - (5/6)x + 1/6 = 0$ **b.** $x^2 - (5/6)x = -1/6$
  c. $x^2 - (5/6)x + 25/144 = -1/6 + 25/144 = 1/144$
  d. $(x - 5/12)^2 = 1/144$ **e.** Roots are $1/2$ and $1/3$.

Exercise Set 2.4.3

1. $-3, -4$ **3.** No roots   **5.** $3/4, -1/2$ **7.** No roots
9. 2   **11.** No roots   **13.** No roots   **15.** $\sqrt{6}$ **17.** $-2, 3$
19. $-1, 1/4$ **21.** $-2, 2$ **23.** $-\pi^2, \pi$
25. **a.** $[x - (-3 + \sqrt{29})/2][x - (-3 - \sqrt{29})/2]$
  b. $[x - (-\pi + \sqrt{\pi^2 - 8})/4][x - (-\pi - \sqrt{\pi^2 - 8})/4]$
  c. This polynomial will not factor into a product of linear factors since it has no roots.
27. $x^2 + 8x + 15$ **29.** $x^2 + (\pi - \sqrt{7})x - \pi\sqrt{7}$
31. $P(x) = 1000x - 2x^2$. If the area is 80,000 ft², the width can be either 100 or 400 ft. If the area is 100,000 ft², the width can be either $250 - 50\sqrt{5}$ or $250 + 50\sqrt{5}$ ft.
33. It will take 10 seconds to fall to ground level. It will take $\sqrt{50}$ seconds to fall halfway to the ground.

Exercise Set 2.5

1. $S(x) = x^3 + 2x, R(x) = 4x + 1$ **3.** $S(x) = 3x^2 + 5x + 6, R(x) = 6$
5. $S(x) = 0, R(x) = x^2 + 2x - 3$ **7.** $S(x) = x^2 + x + 1, R(x) = 2$
9. $S(x) = 3x^3 - 4x^2 + 19x - 54, R(x) = 165x - 5$
11. $x^2(x - 1)(x + 1)$; roots are $-1, 0, 1$
13. $(x + 1)(x + 2)(x^2 + 16)$; roots are $-1, -2$
15. $(x^2 + 6)(x^2 + 4)$; no roots
17. $(x - \sqrt{6})(x + \sqrt{6})(x^2 + 6)$; roots are $-\sqrt{6}, \sqrt{6}$
19. $(x - b)^4$; root is b **21.** Maximum is n, minimum is 0
23. $(x + 1)(x^2 + 1) = x^3 + x^2 + x + 1$ **25.** $(x^2 + 1)^2 = x^4 + 2x^2 + 1$
27. $x^2(x + 1)(x - 1) = x^4 - x^2$ **29.** $a_0 = 0$ **31.** $w = 6, d = 3$
33. $h = 5, r = 4$

Appendix 2.5

3. $1 + i, 1 - i$ **5.** $1 + 2i, 1 - 2i$ **7.** $\dfrac{1}{2} + i, \dfrac{1}{2} - i$

9. $\dfrac{-1 + \sqrt{7}i}{4}, \dfrac{-1 - \sqrt{7}i}{4}$ **11.** $x^2 - 4x + 5$

Chapter 2 Review Exercises

1. $P(x) + Q(x) = x^3 + 4x - 10, P(x) - Q(x) = -x^3 + 6x^2,$
  $P(x)Q(x) = 3x^5 - 7x^4 - 5x^3 + 4x^2 - 20x + 25$
3. $P(x) + Q(x) = \pi x^3 + x^2 + \sqrt{2}x - 2,$
  $P(x) - Q(x) = -\pi x^3 + x^2 + 3\sqrt{2}x - 4,$
  $P(x)Q(x) = \pi x^5 + (2\pi\sqrt{2})x^4 - (3\pi + \sqrt{2})x^3 - 3x^2 + 5\sqrt{2}x - 3$
5. $P(x) + Q(x) = x^3 + 3x^2 + 10x + 24; P(x) - Q(x) = x^3 + 3x^2 + 8x + 30;$
  $P(x)Q(x) = x^4 - 3^4$

7. $x^2 - 10x + 25 = (x - 5)^2$ **9.** $x^2 - \sqrt{2}x + \dfrac{1}{2} = \left(x - \dfrac{\sqrt{2}}{2}\right)^2$

11. $\frac{1}{2}, 3$ **13.** $-\frac{1}{3}, \frac{1}{2}$ **15.** No roots   **17.** $-4, 3/2$

19. $-\sqrt{5}, \sqrt{5}$ **21.** No roots   **23.** No roots

25. $(-7 + \sqrt{61})/2, (-7 - \sqrt{61})/2$ **27.** $-3/5, 2/3$

29. No roots   **31.** $(x - 1)^2(x + 2)$; roots $1, -2$

33. $(x - 2)^2(x + 2)^2$; roots $-2, 2$

35. $(x^2 + x + 1)(x - 3)(x + 3)$; roots $-3, 3$

37. $(x - 3)(x^2 + 3x + 9)$; root 3

39. $x^3(x + 1)(x^2 + 2x + 5)$; roots $-1, 0$

41. $S(x) = x^2 + 1$; $R(x) = 2x - 5$ **43.** $S(x) = 0$; $R(x) = x^2 + 5$

45. $S(x) = -x^3 + 3x^2 - 13x + 40$; $R(x) = 134x - 40$

47. **a.** $P(-2) = -4$; $P(-1) = -11$; $P(0) = -10$; $P(1) = 5$; $P(2) = 40$

 b. $-5, -3, 1$ **c.** 2

49. $A(x) = (17/9)x^2 - (16/3)x + 8$. If the perimeter of the square is 4 ft, then $A(x) = A(1) = 41/9$ ft^2. If $A(x) = 44/9$, then $x = 14/17$ or 2, and the length of wire used to make the square is 56/17 ft or 8 ft, respectively.

Exercise Set 3.2

1. Yes   **3.** No   **5.** Yes   **7.** No   **9.** $(x - 1)/(x + 1)$

11. $(x^3 + x)/(x - 1)$ **13.** $1/(x - 1)$ **15.** $(x^2 - 4)/(x^2 - x + 1)$

19. **a.** $\dfrac{x^3 - x^2 + 2x - 2}{x^2 + 2x - 3}$ **b.** $\dfrac{x^6 - x^5 + x^2 - 2x + 1}{x^4 - x^3 - 3x + 3}$

 c. $(x^2 - 1)/(x - 1)$

21. $R_1(x) = \dfrac{x^4 + x^3 + 5x + 5}{x^3 + x^2 - x - 1}$; $R_2(x) = \dfrac{x^5 - x^4 + 3x^2 - 2x - 1}{x^3 + x^2 - x - 1}$

Exercise Set 3.3

1. $x^3 + x^2 - x - 1$ **3.** $x^3 - 7x + 6$ **5.** $x^6 + x^5 + x^4 - x^2 - x - 1$

7. $30(x - 1)^3(x + 3)^3(x + 4)(x^2 + 4)(x^2 + 5)(x^2 + x + 10)^3$

9. $\dfrac{3x^2 - x - 12}{x^2 - 3x}$ **11.** 0   **13.** $\dfrac{5x + 7}{2x^3 + 8x^2 + 2x - 12}$

15. $\dfrac{3x^2 + 2x + 1}{4(x^3 - 1)}$ **17.** $\dfrac{x^3 - 2x^2 + 2x + 2}{x^3 - 5x^2 + 8x - 4}$

19. $\dfrac{x^4 - 4x^3 + 4x^2 - 3x}{x^5 - 7x^4 + 17x^3 - 20x^2 + 12x - 3}$ **21.** $\dfrac{-x^3 - 5x - 4}{x^3 - x^2 + 4x - 4}$

23. $\dfrac{x + 4}{6x^2 - 18x + 12}$ **25.** $\dfrac{-3x^2 + 10x - 4}{(x - 1)^3(x^2 + x + 1)^2(x - 2)}$

27. $\dfrac{x^4 + 2x^2 + 1}{x^2 - 1}$ **29.** $\dfrac{x^4 - x^3 + 5x^2 - 4x + 4}{x^2 + x - 2}$ **31.** $\dfrac{3x^2 + 2x - 1}{x + 2}$

33. $\dfrac{x - 1}{x^2 + 3x + 9}$ **35.** $\dfrac{x^3 - x^2 - 4x - 6}{x^2 + 3x + 2}$ **37.** $\dfrac{3x^3 - 11x^2 + 24x - 10}{3x^3 - 3x^2 - 3x + 3}$

39. $\dfrac{2x^2 - x + 3}{(x - 1)^3}$ **41.** $\dfrac{-x^2 - 8x + 5}{x^3 - x^2 - 5x - 3}$

43. a. $(x + 3)/(x - 4)$, 0, no value, $-2/5$, $-3/4$, $-4/3$, no value

b. $\dfrac{x^2 - 3x + 2}{x^2 + 4}$, $\dfrac{3}{2}$, no value, $\dfrac{1}{2}$, 0, 0

45. The value is $7 \Leftrightarrow x = \dfrac{1 \pm \sqrt{133}}{6}$. The value is $-7 \Leftrightarrow x = \dfrac{-1 \pm \sqrt{161}}{8}$

47. 20, 5, 0, 1

Chapter 3
Review Exercises

1. Yes **3.** No **5.** Yes **7.** Yes **9.** No **11.** No

13. $x + 2$ **15.** $(x - 3)/(x + 2)$ **17.** $\dfrac{x^2 + x + 1}{x^3 + x^2 + x + 1}$

19. $\dfrac{x - 3}{x + 5}$ **21.** $\dfrac{4x + 2}{x - \pi}$ **23.** $\dfrac{2x - 5}{(x - 2)^2}$ **25.** $\dfrac{4x^2 - 12x + 36}{x^3 - 3x^2 - 4x + 12}$

27. $\dfrac{12x^2 + 26x + 3}{18x^3 + 57x^2 + 32x + 5}$ **29.** $\dfrac{x^4 + x^3 + x^2 + 2x + 1}{x + 1}$

31. $\dfrac{x^4 - 5x^3 - 12x^2 - 17x - 76}{x^5 - 4x^4 + 8x^3 - 32x^2 + 16x - 64}$ **33.** $\dfrac{x^3 - x^2 - 2x - 12}{3x^2 + 8x - 3}$

35. $\dfrac{2x^4 + 9x^3 + 18x^2 + 36x + 40}{x^3 - 9x^2 + 27x - 27}$ **37.** $\dfrac{4x^4 + 6x^3 + 8x^2 + 15x - 5}{x^2 + 6x + 5}$

39. $\dfrac{2x - 10}{x + 3}$ **41.** -3, no value, $-3/2$, $-1/12$, 0 **43.** $3/8, 7/4, 11/8$

45. $-31/3$ **47.** 1 and 2 **49.** -1 and 4

Exercise Set 4.2

1. Yes **3. a.** $P(\pi) = (-1, 0); P(3\pi/2) = (0, -1); P(2\pi) = (1, 0)$
b. $P(\pi/4) = \left(\sqrt{2}/2, \sqrt{2}/2 \right)$
c. $P(3\pi/4) = \left(-\sqrt{2}/2, \sqrt{2}/2 \right)$
7. $(0, +\infty)$

Exercise Set 4.3

1. a. $f(x) = \sqrt[3]{x} + \sqrt{x}$ **b.** $F(x) = \sqrt[3]{x} - \sqrt{x}$ **c.** $S(x) = x^2$
3. a. $V(t) = \pi t^3/12, D_v = $ all $t > 0$;
b. $V(1) = \pi/12, V(10) = 1000\,\pi/12, V(20) = 8000\,\pi/12$
c. $r = 3/\sqrt[3]{\pi}, h = 6/\sqrt[3]{\pi}$
5. a. -2 **b.** -4 **c.** 61 **d.** -30 **e.** $t^3 - 3$
f. $t^6 - 3$ **g.** $t^9 - 3$ **h.** $-t^6 - 3$ **i.** $x^6 - 3$
j. $x^6 + 3x^4 + 3x^2 - 2$ **k.** $\left(\sqrt{x} \right)^3 - 3$
7. a. 6 **b.** $1/38$ **c.** $1/38$ **d.** $1/(x^4 + 4x^2 + 6)$
e. $(2x^2 + 8x + 9)/(x^2 + 4x + 4)$ **f.** $x^4 + 4x^2 + 6$
9. $(-\infty, +\infty)$ **11.** $[2, +\infty)$ **13.** $(-\infty, +\infty)$ **15.** $(1, +\infty)$
17. All real numbers $t \neq -2, -1, 4$

Exercise Set 4.4   **1.** $D_f = (-\infty, +\infty)$ **3.** $D_h = (-\infty, +\infty)$

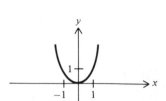

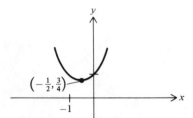

$\left(-\frac{1}{2}, \frac{3}{4}\right)$

5. $D_G = $ all real numbers $z \neq -1$ **7.** $D_T = (-\infty, +\infty)$

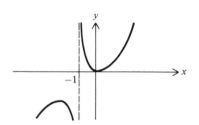

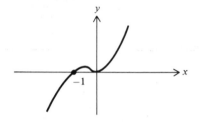

9. $D_g = (-\infty, +\infty)$ **11.** $D_h = $ all real numbers $x \neq 0$

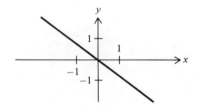

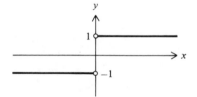

13. **a.** $(-3, 1/2); (-1, 0); (0, -1);$
$(2, 3); (4, 5/3)$

b.

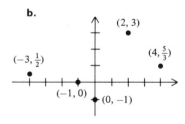

$(2, 3)$

$(4, \frac{5}{3})$

$(-3, \frac{1}{2})$

$(-1, 0)$

$(0, -1)$

c. The range of f consists of the numbers $1/2, 0, -1, 3, 5/3$

15. **a.** D_f consists of the numbers $-2, -1, 0, 1,$ and 2
b. R_f consists of the numbers $3, 4,$ and 5
c. $f(-2) = 4, f(-1) = 4, f(0) = 3, f(1) = 3,$ and $f(2) = 5$

17.

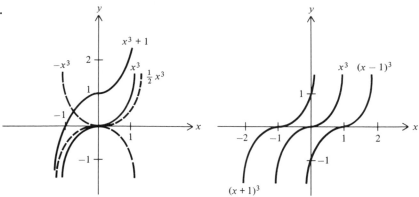

19. Neither
21. Determines y as a function of x and x as a function of y
23. Determines y as a function of x and x as a function of y
25. Determines y as a function of x **27.** Determines y as a function of x

Exercise Set 4.5   **1.** $f^{-1}(x) = (x + 2)/3$ **3.** $g^{-1}(x) = -\sqrt{x}$

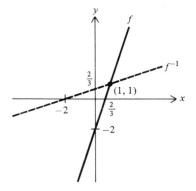

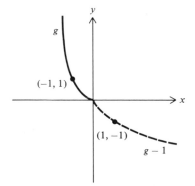

5. g does not have an inverse   **7.** $G^{-1}(x) = -4\sqrt{x}$

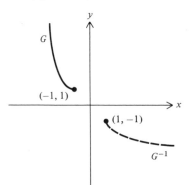

9. $f^{-1}(x) = \sqrt[3]{x+1}$

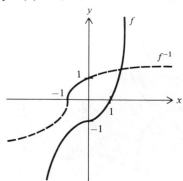

11. $f^{-1}(x) = 1/x$

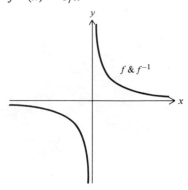

13. T does not have an inverse   **15.** V does not have an inverse
17. g^{-1} exists   **19.** F^{-1} exists

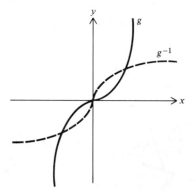

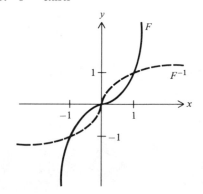

21. $f^{-1}(x) = (4-x)/2$ **23.** $f^{-1}(x) = \sqrt[6]{x}$
25. $f^{-1}(x) = -3 - \sqrt{x+13}$

Exercise Set 4.6.2

1. $(4,0), (0,-12)$ **3.** $(-5,0); (3,0); (0,-15)$ **5.** None
7. $(1,0); \left(-2 + \sqrt{7}, 0\right); \left(-2 - \sqrt{7}, 0\right); (0,3)$ **9.** y-axis   **11.** Origin
13. Origin   **15.** Neither   **17.** Neither
19. $D_f = (-\infty, +\infty)$;   **21.** $D_h =$ all real numbers x such that
intercepts $(0,0), (-1,0), (1,0)$; $x > 1$ or $x < -1$; no intercepts;
symmetry: y-axis   symmetry with respect to the
origin

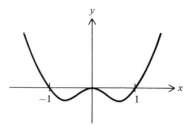

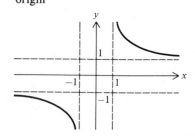

23. $D_G = (-\infty, +\infty)$; intercepts $(0, 0)$; symmetric with respect to the y-axis

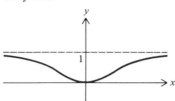

25.

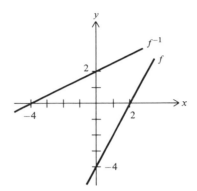

27.

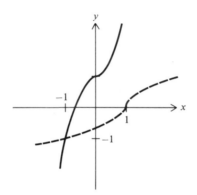

29.

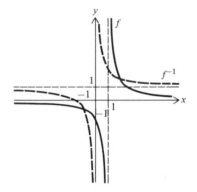

31. **a.** Symmetric with respect to x-axis, y-axis, and origin
c. Symmetric with respect to the y-axis only

b. Not symmetric with respect to x-axis, y-axis, or origin

Exercise Set 4.6.3

1. $(0, -4)$; $(0, 6)$
3. The x-intercepts of the graph of g are 0, 3, and 5; the x-intercepts of the graph of h are -5, -2, and 0
5. The x-intercepts of the graph of g are -19, -5, and 0; the y-intercept is 0

7.

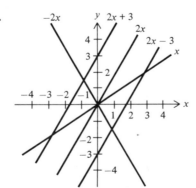

9.

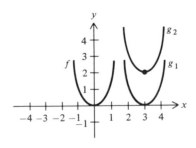

Exercise Set 4.7

1. a. D_{f+g} is $(-\infty, +\infty)$; $(f + g)(x) = x^3 + x^2$
    b. D_{f-g} is $(-\infty, +\infty)$; $(f - g)(x) = x^3 - x^2$
    c. D_{fg} is $(-\infty, +\infty)$; $fg(x) = x^5$ d. $D_{f/g}$ is all $x \neq 0$; $f/g(x) = x$

3. a. D_{f-g} is $(-\infty, 2]$; $(f - g)(x) = x + 1 - \sqrt{2 - x}$
    b. D_{fg} is $(-\infty, 2]$; $fg(x) = (x + 1)\sqrt{2 - x}$
    c. $D_{f/g}$ is $(-\infty, 2)$; $f/g(x) = (x + 1)\sqrt{2 - x}/(2 - x)$
    d. $D_{g/f}$ is all x in $(-\infty, 2]$ except $x = -1$; $g/f(x) = \sqrt{2 - x}/(x + 1)$

5. D_{F-G} is $[6, +\infty)$; $(F - G)(x) = \sqrt{x - 3} - \sqrt{x - 6}$;
    D_{FG} is $[6, +\infty)$; $FG(x) = \sqrt{x^2 - 9x + 18}$;
    $D_{G/F}$ is $[6, +\infty)$; $G/F(x) = \sqrt{x^2 - 9x + 18}/(x - 3)$

7. $D_{f \circ g}$ is $(-\infty, +\infty)$; $f \circ g(x) = x^4 + x^2 - 1$;
    $D_{g \circ f}$ is $(-\infty, +\infty)$; $g \circ f(x) = (x - 2)^4 + (x - 2)^2 + 1$

9. $D_{f \circ g}$ is $(0, +\infty)$; $f \circ g(x) = 1/(\sqrt[4]{x})^3 = \sqrt[4]{x}/x$;
    $D_{g \circ f}$ is $(0, +\infty)$; $g \circ f(x) = 1/(\sqrt[4]{x})^3 = \sqrt[4]{x}/x$

11. $D_{f \circ g}$ is all real numbers x such that $x \geq \sqrt{2}$ or $x \leq -\sqrt{2}$; $f \circ g(x) = \sqrt{x^2 - 2}$;
     $D_{g \circ f}$ is $[2, +\infty)$; $g \circ f(x) = x - 2$

13. $f \circ g(x) = x^3 - 2x^2 + 2x$; $g \circ f(x) = x^3 + x^2 + x$

15. $h \circ g(x) = (x - 4)/(13 - 3x)$; $g \circ h(x) = (x - 3)/(13 - 4x)$

17. $f \circ g(x) = 10$; $g \circ f(x) = -3$

19. $f^{-1}(x) = (x - 4)/2$

21. $h^{-1}(x) = \sqrt[5]{x + 3}$

23. $U^{-1}(x) = \sqrt[4]{x}$

Chapter 4
Review Exercises

1. a. 4   b. 28   c. 28   d. 108   e. $t^8 + 3t^4$
    f. $x^8 + 3x^4$ g. $(x - 2)^4 + 3(x - 2)^2$ h. $1/x^4 + 3/x^2$
    i. 304   j. $(x^4 + 3x^2)^4 + 3(x^4 + 3x^2)^2$

3. a. $2\sqrt{2}$ b. 0   c. 0   d. 2   e. $\sqrt{x^4 - 1}$
    f. $\sqrt{x - 1}$ g. $\sqrt{4x^2 + 4x}$ h. $\sqrt{2}$ i. $\sqrt{u^2 - 2}$
    j. $\sqrt{x - 2}$

5. $[16, +\infty)$ 7. All real numbers u such that $u > 3$ or $u < -3$

9. $[0, +\infty)$ 11. D_{f+g} is $(-\infty, +\infty)$; $(f + g)(x) = x^2 + x + 4$;
                              $D_{f/g}$ is all real numbers $x \neq 3$; $f/g(x) = (x^2 + 7)/(x - 3)$

13. a. D_{g-h} is all real numbers $t \neq 3$;
         $(g - h)(t) = -(t^3 - 8t^2 + 19t - 11)/(t - 3)$
     b. $D_{g/h}$ is all real numbers $t \neq 1, 3$, and 4;
         $g/h(t) = 1/(t^3 - 8t^2 + 19t - 12)$
     c. $D_{h/g}$ is all real numbers $t \neq 3$; $h/g(t) = t^3 - 8t^2 + 19t - 12$

15. a. D_{f-h} is $(-\infty, +\infty)$; $(f - h)(x) = x^3 - x$
     b. D_{fh} is $(-\infty, +\infty)$; $fh(x) = x^4 - x^3 - x + 1$
     c. $D_{f/h}$ is all real numbers $x \neq 1$; $f/h(x) = x^2 + x + 1$

17. a. $2t^6 + 12t^3 + 18$ b. $8t^6 + 3$ c. 32   d. 3
     e. $8(x + 1)^6 + 3$

19. a. $8x^2 - 22x + 20$ b. $4x^2 + 2x + 7$ c. 20   d. 8
     e. $4x^4 + 2x^2 + 7$

21. $D_{f \circ g}$ is $[0, +\infty)$; $f \circ g(x) = x$

23. $D_{f \circ g}$ is all real numbers x such that $x \neq -1$ and $x \neq 0$; $f \circ g(x) = 1/(x^2 + x)$

25. $D_{U \circ V}$ is $[2, +\infty)$; $U \circ V(s) = \left(\sqrt{s-2}\right)^3 + 1$

27. Intercepts: $(0, 0)$, $\left(\sqrt{3}, 0\right)$, $\left(-\sqrt{3}, 0\right)$; symmetric with respect to the origin

29. Intercepts $(1, 0)$, $(-1, 0)$; symmetric with respect to origin

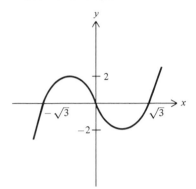

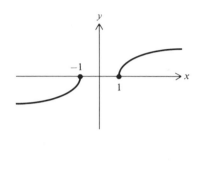

31. Intercepts: $(9, 0)$, $(0, -81)$; no symmetries

33.

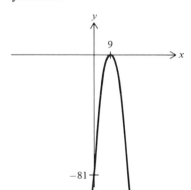

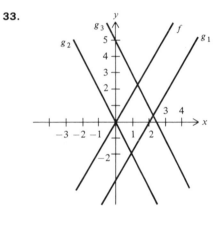

35.

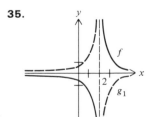

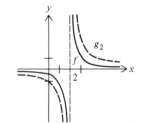

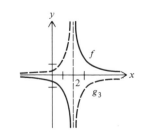

37. $f^{-1}(x) = (x + 5)/7$ **39.** $f^{-1}(x) = \sqrt[3]{x} + 1$

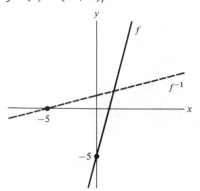   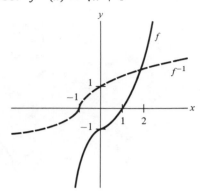

41. $f^{-1}(x) = 1 + \sqrt[3]{2/x}$ **43.** y is a function of x

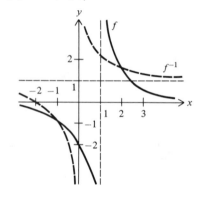

45. x is a function of y **47.** x is a function of y

49. **a.** $f(x) = x^3$ **b.** $g(x) = 6x^2$ **c.** $f^{-1}(V) = \sqrt[3]{V}$
         d. $g \circ f^{-1}(V) = 6\sqrt[3]{V^2}$

51. **a.** $f(x) = 500{,}000 + 1000x$ **b.** $600{,}000$ **c.** 500

53. **a.** $f(r) = 4\pi r^2$ **b.** $g(r) = 2\pi r^3$ **c.** $h(V) = \sqrt[3]{16\pi V^2}$

Exercise Set 5.2   **1.** $m = 2$; intercepts $(0, -6)$, $(3, 0)$ **3.** $m = 3/2$; intercepts $(0, 4)$, $(-8/3, 0)$

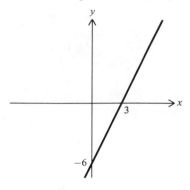   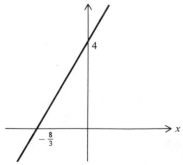

5. No slope; intercept $(5, 0)$

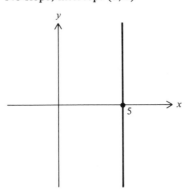

7. $m = 0$; intercept $(0, 3)$

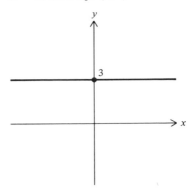

9. $m = -1/3$; intercepts $(0, 2/3)$, $(2, 0)$

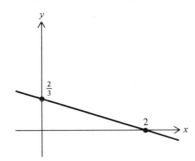

11. $3y - 2x + 1 = 0$

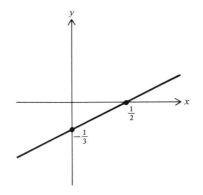

13. $y = -2x + 3$

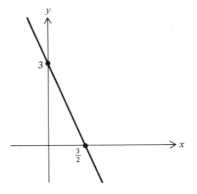

15. $y = 2x - 2$

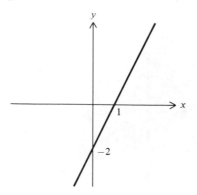

17. $y + 2x = 0$

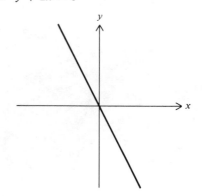

19. $y + 2x = 19$

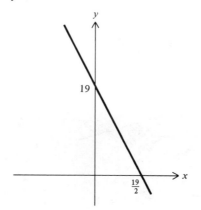

21. $2y + x = -16$

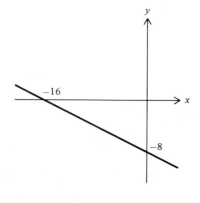

23. $y = -7$

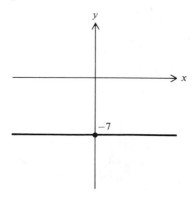

27. $x/5 + y/7 = 1$

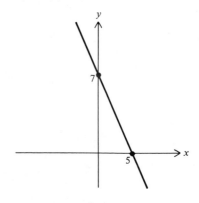

Exercise Set 5.3

1.

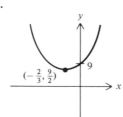

3.

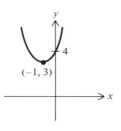

5.

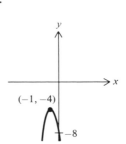

7.

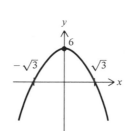

9.

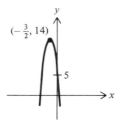

11.

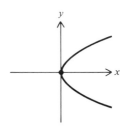

13.

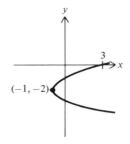

15.

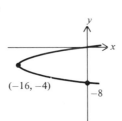

17. $f(x) = \frac{1}{4}x^2$

Exercise Set 5.4

1. $x = 1; y = 3$ **3.** $x = 5; y = 0$
5. $x = 2; y = 6$ and $x = -2; y = -6$ **7.** No solutions
9. $x = 1; y = 5$ **11.** $x = -2; y = 3$ **13.** $x = 1; y = -2$
15. No solutions   **17.** $x = 1; y = 3$ **19.** $x = -2; y = 2$
21. $x = 1; y = 2; z = 0$ **23.** $x = x; y = (19 - 5x)/4; z = (1 + x)/4$
25. $x = -2; y = -1; z = 3$ **27.** No solutions   **29.** $(0, 5)$ and $(3, 4)$
31. 20 by 50   **33.** Ten nickels, 5 dimes, 3 quarters
35. $y = -x^2 + x + 4$

Chapter 5
Review Exercises

1. $y = -6x + 4$ **3.** $2x + 7y = 32$ **5.** $3x + 4y = 34$
7. $-4x + 3y = 13$ **9.** $x/2 + y/(-4) = 1$ **11.** $x = 2$

13. 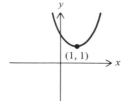   **15.** 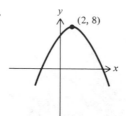   **17.**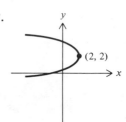

19. 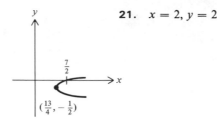   **21.** $x = 2, y = 2$

23. $x = \left(-7 + \sqrt{161}\right)/2, y = 4; x = \left(-7 - \sqrt{161}\right)/2, y = 4$

25. $x = 3, y = -7$ **27.** No solutions

29. $x = 1, y = 0; x = -1, y = 0$

31. $x = 7/2, y = 3/2$ **33.** No solutions   **35.** $x = -1, y = 2, z = -1$

37. $x = z, y = z, z = z$ **39.** No solutions   **41.** $x = 1, y = 0$

43. $x = 1, y = 1 + \sqrt{2}; x = 1, y = 1 - \sqrt{2}$ **45.** 17

47. $\left(\dfrac{3 + \sqrt{41}}{2}, \dfrac{3 - \sqrt{41}}{2}\right)$ and $\left(\dfrac{3 - \sqrt{41}}{2}, \dfrac{3 + \sqrt{41}}{2}\right)$

49. Six nickels, 3 dimes, and 4 quarters

Exercise Set 6.2

1. **a.** $\sin A = 5/13, \cos A = 12/13, \tan A = 5/12, \cot A = 12/5,$
$\sec A = 13/12, \csc A = 13/5$

     b. $\sin B = 12/13, \cos B = 5/13, \tan B = 12/5, \cot B = 5/12,$
$\sec B = 13/5, \csc B = 13/12$

3. **a.** $\sin G = \sqrt{2}/2, \cos G = \sqrt{2}/2, \tan G = 1, \cot G = 1, \sec G = \sqrt{2},$
$\csc G = \sqrt{2}$

     b. $\sin H = \sqrt{2}/2, \cos H = \sqrt{2}/2, \tan H = 1, \cot H = 1, \sec H = \sqrt{2},$
$\csc H = \sqrt{2}$

5. **a.** $\sin M = \sqrt{5}/5, \cos M = 2\sqrt{5}/5, \tan M = 1/2, \cot M = 2, \sec M = \sqrt{5}/2,$
$\csc M = \sqrt{5}$

     b. $\sin N = 2\sqrt{5}/5, \cos N = \sqrt{5}/5, \tan N = 2, \cot N = 1/2, \sec N = \sqrt{5},$
$\csc N = \sqrt{5}/2$

7. $10/3$ **9.** **b.** $14/3$ **11.** 226.8 feet   **13.** 0.707 feet

15. 31.06 feet   **17.** One mile

Exercise Set 6.3

1. $3\pi/4$ **3.** $\pi/12$ **5.** $5\pi/4$ **7.** $23\pi/180$ **9.** $7\pi/6$

11. $(900/\pi)° \doteq 286.5°$ **13.** $(180/\pi)° \doteq 57.3°$ **15.** 270°

17. 225°   **19.** 120°   **21.** $25\pi/6$ **23.** $2\pi/3$

Exercise Set 6.4
1. -2π 3. 2π 5. 4π 7. $\dfrac{\pi}{4} + 2n\pi,\ n = 0,\ \pm 1,\ \pm 2,\ldots$

9. $(1, 0)$ 11. $(-1, 0)$ 13. $(1, 0)$ 15. $(-1, 0)$ 17. $(1, 0)$
19. 1   21. -1 23. Undefined   25. -1 27. 1

29.

Quadrant	I	II	III	IV
Sine	+	+	−	−
Cosine	+	−	−	+
Tangent	+	−	+	−
Cotangent	+	−	+	−
Secant	+	−	−	+
Cosecant	+	+	−	−

31. Negative   33. Negative   35. Positive   37. Positive
39. a. $(0, -100)$ b. $(100, 0)$ c. $(0, 100)$ d. $\left(50\sqrt{2}, 50\sqrt{2}\right)$

Exercise Set 6.5
1. $\pi/3$ 3. $\pi/4$ 5. $\pi/6$ 7. $\pi/6$ 9. $\pi/6$ 11. $\pi/12$
13. $\pi/3$ 15. $4 - \pi \doteq 0.8584$ 17. $\sqrt{3}/2$ 19. $\sqrt{2}$ 21. -1
23. $-\sqrt{2}/2$ 25. -7.02 27. $-\sqrt{3}$ 29. -2

31. $x = \dfrac{\pi}{6} + 2n\pi$ or $x = \dfrac{5\pi}{6} + 2n\pi;\ n = 0,\ \pm 1,\ \pm 2,\ldots$

33. $x = \dfrac{3\pi}{4} + 2n\pi$ or $x = \dfrac{7\pi}{4} + 2n\pi;\ n = 0,\ \pm 1,\ \pm 2,\ldots$

35. $x = \dfrac{2\pi}{3} + 2n\pi$ or $x = \dfrac{4\pi}{3} + 2n\pi;\ n = 0,\ \pm 1,\ \pm 2,\ldots$

37. $x = -2 + \dfrac{4\pi}{3} + 2n\pi$ or $x = -2 + \dfrac{5\pi}{3} + 2n\pi;\ n = 0,\ \pm 1,\ \pm 2,\ldots$

39. $\pm\sqrt{\pi/4 + n\pi};\ n = 0,\ \pm 1,\ \pm 2,\ldots$

Exercise Set 6.6
1.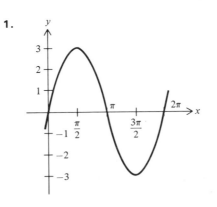

3.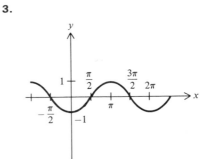

5.

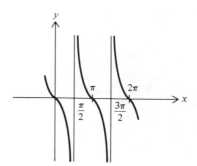

7.

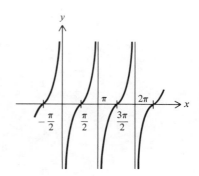

9.

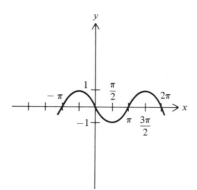

11.

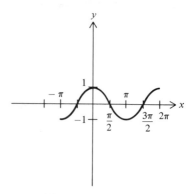

13.

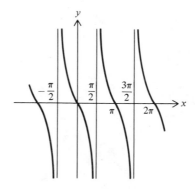

15.

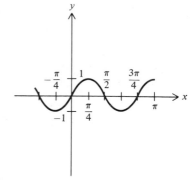

17.

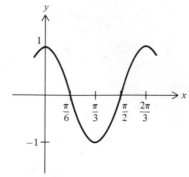

19. The graph of the cotangent can be obtained by translating the graph of the tangent $\pi/2$ units to the left and then reflecting in the x-axis.
21. **a.** $2, 0, 2 \cos 11 \doteq 0.0216$
    b. When $(2n - 1)\pi \leq t \leq 2n\pi$ $\quad$ $n = 0, \pm 1, \pm 2, \ldots$
    c. When $2n\pi \leq t \leq (2n + 1)\pi$ $\quad$ $n = 0, \pm 1, \pm 2, \ldots$

Exercise Set 6.7

1. $\pm 2\sqrt{2}/3$ $\quad$ **3.** $-\sqrt{15}$ $\quad$ **5.** $-3\sqrt{10}/10$ $\quad$ **7.** $\dfrac{\sqrt{2} + \sqrt{6}}{4}$

11. $\tan t = \pm \sin t / \sqrt{1 - \sin^2 t}$ $\quad$ **13.** $\cos t = \pm \cot t / \sqrt{1 + \cot^2 t}$
15. $AB = BC = 6.2$ $\quad$ **17.** **a.** Symmetry with respect to the x-axis
    $\quad$ **b.** $P(x, -y)$ $\quad$ **c.** $\sin A = -\sin(-A)$
    $\quad$ **d.** $\cos A = \cos(-A)$
19. $\sin 2A = 2 \sin A \cos A$

Exercise Set 6.8

1. $\dfrac{\sqrt{6} + \sqrt{2}}{4}$ $\quad$ **3.** $\dfrac{\sqrt{3} - 1}{1 + \sqrt{3}}$ $\quad$ **5.** $-1/2$ $\quad$ **7.** $-1/2$ $\quad$ **9.** -1

11. $\sqrt{3}/2$ $\quad$ **13.** $1/2$ $\quad$ **15.** $\dfrac{\sqrt{6} + \sqrt{2}}{4}$ $\quad$ **17.** $-2 - \sqrt{3}$

19. $-\sqrt{3}/2$ $\quad$ **21.** **a.** $\dfrac{\sqrt{6} - \sqrt{2}}{4}$ $\quad$ **b.** $\dfrac{\sqrt{2 - \sqrt{3}}}{2}$
27. $\sin A \cos B = [\sin(A + B) + \sin(A - B)]/2$ $\quad$ **29.** $1/2$
31. Undefined $\quad$ **33.** **a.** $g(x + \pi/2) = \sin(x + \pi/2)$
    $\quad$ **b.** The graph of $y = g(x + \pi/2)$ is found by translating the graph of $y = g(x)$ $\pi/2$ units to the left

Exercise Set 6.9

1. $-\pi/6$ $\quad$ **3.** $\pi/6$ $\quad$ **5.** $\pi/3$ $\quad$ **7.** π $\quad$ **9.** 0 $\quad$ **11.** $\pi/3$
13. $-\pi/4$ $\quad$ **15.** $-\pi/6$ $\quad$ **17.** $\pi/6$ $\quad$ **19.** $\pi/2$ $\quad$ **21.** $1/2$
23. $\sqrt{2}$ $\quad$ **25.** 1 $\quad$ **27.** $\pi/3$ $\quad$ **29.** $3\pi/4$

Chapter 6
Review Exercises

1. $38.66°$ $\quad$ **3.** 223.9 $\quad$ **5.** $7\pi/4$ $\quad$ **7.** $2\pi/3$ $\quad$ **9.** $25\pi/4$
11. $7\pi/180$ $\quad$ **13.** $20\pi/9$ $\quad$ **15.** π $\quad$ **17.** $1260°$ $\quad$ **19.** $-270°$
21. $150°$ $\quad$ **23.** $3600°$ $\quad$ **25.** $57.30°$ $\quad$ **27.** $25\pi/8$
29. First and third $\quad$ **31.** First and fourth $\quad$ **33.** $\sqrt{3}/2$
35. Undefined $\quad$ **37.** $-\sqrt{3}/3$ $\quad$ **39.** $\pm\sqrt{1 - \cos^2 x}$ $\quad$ **41.** $\dfrac{\sqrt{6} - \sqrt{2}}{4}$
43. $\pi/4$ $\quad$ **45.** Because $5\pi/6$ is not in the interval $[-\pi/2, \pi/2]$
47. $x = \dfrac{\pi}{6} + 2n\pi$ $\quad$ or $\quad$ $x = \dfrac{5\pi}{6} + 2n\pi$; $n = 0, \pm 1, \pm 2, \ldots$

Exercise Set 7.1

$100/(2^{50})$; since $2^7 = 128$, $100/(2^{50}) < (2^7/2^{50}) = 1/(2^{43})$

Exercise Set 7.2

1. 4   **3.** 5   **5.** 1/5   **7.** 1/8   **9.** 1/15   **11.** 1/2
13. 4   **15.** -2 **17.** 9   **19.** 1024   **21.** 2^{15} **23.** 4^{7}
25. 2^{2} **27.** $x = -2$
29. **a.** $x = 3$
    b. Not unless we can write 9 and 2 as exponentials with the same base
31. 6   **33.** -4 **35.** 60   **37.** 14   **39.** 0   **41.** b^{2}/a^{5}
43. s^{2}/t^{5} **45.** $a^{5}d^{3}/b^{3}c^{4}$ **47.** $b^{3}c^{6}/a^{9}$ **49.** x^{2}/y^{6}
51. $ab^{-2}c^{-1}$ **53.** $a^{-5}b^{-4}$ **55.** **a.** $10^{0.7781}$ **b.** $10^{1.1761}$ **c.** $10^{1.4771}$
    d. $10^{0.9030}$ **e.** $10^{1.0791}$ **f.** $10^{1.8751}$

Exercise Set 7.3

5. $f(x) = (3^{-1})^{x}$ or $f(x) = (\frac{1}{3})^{x}$

7.

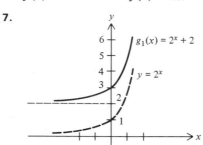

9.

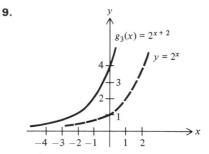

13. **a.** $2^{x}(x^{2} + 1)$ **b.** $2^{x}/(x^{2} + 1)$ **c.** $2^{x^{2}+1}$ **d.** $2^{2x} + 1$
    e. 4   **f.** 5
15. **a.** $x2^{x}$ **b.** $x/2^{x}$ **c.** 2^{x} **d.** 2^{x}
17. **a.** 4
    b. It gets smaller, approaching zero
    c. It gets larger, approaching 5   **e.**
    d. No

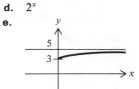

19. **a.** 25 grams   **b.** 5568 years

Exercise Set 7.4

1. $10^{3} = 1000$ **3.** $25^{1} = 25$ **5.** $4^{-2} = 1/16$ **7.** $5^{4} = 625$
9. $6^{3} = 216$ **11.** $\log_{3} 27 = 3$ **13.** $\log_{4} 256 = 4$
15. $\log_{2} (1/32) = -5$ **17.** $\log_{8} 4 = 2/3$ **19.** $\log_{10} 10{,}000 = 4$
21. 2   **23.** -2 **25.** 0   **27.** 3/4   **29.** -3
33.

x	1	2	3	4	5	6	7	8	9	10
$\log_{10} x$	0	0.3010	0.4771	0.6021	0.6990	0.7781	0.8451	0.90301	0.9542	1

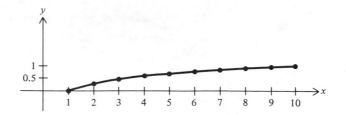

Exercise Set 7.5

1. 1.2040   **3.** 0.7781   **5.** 1.8751   **7.** -2.2219 **9.** -0.7781
11. 1.6990   **13.** 1.4313   **15.** 0.1761   **17.** $1 + B + C$
19. $2A - B - C$ **21.** $2(A + B) - 3C$ **23.** $3A + 3B + 6C$
25. $\log 2/\log 3 = 0.6310$ **27.** $\log 12/\log 7 = 1.2770$ **29.** 3
31. -0.3713 and 2.694 **33.** **a.** 5   **b.** 12
35. **a.** $\log_k a = x \Leftrightarrow k^x = a;\ \log_k b = y \Leftrightarrow k^y = b$
    b. $ab = k^{x+y}$ **c.** $ab = k^{x+y} \Leftrightarrow \log_k ab = x + y = \log_k a + \log_k b$
37. **a.** $N = 5000$ **b.** $k \doteq 0.3495$ **c.** Approximately 279,500

Appendix 7.5

1. 2.9253   **3.** 3.4871   **5.** 4.2253   **7.** -1.3270 **9.** -3.5017
11. 230   **13.** 46,700   **15.** 632   **17.** 0.704   **19.** 9530
21. 2.3530

Exercise Set 7.6

1. **a.**

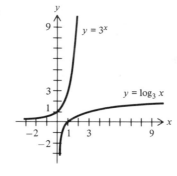

b. They are mirror images of each other in the line $y = x$

3.

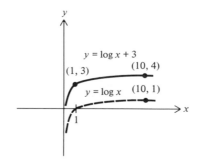

5.

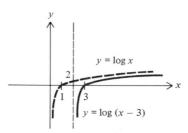

7.

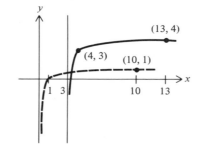

9. $\log_5 x = \dfrac{\log x}{\log 5} \doteq 1.4306 \log x;$ **11.** 10^{1000} **13.** 10^{-50}

$\log_{1/5} x = \dfrac{\log x}{\log 1/5} \doteq -1.4306 \log x$

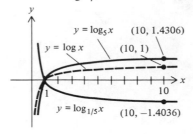

15.   **17.**   **19.** $t = \dfrac{1}{k} \log\left(\dfrac{P}{N}\right)$

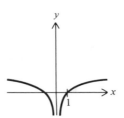   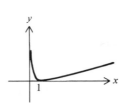

Exercise Set 7.7

1. a. $2 = 10^{0.3010}; \ 3 = 10^{0.4771}; \ 5 = 10^{0.6990}$
   b. $3 \cdot 10^{1.5050x}; \ 1.3 \cdot 10^{0.9542x}; \ 4 \cdot 10^{0.3495x}$
5. The relationship is probably exponential. The fact that the points do not lie exactly on an exponential curve may be due to experimental error.
7. $k = 5/4, \ c = \log 8, \ a = 10, \ n = 3$

Chapter 7
Review Exercises

1. $1/4$ **3.** $1/4$ **5.** $1/5$ **7.** 1 **9.** $1/2$ **11.** 4
13. -1 **15.** 7 **17.** 5 **19.** $9/2$

21.

x	1	2	3	4	5
$\log y$	-0.3010	0.6021	1.1303	1.5051	1.7959

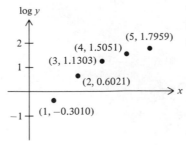

The graph suggests the points (x, y) do not lie on the graph of an exponential function.

23. 2.1761 **25.** -0.3979 **27.** 1.4651 **29.** 1.9084

Exercise Set 8.1 1. It is likely that the per-unit cost will decrease as the number of cars manufactured increases.

Exercise Set 8.2 1. $x \leq -13/8$ 3. $x \geq 9$ 5. $x < -12$

      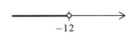

7. $x \geq -7/2$ 9. $x \leq 5$ 11. True for all values of x

13. $x \geq 11/2$ 15. $x \leq 6$

17. Because x may be negative, in which case $6/x < 12 \not\Rightarrow 6 < 12x$

19. **a.** $x < -2; -2 < x < 2; x > 3$ **b.** $2 < x < 3$
    c. $x = -2; x = 2; x = 3$

21. **a.** 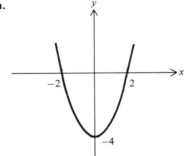   **b.** $-2 \leq x \leq 2$

25. **a.** $f(x) = 2500x - 5,000,000$ **b.** 2000   **c.** 2100   **d.** 2001

27. 2000/3 gallons

Exercise Set 8.3 1. $-10 < x < -1$ 3. True for all values of x 5. $z < 0$ or $z > 1$

   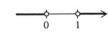

7. $\dfrac{-1 - \sqrt{5}}{2} < x < \dfrac{-1 + \sqrt{5}}{2}$ 9. $\dfrac{-7 - \sqrt{73}}{2} < x < \dfrac{-7 + \sqrt{73}}{2}$

11. $x < -2$ or $x > 7$ **13.** $x < -2$ or $x > 0$ **15.** $-10 \leq z \leq -2$

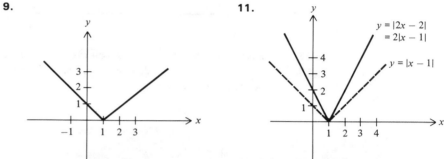

17. $1 \leq x \leq 3$ **19.** $1 \leq x \leq \sqrt{3}$ or $-\sqrt{3} \leq x \leq -1$

21. $x \geq 3$ **23.** $-\pi/4 + n\pi \leq x \leq \pi/4 + n\pi; \; n = 0, \pm 1, \ldots$
25. $\pi/2 + 2n\pi < x < 3\pi/2 + 2n\pi; \; n = 0, \pm 1, \pm 2, \ldots$ **27.** 3/4 inches

Exercise Set 8.4   **1.** 7   **3.** 5   **5.** 6   **7.** 1

9.

11.

13. **a.** $y = |x| + 1$ **b.** $y = |x - 3|$ **c.** $y = |x - 3| + 1$
15. **a.** x and y must be either both non-negative or both non-positive
        b. Choose one of the numbers positive and one of the numbers negative
        c. $|x + y| \leq |x| + |y|$

17.

19.

21. **a.** All $x \neq 0$
        b. $f(x) = 1$
        c. $f(x) = -1$
        d.

23.

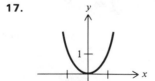

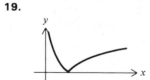

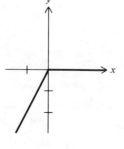

Exercise Set 8.5

1. $x = 1/4$ or $x = -1$ **3.** $-1 < x < 7$ **5.** $14 < x < 16$

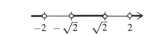

7. All $y \neq 5$ **9.** $-1/4 < x < 3/4$ **11.** $x > 8$ or $x < 2$

13. $-17/7 \leq x \leq -1$ **15.** $-1 < x < 3$

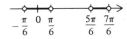

17. $y < -2, -\sqrt{2} < y < \sqrt{2}$ or $y > 2$ **19.** $x > 0$

21. $-\pi/6 + n\pi < x < \pi/6 + n\pi; n = 0, \pm 1, \pm 2, \ldots$

Chapter 8
Review Exercises

1. $x \leq 1$ **3.** $x > -1$ or $x < -7$ **5.** All real numbers
7. $x \leq 27$ **9.** $-4 \leq x \leq 4$ **11.** All real numbers
13. $0 \leq x \leq 5$ **15.** $x > 15$ or $x < -9$ **17.** $x > 3/5$ or $x < 0$
19. All real numbers   **21.** $x = -1$ or $x = -7/3$ **23.** $x = 3$

25. Graph of $\cos x$ **27.**

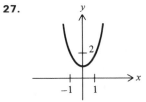

Exercise Set 9.1

1. **a.** $f(1) = 1; f(2) = 1.5; f(5) = 1.8; f(10) = 1.9; f(1000) = 1.999$
    b. Larger   **c.** No   **d.** 2   **e.** No
3. **a.** $h(1) = -2; h(2) = -2.5; h(5) = -2.8; h(10) = -2.9;$
    $h(1000) = -2.999$
    b. Smaller   **c.** -3 **d.** No   **e.** It is shifted 3 units down
5. **a.** $\lim\limits_{x \to \frac{\pi}{2}^+} \tan x = -\infty; \lim\limits_{x \to \frac{\pi}{2}^-} \tan x = +\infty$ **7.** $\lim\limits_{x \to -\infty} (3^x) = 0$
    b. $\lim\limits_{x \to -\frac{\pi}{2}^+} \tan x = -\infty; \lim\limits_{x \to -\frac{\pi}{2}^-} \tan x = +\infty$
9. **a.** $f(1) = 1; f(0.1) = 0.01; f(0.01) = 0.0001; f(0.001) = 0.000001$
    b. 0
    c. $f(-1) = 1; f(-0.1) = 1.9; f(-0.01) = 1.99; f(-0.001) = 1.999$
    d. $\lim\limits_{x \to 0^-} f(x) = 2$

11. **a.** $f(1) = 1; f(0.1) = 0.01; f(0.01) = 0.0001; f(0.001) = 0.000001$
  b. $f(-1) = 1; f(-0.1) = 0.01; f(-0.01) = 0.0001; f(-0.0001) = 0.000001$
  c. 0

Exercise Set 9.2   **1.** 70   **5.** **a.** It is likely that our score when $t = 0$ would be very low.

Exercise Set 9.3   **1.** 1/3   **3.** π **5.** 1/6   **7.** 0   **9.** 0   **11.** 0   **13.** 0
  15. $\sin(1°) = 0.0174524; \dfrac{\sin(0.5°)}{0.5} = 0.0174531;$

  $\dfrac{\sin(0.1°)}{0.1} = 0.0174533; \dfrac{\sin(0.01°)}{0.01} = 0.0174533$

Exercise Set 9.5   **1.** **a.** $\left(-7 - \sqrt{61}\right)/2; \left(-7 + \sqrt{61}\right)/2$ **b.** 0.41   **3.** 6.918
  5. 0.38   **7.** 0.7391   **9.**

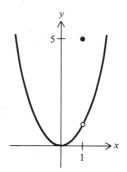

**Chapter 9
Review Exercises**   **1.** **a.** 6   **b.** $+\infty$ **3.** **a.** 4   **5.** $f(x) = 3 + 1/x$
  7. **a.**

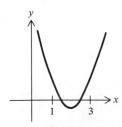

  b. $f(1.1) = 1.21; f(1.01) = 1.0201;$ **c.** 1   **d.** $\lim\limits_{x \to 1} f(x) \neq f(1)$
  $f(1.001) = 1.002001$

  11. 1/6   **13.** $1/\pi$ **15.** 0   **17.** Approximately 0.2027 square units
  19. 1.193   **21.** -1.105 **23.** 0.566   **25.** 0.786   **27.** 1.732

Index